A Course in Behavioral Economics

A Course in Behavioral Economics

ERIK ANGNER

Third edition

BLOOMSBURY ACADEMIC
NEW YORK • LONDON • OXFORD • NEW DELHI • SYDNEY

BLOOMSBURY ACADEMIC
Bloomsbury Publishing Plc
50 Bedford Square, London, WC1B 3DP, UK
1385 Broadway, New York, NY 10018, USA
29 Earlsfort Terrace, Dublin 2, Ireland

BLOOMSBURY, BLOOMSBURY ACADEMIC and the Diana logo
are trademarks of Bloomsbury Publishing Plc

Reprinted by Bloomsbury Academic 2021

Previous editions published under the imprint PALGRAVE

ISBN 978-1-352-01080-0 paperback

This book is printed on paper suitable for recycling and made from fully managed and sustained forest sources. Logging, pulping and manufacturing processes are expected to conform to the environmental regulations of the country of origin.

A catalogue record for this book is available from the British Library.

A catalog record for this book is available from the Library of Congress.

To find out more about our authors and books visit
www.bloomsbury.com and sign up for our newsletters.

To Iris

The master-piece of philosophy would be to develop the means which Providence employs to attain the ends it proposes over man, and to mark out accordingly a few lines of conduct which might make known to this unhappy biped individual the way in which he must walk within the thorny career of life, that he might guard against the whimsical caprices of this fatality to which they give twenty different names, without having as yet come to understand or define it.

The Marquis de Sade, *Justine*

BRIEF CONTENTS

CONTENTS

LIST OF FIGURES

LIST OF TABLES

PREFACE

As a Ph.D. student in economics, behavioral economics struck me as the most exciting field of study by far. Even with the benefit of some spectacular teachers, though, I felt that existing literature failed to convey an adequate understanding of the nature and significance of the project, and how the many different concepts and theories described as "behavioral" were tied together. When as an assistant professor I was offered the opportunity to teach my own course, I discovered that there were few texts available in the sweet spot between popular-science treatments, which do not contain enough substance for a university-level course, and scientific papers and more advanced textbooks, which are not easily readable and typically fail to provide sufficient background to be comprehensible to a novice reader.

This introduction to behavioral economics was written to be the book I wish I had had as a student, and the book that I want to use as a teacher. It aspires to situate behavioral economics in historical context, seeing it as the result of a coherent intellectual tradition; it offers more substance than popular books but more context than original articles; and while behavioral economics is a research program as opposed to a unified theory, the book not only describes individual concepts and theories but also tries to show how they hang together. The book was designed as a user-friendly, self-contained, freestanding textbook suitable for a one-semester course at the undergraduate level, but can easily be used in conjunction with books or articles in a variety of higher-level courses and programs.

In recognition of the fact that many students of behavioral economics come from outside traditional economics, the exposition was developed to appeal to advanced undergraduates across the social and behavioral sciences, humanities, business, public health, etc. The book contains no advanced mathematics and presupposes no knowledge of standard economic theory. If you are sufficiently interested to pick up a copy of this book and read this far, you have what it takes to grasp the material. Thorough battle-testing at two medium-sized state universities in the US over the course of several years has confirmed that the treatment is accessible to diverse audiences – including to economics majors and non-majors alike.

Serious economics does not need to be intimidating, and this book aims to prove it. Abstract, formal material is introduced in a progressively more difficult manner, which serves to build confidence in students with limited previous exposure. A wealth of examples and exercises help make the underlying intuitions as clear as possible. (Answers to the exercises are provided in an appendix.) In order to sustain the interest of readers with different backgrounds, and to illustrate the vast applicability of economic analysis, examples are drawn from economics, business, marketing, medicine, philosophy, public health, political science, public policy, and elsewhere. More open-ended problems encourage students to apply the ideas and theories presented here to decision problems they might come across outside the classroom.

The book is arranged in six main parts. The first five cover (1) choice under certainty, (2) judgment under risk and uncertainty, (3) choice under risk and uncertainty, (4) intertemporal choice, and (5) strategic interaction. Each of these parts contains two chapters: an even-numbered one outlining standard neoclassical theory and an odd-numbered one discussing behavioral alternatives. The unique structure makes it easy for instructors to teach the book at a more advanced level, as they can easily assign even-numbered chapters as background reading and supplement the odd-numbered chapters with more advanced material of their choosing. A final part (6) explores policy applications – including libertarian paternalism and the nudge agenda – and concludes. Additional material for general readers, students, and instructors is available via the companion website macmillanihe.com/angner-behavioral-economics-3e.

The non-trivial amount of neoclassical theory in this book may warrant explanation. First, because behavioral economics was developed in response to neoclassical economics, large portions of behavioral economics can only be understood against this background. Second, while behavioral economists reject the standard theory as a descriptive theory, they often accept it as a normative theory. Third, much of behavioral economics is a modification or extension of neoclassical theory, which remains useful under a wide range of conditions. Finally, to assess the relative merits of neoclassical and behavioral economics, it is necessary to understand both. Just as the study of a foreign language teaches you a great deal about your native tongue, so the study of behavioral economics can teach you a lot about standard economics.

As a textbook rather than an encyclopedia, this book does not aspire to be a complete record of contemporary theorizing in behavioral economics. Instead, it explores a selection of the most important ideas in behavioral economics and their interrelations. Many fascinating ideas, developments, and avenues of research have deliberately been omitted. No doubt every behavioral economist will disagree with some of my decisions about the things that were left out. But I think most will agree about the things that were included. The material presented in this book is, on the whole, uncontroversially part of the canon, and as such should be familiar to anyone who wishes to have a basic grasp of behavioral economics.

Like other introductory textbooks across the sciences, this book does not purport to describe the evidence supporting the theory in any detail. To keep the focus on theory and applications, the exposition is intentionally uncluttered by extensive discussion of data, standards of evidence, empirical (including experimental) methodology, and statistical techniques. Instead, theories are illustrated by reference to "stylized facts" and stories intended to elicit the intuition underlying the theory and to demonstrate that it is not entirely implausible. In this respect, the present book is no different from any of the standard introductions to microeconomics, to take one example.

For each new edition of the book, the publisher sought extensive feedback from current and prospective readers, including students and instructors, from across the world. As a result, the book has been updated with a wealth of new material, including entire new sections on Rabin's calibration theorem and the economics of happiness. Since the exercises (and answer key) turned out to be one of the most appreciated features of the original edition, I have added even more – and of a wider range of difficulty levels. Meanwhile, I have done my best to keep the book readable and to-the-point (and affordable too).

For readers who wish to continue their study of behavioral economics, or who want to know more about its evidential support, methodology, history, and philosophy, every chapter ends with a further reading section, which offers a selection of citation classics, review articles, and advanced textbooks. While this may be the first book in behavioral economics for many readers, my hope is that it will not be the last.

Erik Angner
Stockholm, Sweden

ACKNOWLEDGEMENTS

This book draws on many different sources of inspiration. I am particularly grateful to the teachers who set me off on this path (among them Cristina Bicchieri, Robyn Dawes, Baruch Fischhoff, George Loewenstein, Philip Reny, and Alvin Roth) and to the students (too numerous to name) who have kept me on the straight and narrow by catching errors and challenging me to continually improve the presentation.

This book project was originally conceived in conversation with Jaime Marshall and Aléta Bezuidenhout at what was then Palgrave Macmillan. I continue to be impressed with their wisdom and foresight in helping me articulate the details of the project. Kirsty Reade, Elizabeth Stone, and Jared Sutton played important roles in realizing the vision, just like Melanie Birdsall, Georgia Walters, and Lillie Flowers did with the second edition. Mallick Hossain kindly assisted me with some of the ancillary materials. I am grateful for all the careful work they put into this project, without which the text would have been far less effective. Luke Block, Amy Brownbridge, Jon Finch, Lloyd Langman, and Verity Rimmer at Red Globe Press worked with me on this third edition, offering invaluable suggestions and gentle encouragement. I am delighted about having the opportunity to refresh the text once more, and have benefited greatly from their thoughtful advice. A most sincere thank you to all.

Many others – including friends, colleagues, and anonymous reviewers – have offered invaluable feedback on earlier versions of this text. I am particularly grateful to Jörg Franke, Nick Huntington–Klein, Ramzi Mabsout, George MacKerron, Ivan Moscati, Norris Peterson, Mark Raiffa, Daniel Wood, and others who got in touch unprompted to share their thoughts on the text and its ancillary materials. Being able to leverage the intelligence and professional judgment of so many fantastically experienced colleagues is gratifying indeed, and I am most thankful.

The image of the mosaic of Ulysses and the Sirens is used with photographer Dennis Jarvis's kind permission. The Swedish Transport Agency (*Transportstyrelsen*) graciously granted the permission to use their images of traffic signs. Original illustrations by Cody Taylor are used with the artist's kind permission. The epigraph appears in Opus Sadicum (de Sade, 1889 [1791], p. 7). Every effort has been made to trace copyright holders, but if any have been inadvertently overlooked we will be pleased to make the necessary arrangements at the first opportunity. I gratefully acknowledge support from a Quality Enhancement Plan Development Grant from the University of Alabama at Birmingham. Most importantly, Elizabeth Blum's love and support every step along the way were essential to the completion of the project.

I continue to welcome suggestions for improvement via the companion website macmillanihe.com/angner-behavioral-economics-3e. As always, errors remain my own.

ABOUT THE AUTHOR

Erik Angner is Professor of Practical Philosophy at Stockholm University and Researcher at the Institute for Futures Studies in Stockholm, Sweden. He is affiliated with the Interdisciplinary Center for Economic Science at George Mason University, USA, where he previously taught Philosophy, Economics, and Public Policy. As a result of serious mission creep, he holds two PhDs – one in Economics and one in History and Philosophy of Science – both from the University of Pittsburgh. He is the author of a scholarly book and multiple journal articles and book chapters on behavioral and experimental economics, the economics of happiness, and the history, philosophy, and methodology of contemporary economics. He lives in Stockholm with his wife and their three children.

1 INTRODUCTION

Learning objectives

After studying this chapter you will:

- Know the difference between descriptive and normative theories of decision
- Understand how behavioral economics differs from standard (neoclassical) economics – and why
- Appreciate the variety of methods used by behavioral economists

1.1 Economics: Neoclassical and behavioral

This is a book about **theories of decision**. To use the language of the epigraph, such theories are about the negotiation of "the thorny career of life": they tell us how we make, or how we should make, decisions. Not that the Marquis de Sade would have spoken in these terms, living as he did in the eighteenth century, but the theory of decision seems to be exactly what he had in mind when he imagined "the master-piece of philosophy."

Developing an acceptable theory of decision would be an achievement. Most human activity – finance, science, medicine, arts, and life in general – can be understood as a matter of people making certain kinds of decisions. Consequently, an accurate theory of decision would cover a lot of ground. Maybe none of the theories we will discuss is the masterpiece of which de Sade thought so highly. Each theory can be, has been, and perhaps should be challenged on various grounds. However, decision theory has been an active area of research in recent decades, and it may have generated real progress.

Modern theories of decision (or **theories of choice** – I will use the terms interchangeably) say little about what goals people will or should pursue. Goals may be good or evil, mean-spirited or magnanimous, altruistic or egoistic, short-sighted or far-sighted; they may be Mother Teresa's or the Marquis de Sade's. Theories of decision simply take a set of goals as given. Provided a set of goals, however, the theories have much to say about how people will or should pursue those goals.

Theories of decision are variously presented as descriptive or normative. A **descriptive** theory describes how people *in fact* make decisions. A **normative** theory captures how people *should* make decisions. It is at least theoretically possible that people make the decisions that they should make. If so, one and the same theory can simultaneously be descriptively adequate and normatively correct. However, it is possible that people fail to act in the manner in which they should. If so, no one theory can be both descriptively adequate and normatively correct.

Exercise 1.1 Descriptive vs. normative Which of the following claims are descriptive and which are normative? (Answers to this and other exercises can be found in the Appendix.)
(**a**) On average, people save less than 10 percent of their income for retirement.
(**b**) People do not save as much for retirement as they should.
(**c**) Very often, people regret not saving more for retirement.

It can be unclear whether a claim is descriptive or normative. "People save too little" is an example. Does this mean that people do not save as much as they should? If so, the claim is normative. Does this mean that people do not save as much as they wish they did? If so, the claim is descriptive.

Example 1.2 Poker Suppose that you are playing poker, and that you are playing to win. Would you benefit from having an adequate descriptive theory, a correct normative theory, or both?

A descriptive theory would give you information about the actions of the other players. A normative theory would tell you how you should behave in light of what you know about the nature of the game, the expected actions of the other players, and your ambition to win. All this information is obviously useful when playing poker. You would benefit from having both kinds of theory.

Some theories of decision are described as **theories of rational choice**. In everyday speech, the word "rationality" is used loosely; frequently it is used simply as a mark of approval. For our purposes, a theory of rational decision is best seen as a **definition** of rationality, that is, as specifying what it means to be rational. Every theory of rational decision serves to divide decisions into two classes: rational and irrational. Rational decisions are those that are in accordance with the theory; irrational decisions are those that are not. A theory of rational choice can be thought of as descriptive or normative (or both). To say that a theory of rational decision is descriptive is to say that people in fact act rationally. To say that a theory of rational decision is normative is to say that people should act rationally. To say that a theory of rational decision is simultaneously descriptive and normative is to say that people act and should act rationally. Typically, the term **rational-choice theory** is reserved for theories that are (or that are thought to be) normatively correct, whether or not they are simultaneously descriptively adequate.

For generations now, economics has been dominated by an intellectual tradition broadly referred to as **neoclassical economics**. If you have studied economics but do not know whether or not you were taught in the neoclassical tradition, it is almost certain that you were. Neoclassical economics is characterized by its commitment to a theory of rational choice that is simultaneously presented as descriptively adequate and normatively correct. This approach presupposes that people by and large act in the manner that they should. Neoclassical economists do not need to assume that all people act rationally all the time, but they insist that deviations from perfect rationality are so small or so unsystematic as to be negligible. Because of its historical dominance, I will refer to neoclassical economics as standard economics, and to neoclassical economic theory as standard theory.

This is an introduction to **behavioral economics**: the attempt to increase the explanatory and predictive power of economic theory by providing it with more psychologically plausible foundations, where "psychologically plausible" means consistent with the best available psychology. Behavioral economists share neoclassical economists' conception of **economics** as the study of people's decisions under conditions of scarcity and of the results of those decisions for society. But behavioral economists reject the idea that people by and large behave in the manner that they should. While behavioral economists certainly do not deny that some people act rationally some of the time, they believe that the deviations from rationality are large enough, systematic enough, and consequently predictable enough to warrant the development of new descriptive theories of decision. If this is right, a descriptively adequate theory cannot at the same time be normatively correct, and a normatively correct theory cannot at the same time be descriptively adequate.

1.2 The origins of behavioral economics

Behavioral economics can be said to have a short history but a long past. Only in the last few decades has it emerged as an independent subdiscipline of economics. By now, top departments of economics have behavioral economists on their staff. Behavioral economics gets published in mainstream journals. Traditional economists incorporate insights from behavioral economics into their work. In 2002, Daniel Kahneman (one of the most famous behavioral economists) won the Nobel Memorial Prize "for having integrated insights from psychological research into economic science, especially concerning human judgment and decision-making under uncertainty." And then, in 2017, Richard Thaler (another leading figure) won the Prize for his contributions to behavioral economics. In spite of its short history, however, efforts to provide economics with plausible psychological foundations go back a long way.

The establishment of modern economics is marked by the publication in 1776 of Adam Smith's *The Wealth of Nations*. Classical economists such as Smith are often accused of having a particularly simple-minded (and false) picture of human nature, according to which people everywhere and always, in hyper-rational fashion, pursue their narrowly construed self-interest. This accusation, however, is unfounded. Smith did not think people were rational:

> How many people ruin themselves by laying out money on trinkets of frivolous utility? What pleases these lovers of toys is not so much the utility, as the aptness of the machines which are fitted to promote it. All their pockets are stuffed with little conveniences ... of which the whole utility is certainly not worth the fatigue of bearing the burden.

Smith wrote these words 200 years before the era of pocket calculators, camera phones, iPads, and smartwatches. Nor did Smith think people were selfish: "[There] are evidently some principles in [man's] nature, which interest him in the fortune of others, and render their happiness necessary to him, though he derives nothing from it except the pleasure of seeing it." Smith and the other classical economists had a conception of human nature that was remarkably multi-faceted; indeed, they did not draw a sharp line between psychology and economics the way we do.

Early neoclassical economics was built on the foundation of **hedonic psychology**: an account of individual behavior according to which individuals seek to maximize pleasure and minimize pain. In W. Stanley Jevons's words: "Pleasure and pain are undoubtedly the ultimate objects of the Calculus of Economics. To satisfy our wants to the utmost with the least effort ... in other words, to *maximise pleasure*, is the problem of Economics." The early neoclassical economists were inspired by the philosopher Jeremy Bentham, who wrote: "Nature has placed mankind under the governance of two sovereign masters, *pain* and *pleasure* ... They govern us in all we do, in all we say, in all we think." Because it was assumed that individuals have direct access to their conscious experience, some economists defended the principles of hedonic psychology on the basis of their introspective self-evidence alone.

After World War II, however, many economists were disappointed with the meager results of early neoclassicism in terms of generating theories with predictive power and so came to doubt that introspection worked. Similar developments took place in other fields: behaviorism in psychology, verificationism in philosophy, and operationalism in physics can all be seen as expressions of the same intellectual trend. Postwar neoclassical economists aimed to improve the predictive power of their theories by focusing on what can be publicly observed rather than on what must be experienced. Instead of taking a theory about pleasure and pain as their foundation, they took a theory of preference. The main difference is that people's feelings of pleasure and pain are unobservable, whereas their choices can be directly observed. On the assumption that choices reflect personal preferences, we can have direct observable evidence about what people prefer. Thus, postwar neoclassical economists hoped to completely rid economics of its ties to psychology – hedonic and otherwise.

In spite of the relative hegemony of neoclassical economics during the second half of the twentieth century, many economists felt that their discipline would benefit from closer ties to psychology and other neighboring fields. What really made a difference, however, was the cognitive revolution. In the 1950s and 1960s, researchers in psychology, computer science, linguistics, anthropology, and elsewhere rejected the demands that science focus on the observable and that all methods be public. Instead, these figures advocated a "science of cognition" or **cognitive science**. The cognitive scientists were skeptical of naive reliance on introspection, but nevertheless felt that a scientific psychology must refer to things "in the head," including beliefs and desires, symbols, rules, and images. Behavioral economics is a product of the cognitive revolution. Like cognitive scientists, behavioral economists – though skeptical of the theories and methods of the early neoclassical period – are comfortable talking about beliefs, desires, rules of thumb, and other things "in the head." Below, we will see how these commitments get played out in practice.

To some, the fact that behavioral economists go about their work in such a different way means that they have become economists in name only. But notice that behavioral economics is still about the manner in which people make choices under conditions of scarcity and the results of those choices for society at large – which is the very definition of economics. **Behavioral science** refers to the scientific study of behavior, which makes behavioral economics a kind of behavioral science.

Psychology and economics is also a broader category, referring to anything that integrates the two disciplines, and which therefore does not need to be about choice at all.

1.3 Methods

Before we explore in earnest the concepts and theories developed by behavioral economists in the last few decades, I want to discuss the data that behavioral economists use to test their theories and the methods they use to generate such data. I also want to assuage some skepticism that people may have about those methods.

Some of the earliest and most influential papers in behavioral economics relied on participants' responses to hypothetical choices. In such studies, participants were asked to imagine that they found themselves in a given choice situation and to indicate what decision they would make under those conditions. Here is one such question: "Which of the following would you prefer? A: 50% chance to win 1,000, 50% chance to win nothing; B: 450 for sure." Other early papers relied on readers' intuitions about how people might behave under given conditions. Thus, they offered scenarios such as: "Mr S. admires a $125 cashmere sweater at the department store. He declines to buy it, feeling that it is too extravagant. Later that month he receives the same sweater from his wife for a birthday present. He is very happy. Mr and Mrs S. have only joint bank accounts." These thought experiments were apparently inspired in part by the author's observations of the behavior of fellow economists, who argued that people were always rational but at times behaved irrationally in their own lives.

Soon enough, hypothetical choice studies were almost completely displaced by **laboratory experiments** in which laboratory participants make real choices involving real money. Such experiments have been run for decades. In the early 1970s, for example, psychologists Sarah Lichtenstein and Paul Slovic ran experiments at a Las Vegas casino, where a croupier served as experimenter, professional gamblers served as participants, and winnings and losses were paid in real money. More frequently, behavioral economists use college undergraduates or other easily accessible participants. When behavioral economists engage in experimental studies, they can be hard to distinguish from neoclassical experimental economists, that is, neoclassical economists who use experiments to explore how people make decisions. Experimentalists agree that decisions performed by laboratory participants must be real, and that actual winnings must be paid out.

Behavioral economists, during the last two decades, have increasingly relied on data gathered "in the field." In one famous **field study**, Colin F. Camerer and colleagues studied the behavior of New York City cab drivers by using data from "trip sheets" – forms that drivers use to record the time passengers are picked up and dropped off as well as the amount of the fares – and from the cabs' meters, which automatically record the fares. Researchers in this study simply observed how participants behaved under different conditions. In **field experiments**, researchers randomly assign participants to test and control groups, and then note how (if at all) the behavior of individuals in the two groups differs. In one prominent field experiment, Jen Shang and Rachel Croson tracked how voluntary donations to a public radio station varied when prospective donors were given different social information, that is, information about how much other people had given.

To some extent, behavioral economists use what psychologists call **process measures**, that is, methods that provide hints about cognitive and emotional processes underlying decision-making. Some rely on **process-tracing** software to assess what information people use when making decisions in games. Others employ brain scans, typically functional magnetic resonance imaging (fMRI), which allows researchers to examine, albeit crudely, which parts of an individual's brain are activated in response to a task or decision. Imaging methods have already been applied to a diversity of economic tasks, including decision-making under risk and uncertainty, intertemporal choice, buying and selling behavior, and strategic behavior in games. Even more exotic neuroscience methods are sometimes employed. For example, a tool called transcranial magnetic stimulation can be used to temporarily disable a part of participants' brains as they make decisions. The increasing use of methods borrowed from neuroscience is, not coincidentally, connected to the rise of **neuroeconomics**, which integrates economics with neuroscience.

The use of multiple methods to generate evidence raises interesting methodological problems. This is particularly true when evidence from different sources points in slightly different directions. Sometimes, however, evidence from multiple sources points in the same direction. When this is true, behavioral economists have more confidence in their conclusions. It can be argued that part of the reason why behavioral economics has turned into such a vibrant field is that it successfully integrates evidence of multiple kinds, generated by a variety of methods.

Recently, social and behavioral science has been thrown into something called the "replication crisis," as several well-known empirical results have proven difficult to replicate. It may turn out that these findings were mere experimental artifacts all along. The lack of reproducibility is obviously unwelcome news for the researchers invested in the results, and has fueled skepticism about the methods of social and behavioral science – and perhaps the entire enterprise of trying to understand human behavior with scientific methods. But it is important to note that (at least within bounds) the fact that some alleged findings are revised in light of new evidence is not as such devastating for social and behavioral science. In fact, what makes science different from other kinds of human activity is that *it is supposed to be* open to revision in light of new data. On statistical grounds alone, we should expect that some of the results generated by behavioral economists – and consequently some of the results discussed in the below – will not hold up. That said, systematic studies of reproducibility in psychology and economics suggest that economics is doing reasonably well by comparison. A 2016 report in the prestigious journal *Science* concludes that results from laboratory experiments in economics are at least as robust (and maybe more robust) than any other empirical result in economics, and moreover that laboratory experiments published in top economic journals have relatively high rates of replicability. The authors conclude on a positive note: "There is every reason to be optimistic that science in general, and social science in particular, will emerge much improved after the current period of critical self-reflection."

1.4 Looking ahead

As stated in the Preface, this book is arranged in six main parts: (1) choice under certainty, (2) judgment under risk and uncertainty, (3) choice under risk and uncertainty, (4) intertemporal choice, (5) strategic interaction, and (6) policy applications and conclusions. As suggested in Section 1.1, the ultimate goal of behavioral economics is to generate novel insights into people's decisions under conditions of scarcity and the results of those decisions for society. Behavioral and neoclassical economists alike try to attain this goal by building abstract, formal theories. In this book we will explore increasingly general theories, both neoclassical and behavioral.

Studying behavioral economics is a non-trivial enterprise. For one thing, the level of abstraction can pose an initial challenge. But as we will see below, it is the very fact that economics is so abstract that makes it so very useful: the more abstract the theory, the wider its potential application. Some readers may be prone to putting down a book like this as soon as they notice that it contains mathematics. Please do not. There is no advanced math in the book, and **numeracy** – the ability with or knowledge of numbers – is incredibly important, even to people who think of themselves as practically oriented.

Exercise 1.3 Numeracy In a 2010 study on financial decision-making, people's answers to three quick mathematics questions were strong predictors of their wealth: households where both spouses answered all three questions correctly were *more than eight times* as wealthy as households where neither spouse answered any question correctly. So if you have ever struggled with math, be glad that you did. You can try answering the three questions for yourself:

(**a**) If the chance of getting a disease is 10 percent, how many people out of 1000 would be expected to get the disease?

(**b**) If five people all have the winning numbers in the lottery, and the prize is 2 million dollars, how much will each of them get?

(**c**) Let us say you have $200 in a savings account. The account earns 10 percent interest per year. How much would you have in the account at the end of two years?

You will find the correct answers in the answer key at the end of the book.

There's also evidence that people who fall prey to the specific fallacies and mistakes that behavioral economists study are more likely to experience poor outcomes in their own lives. In a widely cited 2007 study, researchers assessed people's decision-making competence by checking to what extent they make mistakes such as honoring sunk costs (see Section 3.3) in pen-and-paper questionnaires. The study found that people with low decision-making competence were more likely to report poor real-world decision outcomes, such as having gotten a divorce, declared bankruptcy, lost one's driver's license, gotten oneself kicked out of a bar, and so on. The authors suggest that decision-making competence should be considered a separate cognitive skill that helps us avoid negative real-world outcomes.

To underscore the usefulness of behavioral economics, the book discusses a variety of applications. Among other things, you will learn how to choose a wingman or wingwoman, how to design a marketing scheme that works, how not to fall for such marketing schemes, how to compute the probability that your love interest is seeing somebody else, how to sell tires, and how to beat anyone at rock-paper-scissors. Ultimately, behavioral economics sheds light on human beings living in society – the way they really are, as opposed to the way great thinkers of the past have thought they should be – and on the nature of the human condition. Behavioral economics helps us live better lives – and to improve the world to boot.

Further reading

Kahneman's *Thinking, Fast and Slow* (2011) and Thaler's *Misbehaving: The Making of Behavioral Economics* are must-reads for anyone interested in behavioral economics, both for their unparalleled understanding of the theory and for their illuminating personal reminiscences. Angner and Loewenstein (2012) and Heukelom (2014) discuss the nature, historical origins, and methods of behavioral economics; Angner (2015a, 2019) explores further the relationship between behavioral and neoclassical economics. *The Wealth of Nations* is Smith (1976 [1776]); the quotations in the history section are from Smith (2002 [1759], p. 211) and Smith (2002 [1759], p. 11), Jevons (1965 [1871], p. 37), and Bentham (1996 [1789], p. 11). The sample questions in the methods section come from Kahneman and Tversky (1979, p. 264) and Thaler (1985, p. 199). The psychologists who went to Vegas are Lichtenstein and Slovic (1973). The study of NYC cabdrivers is Camerer et al. (1997); the one about social information is Shang and Croson (2009). Camerer et al. (2005) provide a widely cited overview of neuroeconomics, and Camerer, Dreber, et al. (2016, pp. 1435–6) examine the reproducibility of economics. The study on financial decision-making is Smith et al. (2010); the three numeracy questions were adapted from the University of Michigan Health and Retirement Study.

PART 1

CHOICE UNDER CERTAINTY

2 RATIONAL CHOICE UNDER CERTAINTY

Learning objectives

After studying this chapter you will:

- Know the theory of choice under certainty
- Understand the concept of rationality built into this theory
- Be able to prove theorems on the basis of axioms and definitions

2.1 Introduction

As promised, we begin by discussing the theory of rational choice. This theory forms the foundation of virtually all modern economics and is one of the first things you would learn in a graduate-level microeconomics class. As a theory of rational choice (see Section 1.1), the theory specifies what it means to make rational decisions – in short, what it means to be rational.

In this chapter, we consider **choice under certainty**. The phrase "under certainty" simply means that there is no doubt as to which outcome will result from a given act. For example, if the staff at your local gelato place is minimally competent, so that you actually get vanilla every time you order vanilla and stracciatella every time you order stracciatella, you are making a choice under certainty. (We will discuss other kinds of choice in future chapters.) Before discussing what it means to make rational choices under conditions of certainty, however, we need to talk about what preferences are and what it means to have rational preferences.

The theory of rational choice under certainty is an **axiomatic** theory. This means that the theory consists of a set of **axioms**: basic propositions that cannot be proven using the resources offered by the theory, and which will simply have to be taken for granted. When studying the theory, the first thing we want to do is examine the axioms. As we go along, we will also introduce new terms by means of definitions. Axioms and definitions have to be memorized. Having introduced the axioms and definitions, we can prove many interesting claims. Thus, much of what we will do below involves proving new propositions on the basis of axioms and definitions.

2.2 Preferences

The concept of **preference** is fundamental in modern economics, neoclassical and behavioral. Formally speaking, a preference is a **relation**. The following are examples of relations: "Alf is older than Betsy," "France is bigger than Norway," and "Bill is worried he may not do as well on the exam as Jennifer." Notice that each of these sentences expresses a relationship between two entities (things, individuals). Thus, "Alf is older than Betsy" expresses a relationship between Alf and Betsy, namely, that the former is older than the latter. Because these examples express a relation between two entities, they are called **binary** relations. The following relation is not binary: "Mom stands between Bill and Bob." This relation is **ternary**, because it involves three different entities; in this case, people.

For convenience, we often use small letters to denote entities or individuals. We may use a to denote Alf and b to denote Betsy. Similarly, we often use capital letters to denote relations. We may use R to denote the relation "is older than." If so, we can write aRb for "Alf is older than Betsy." Sometimes we write Rab. Notice that the order of the terms matters: aRb is not the same thing as bRa. The first says that Alf is older than Betsy, and the second that Betsy is older than Alf. Similarly, Rab is not the same thing as Rba.

Exercise 2.1 Relations Assume that f denotes France and n denotes Norway, and that B means "is bigger than."

(**a**) How would you write that France is bigger than Norway?
(**b**) How would you write that Norway is bigger than France?
(**c**) How would you write that Norway is bigger than Norway?

In order to speak clearly about relations, we need to specify what sort of entities may be related to one another. When talking about who is older than whom, we may be talking about people. When talking about what is bigger than what, we may be talking about countries, houses, people, dogs, or many other things. Sometimes it matters what sort of entities we have in mind. When we want to be careful, which is most of the time, we define a **universe** U. The universe is the set of all things that can be related to one another. Suppose we are talking about Donald Duck's nephews Huey, Dewey, and Louie. If so, that is our universe. The convention is to list all members of the universe separated by commas and enclosed in curly brackets, like so: {Huey, Dewey, Louie}. Here, the order does not matter. So, the same universe can be written like this: {Louie, Dewey, Huey}. Thus: U = {Huey, Dewey, Louie} = {Louie, Dewey, Huey}.

Exercise 2.2 The universe Suppose we are talking about all countries that are members of the United Nations. How would that be written?

A universe may have infinitely many members, in which case simple enumeration is inconvenient. This is true, for instance, when you consider the time at which you entered the space where you are reading this. There are infinitely many points in time between 11:59 am and 12:01 pm, for example, as there are between 11:59:59 am and 12:00:01 pm. In such cases, we need to find another way to describe the universe.

One relation we can talk about is this one: "is at least as good as." For example, we might want to say that "coffee is at least as good as tea." The "at least as good as" relation is often expressed using this symbol: $\succcurlyeq$. If c denotes coffee and t denotes tea, we can write this sentence as $c \succcurlyeq t$. This is the **(weak) preference relation**. People may have, and often will have, their own preference relations. If we wish to specify whose preferences we are talking about, we use subscripts to denote individuals. If we want to say that for Alf coffee is at least as good as tea, and that for Betsy tea is at least as good as coffee, we say that c $\succsim_{\text{Alf}}$t and t $\succsim_{\text{Betsy}}$c, or that c $\succsim_{\text{A}}$t and t $\succsim_{\text{B}}$c.

Exercise 2.3 Preferences Suppose d denotes "enjoying a cool drink on a hot day" and r denotes "getting roasted over an open fire."
(**a**) How would you state your preference over these two options?
(**b**) How would you express a masochist's preference over these two options?

In economics, we are typically interested in people's preferences over **consumption bundles**, which are collections of goods. You face a choice of commodity bundles when choosing between the #1 Big Burger meal and the #2 Veggie Burger meal at your local hamburger restaurant. In order to represent commodity bundles, we think of them as collections of individual goods along the following lines: three apples and two bananas, or two units of guns and five units of butter. When talking about preference relations, the universe can also be referred to as the **set of alternatives**. If bundles contain no more than two goods, it can be convenient to represent the set of alternatives on a plane, as in Figure 2.1. When bundles contain more than two goods, it is typically more useful to write $\langle 3, 2\rangle$ for three apples and two bananas; $\langle 6, 3, 9\rangle$ for six apples, three bananas, and nine coconuts; and so on.

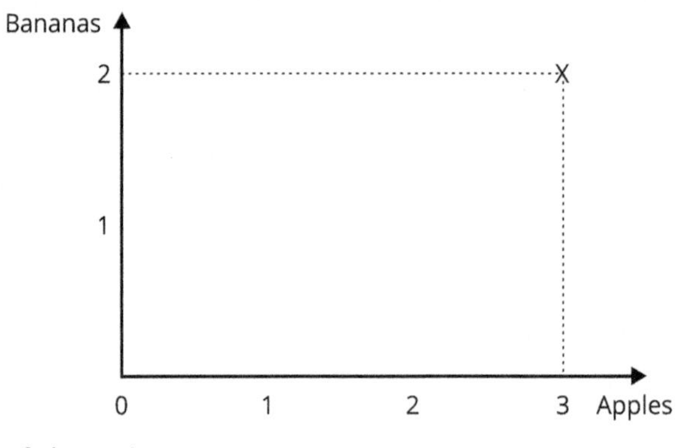

Figure 2.1 Set of alternatives

2.3 Rational preferences

We begin building our theory of rational choice by specifying what it means for a preference relation to be rational. A **rational** preference relation is a preference relation that is transitive and complete.

A relation R is **transitive** just in case the following condition holds: for all x, y, and z in the universe, if x bears relation R to y, and if y bears relation R to z, then x must bear relation R to z. Suppose the universe is the set of all the Marx brothers. If so, "is taller than" is a transitive relation: if Zeppo is taller than Groucho, and Groucho is taller than Harpo, then Zeppo must be taller than Harpo (Figure 2.2).

Figure 2.2 The Marx brothers. Illustration by Cody Taylor

Example 2.4 ***30 Rock*** Consider the following exchange from the TV show *30 Rock*. Tracy, Grizz, and Dot Com are playing computer games. Tracy always beats Grizz and Dot Com. When Kenneth beats Tracy but gets beaten by Grizz, Tracy grows suspicious.

Tracy: *"How were you beating Kenneth, Grizz?"*
Grizz: *"I don't know."*
Tracy: *"If Kenneth could beat me and you can beat Kenneth, then by the transitive property, you should beat me too! Have you been letting me win?"*
Dot Com: *"Just at some things."*
Tracy: *"Things? Plural?"*

Now you are the first kid on the block who understands *30 Rock*. You also know that the show had a former economics or philosophy student on its staff.

If the universe consists of all people, examples of **intransitive** relations include "is in love with." Just because Sam is in love with Pat, and Pat is in love with Robin, it is not necessarily the case that Sam is in love with Robin. Sam *may* be in love with Robin. But Sam may have no particular feelings about Robin, or Sam may resent Robin for attracting Pat's attention. It may also be the case that Robin is in love with Sam. This kind of intransitivity is central to the play *No Exit*, by the French existentialist philosopher Jean-Paul Sartre. In the play, which takes place in a prison cell, a young woman craves the affection of a man who desires the respect of an older woman, who in turn is in love with the young woman. Hence the most famous line of the play: "Hell is other people." To show that a relation is intransitive, it is sufficient to identify three members of the universe such that the first is related to the second, and the second is related to the third, but the first is not related to the third.

Formally speaking, a preference relation $\succsim$ is transitive just in case the following is true:

Axiom 2.5 Transitivity of $\succsim$ *If $x \succsim y$ and $y \succsim z$, then $x \succsim z$ (for all x, y, z).*

There are other ways of expressing the same thing. We might write: If $x \succsim y \succsim z$, then $x \succsim z$ (for all x, y, z). Using standard logic symbols, we might write: $x \succsim y \,\&\, y \succsim z \rightarrow x \succsim z$ (for all x, y, z). See the text box below for a useful list of logical symbols. Either way, transitivity says that if you prefer coffee to tea, and tea to root beer, you must prefer coffee to root beer; that is, you cannot prefer coffee to tea and tea to root beer while failing to prefer coffee to root beer.

A relation R is **complete** just in case the following condition holds: for any x and y in the universe, either x bears relation R to y, or y bears relation R to x (or both). If the universe consists of all people – past, present, and future – then "is at least as tall as" is a complete relation. You may not know how tall Caesar and Brutus were, but you do know this: either Caesar was at least as tall as Brutus, or Brutus was at least as tall as Caesar (or both, in case they were equally tall).

Given the universe of all people, examples of **incomplete** relations include "is in love with." For any two randomly selected people – your landlord and the current President of the US, for example – it is not necessarily the case that either one is in love with the other. Your landlord may have a crush on the President, or the other way around. But this need not be the case, and it frequently will not be. To show that a relation is incomplete, then, it is sufficient to identify two objects in the universe such that the relation does not hold either way.

Formally speaking, a preference relation $\succsim$ is complete just in case the following is true:

Axiom 2.6 Completeness of $\succsim$ *Either $x \succsim y$ or $y \succsim x$ (or both) (for all x, y).*

Completeness means that you must prefer tea to coffee or coffee to tea (or both); though your preference can go both ways, you cannot fail to have a preference between the two. The use of the phrase "(or both)" in the formula above is, strictly speaking, redundant: we use the "inclusive or," which is equivalent to "and/or" in everyday language. Using standard logical symbols, we might write: $x \succsim y \vee y \succsim x$ (for all x, y). If both $x \succsim y$ and $y \succsim x$, we say that there is a tie (see Section 2.4).

Logical symbols

Here is a list of the most common logical symbols:

$x \,\&\, y$	x and y
$x \vee y$	x or y
$x \rightarrow y$	if x then y; x only if y
$x \leftrightarrow y$	x if and only if y; x just in case y
$\neg p$	not p

The following exercise serves to illustrate the concepts of transitivity and completeness.

Exercise 2.7 Assuming the universe is the set of all people – past, present, and future – are the following relations transitive? Are they complete?
(**a**) "is the mother of"
(**b**) "is an ancestor of"
(**c**) "is the sister of"
(**d**) "detests"
(**e**) "weighs more than"
(**f**) "has the same first name as"
(**g**) "is taller than"

When answering questions such as these, ambiguity can be a problem. A word such as "sister" is ambiguous, which means that answers might depend on how it is used. As soon as the word is defined, however, the questions have determinate answers.

Exercise 2.8 The enemy of your enemy Suppose it is true, as people say, that the enemy of your enemy is your friend. What does this mean for the transitivity of "is the enemy of"? (Assume there are no true frenemies: people who are simultaneously friends and enemies.)

Exercise 2.9 Assuming the universe is the set of all natural numbers, meaning that U = {1, 2, 3, 4, ...}, are the following relations transitive? Are they complete?
(**a**) "is at least as great as" (≥)
(**b**) "is equal to" (=)
(**c**) "is strictly greater than" (>)
(**d**) "is divisible by" (|)

Exercise 2.10 Preferences and the universe Use your understanding of transitivity and completeness to answer the following questions:
(**a**) If the universe is {apple, banana, starvation}, what does the transitivity of the preference relation entail?
(**b**) If the universe is {apple, banana}, what does the completeness of the preference relation entail?

As the last exercise suggests, the completeness of the preference relation implies that it is **reflexive**, meaning that $x \succsim x$ (for all x). This result might strike you as surprising. But recall that completeness says that, any time you pick two elements from the universe, the relation must hold one way or the other. The axiom does not say that the two elements must be different. If you pick the same element twice, which you may, completeness requires that the thing stands in the relation to itself.

The choice of a universe might determine whether a relation is transitive or intransitive, complete or incomplete. If the universe were U = {Romeo, Juliet}, the relation "is in love with" would be complete, since for any two members of the universe, either the one is in love with the other, or the other is in love with the one. (This assumes that Romeo and Juliet are both in love with themselves, which might perhaps not be true.) Perhaps more surprisingly, the relation would also be transitive: whenever $x \succsim y$ and $y \succsim z$, it is in fact the case that $x \succsim z$.

The assumption that the weak preference relation is rational (transitive and complete) might seem fairly modest. Yet, in combination with a couple of definitions, this assumption is in effect everything necessary to build a theory of choice under certainty. This is a wonderful illustration of how science works: based on a small number of assumptions, we will build an extensive theory, whose predictions will then be confronted with actual evidence. The rest of this chapter spells out the implications of the assumption that the weak preference relation is rational.

2.4 Indifference and strict preference

As the previous section shows, the (weak) preference relation admits ties. When two options are tied, we say that the first option is **as good as** the second or that the agent is **indifferent** between the two options. That is, a person is indifferent between two options just in case, to her, the first option is at least as good as the second and the second is at least as good as the first. We use the symbol ~ to denote indifference. Formally speaking:

Definition 2.11 Definition of indifference *$x \sim y$ if and only if $x \succcurlyeq y$ and $y \succcurlyeq x$.*

Using logical symbols, we might write: $x \sim y \Leftrightarrow x \succcurlyeq y \;\&\; y \succcurlyeq x$.

Assuming that the "at least as good as" relation is rational, the indifference relation is both reflexive and transitive. It is also **symmetric**: if x is as good as y, then y is as good as x. These results are not just intuitively plausible; they can be established by means of **proofs**. (See the text box on page 20 for more about proofs.) Properties of the indifference relation are established by the following proposition.

Proposition 2.12 Properties of indifference *The following conditions hold:*

(i) $x \sim x$ *(for all x)*

(ii) $x \sim y \rightarrow y \sim x$ *(for all x, y)*

(iii) $x \sim y \;\&\; y \sim z \rightarrow x \sim z$ *(for all x, y, z)*

Proof.
Each part of the proposition requires a separate proof:

(i)	1. $x \succcurlyeq x$	by Axiom 2.6
	2. $x \succcurlyeq x \;\&\; x \succcurlyeq x$	from (1), by logic
	$\therefore$ $x \sim x$	from (2), by Definition 2.11 □

(ii)
1. $x \sim y$ — by assumption
2. $x \succcurlyeq y \,\&\, y \succcurlyeq x$ — from (1), by Definition 2.11
3. $y \succcurlyeq x \,\&\, x \succcurlyeq y$ — from (2), by logic
4. $y \sim x$ — from (3), by Definition 2.11

$\therefore\ x \sim y \rightarrow y \sim x$ — from (1)–(4), by logic □

(iii)
1. $x \sim y \,\&\, y \sim z$ — by assumption
2. $x \succcurlyeq y \,\&\, y \succcurlyeq x$ — from (1), by Definition 2.11
3. $y \succcurlyeq z \,\&\, z \succcurlyeq y$ — from (1), by Definition 2.11
4. $x \succcurlyeq z$ — from (2) and (3), by Axiom 2.5
5. $z \succcurlyeq x$ — from (2) and (3), by Axiom 2.5
6. $x \sim z$ — from (4) and (5), by Definition 2.11

$\therefore\ x \sim y \,\&\, y \sim z \rightarrow x \sim z$ — from (1)–(6), by logic □

These are the complete proofs. In what follows, I will often outline the general shape of the proof rather than presenting the whole thing.

The indifference relation is not complete. To show this, it is enough to give a single counterexample. Any rational preference relation according to which the agent is not indifferent between all options will do (see, for instance, Figure 2.3).

Exercise 2.13 Prove the following principle: $x \succcurlyeq y \,\&\, y \sim z \rightarrow x \succcurlyeq z$.

In your various proofs, it is always acceptable to rely on propositions you have already established. The following exercise shows how useful this can be.

Exercise 2.14 Iterated transitivity In this exercise you will prove the following principle in two different ways: $x \sim y \,\&\, y \sim z \,\&\, z \sim p \rightarrow x \sim p$.

(**a**) First prove it by applying the transitivity of indifference (Proposition 2.12(iii)).

(**b**) Then prove it without assuming the transitivity of indifference. (You may still use the transitivity of weak preference, since it is an axiom.)

If you have difficulty completing the proofs, refer to the text box on page 20 for hints.

Heavenly Bliss

$\curlyvee$

Coke $\sim$ Pepsi

$\curlyvee$

Eternal Suffering

Figure 2.3 Preference ordering with tie

When a first option is at least as good as a second, but the second is not at least as good as the first, we say that the first option is **better than** the second or that the agent **strictly** or **strongly** prefers the first over the second. We use the symbol $\succ$ to denote **strict** or **strong preference**. Formally speaking:

Definition 2.15 Definition of strict preference *$x \succ y$ if and only if $x \succsim y$ and it is not the case that $y \succsim x$.*

Using logical notation, that is to say: $x \succ y \Leftrightarrow x \succsim y \;\&\; \neg y \succsim x$. For clarity, sometimes the "is at least as good as" relation will be called **weak preference**.

Assuming (still) that the weak preference relation is rational, it is possible to prove logically that the strict preference relation will have certain properties. The following proposition establishes some of them.

Proposition 2.16 Properties of strict preference *The following conditions hold:*

(i) $x \succ y \;\&\; y \succ z \rightarrow x \succ z$ *(for all x, y, z)*
(ii) $x \succ y \rightarrow$ *not* $y \succ x$ *(for all x, y)*
(iii) *not* $x \succ x$ *(for all x)*

Proof.

(i) Suppose that $x \succ y \;\&\; y \succ z$. In order to establish that $x \succ z$, Definition 2.15 tells us that we need to show that $x \succsim z$ and that it is not the case that $z \succsim x$. The first part is Exercise 2.17. The second part goes as follows: suppose for a **proof by contradiction** that $z \succsim x$. From the first assumption and the definition of strict preference, it follows that $x \succsim y$. From the second assumption and Axiom 2.5, it follows that $z \succsim y$. But from the first assumption and the definition of strict preference, it also follows that $\neg z \succsim y$. We have derived a contradiction, so the second assumption must be false, and therefore $\neg z \succsim x$.
(ii) Begin by assuming $x \succ y$. Then, for a proof by contradiction, assume that $y \succ x$. Given the first assumption, Definition 2.15 implies that $x \succsim y$. Given the second assumption, the same definition implies that $\neg x \succsim y$. But this is a contradiction, so the second assumption must be false, and therefore $\neg y \succ x$
(iii) See Exercise 2.19. □

Proposition 2.16(i) says that the strict preference relation is transitive, 2.16(ii) that it is **anti-symmetric**, and 2.16(iii) that it is **irreflexive**.

Exercise 2.17 Using the definitions and propositions discussed so far, complete the first part of the proof of Proposition 2.16(i).

Notice that the proofs of Proposition 2.16(i) and (ii) involve constructing proofs by contradiction. Such proofs are also called **indirect proofs**. This mode of reasoning might look weird, but it is actually quite common in mathematics, science, and everyday thinking. For example, when mathematicians prove that $\sqrt{2}$ is an irrational number, they can proceed by assuming (for a proof by contradiction) that $\sqrt{2}$ is a rational number (meaning that $\sqrt{2}$ can be expressed as a fraction p/q of natural numbers p and q) and then use this assumption to derive a contradiction.

Exercise 2.18 The enemy of your enemy, cont. Use a proof by contradiction to establish that "is the enemy of" is not transitive, as in Exercise 2.8 on page 15.

In future exercises, you will see just how useful proofs by contradiction can be.

Exercise 2.19 Prove Proposition 2.16(iii). Prove it by contradiction, by first assuming that there is an x such that $x \succ x$.

Exercise 2.20 Prove the following principle: $x \succ y \;\&\; y \succcurlyeq z \rightarrow x \succ z$ (for all x, y, z). Notice that this proof has two parts. First, prove that $x \succcurlyeq z$; second, prove that $\neg z \succcurlyeq x$.

Exercise 2.21 Establish the following important and intuitive principles. (For the record, some of them are logically equivalent.)

(a) If $x \succ y$ then $x \succcurlyeq y$
(b) If $x \succ y$ then $\neg y \succcurlyeq x$
(c) If $x \succcurlyeq y$ then $\neg y \succ x$
(d) If $x \succ y$ then $\neg x \sim y$
(e) If $x \sim y$ then $\neg x \succ y$
(f) If $\neg x \succcurlyeq y$ then $y \succcurlyeq x$
(g) If $\neg x \succcurlyeq y$ then $y \succ x$
(h) If $\neg x \succ y$ then $y \succcurlyeq x$

If you run into trouble with parts (f) and (g), note that you can always play the completeness card and throw in the expression $x \succcurlyeq y \vee y \succcurlyeq x$ any time. Also note that $p \vee q$ and $\neg p$ implies that q. If you find part (h) difficult, feel free to invoke the principle known as **de Morgan's law**, according to which $\neg (p \;\&\; q)$ is logically equivalent to $\neg p \vee \neg q$. Also note that $p \vee q$ and $p \rightarrow q$ implies that q.

For the next exercise, recall that it is acceptable to rely on propositions already established.

Exercise 2.22 Prove that if $x \sim y$ and $y \sim z$, then $\neg x \succ z$.

Exercise 2.23 Negative transitivity Prove the following two principles. You might already have been tempted to invoke these two in your proofs. But remember that you may not do so before you have established them.

(a) If $\neg x \succcurlyeq y$ and $\neg y \succcurlyeq z$, then $\neg x \succcurlyeq z$
(b) If $\neg x \succ y$ and $\neg y \succ z$, then $\neg x \succ z$

The last two exercises illustrate some potentially problematic implications of the theory that we have studied in this chapter. Both are classics.

Exercise 2.24 Vacations Suppose that you are offered two vacation packages, one to California and one to Florida, and that you are perfectly indifferent between the two. Let us call the Florida package f and the California package c. So f ~ c. Now, somebody improves the Florida package by adding an apple to it. You like apples, so the enhanced Florida package f^+ improves the original Florida package, meaning that $f^+ \succ f$. Assuming that you are rational, how do you feel about the enhanced Florida package f^+ compared with the California package c? Prove it.

How to do proofs

The aim of a **proof** of a proposition is to establish the truth of the proposition with logical or mathematical certainty (see the proofs of Proposition 2.12(i)–(iii) for examples). A proof is a sequence of propositions, presented on separate lines of the page. The last line of the proof is the proposition you intend to establish, that is, the conclusion; the lines that come before it establish its truth. The conclusion is typically preceded by the symbol ∴. All other lines are numbered using Arabic numerals. The basic rule is that each proposition in the proof must follow logically from (a) a proposition on a line above it, (b) an axiom of the theory, (c) a definition that has been properly introduced, and/or (d) a proposition that has already been established by means of another proof. Once a proof is concluded, logicians like to write "QED" – Latin for "quod erat demonstrandum," meaning "that which was to be shown" – or add a little box. □

There are some useful hints, or rules of thumb, that you may want to follow when constructing proofs. **Hint one**: if you want to establish a proposition of the form $x \rightarrow y$, you typically want to begin by assuming what is to the left of the arrow; that is, the first line will read "1. x by assumption." Then, your goal is to derive y, which would permit you to complete the proof. If you want to establish a proposition of the form $x \leftrightarrow y$, you need to do it both ways: first, prove that $x \rightarrow y$, and second, that $y \rightarrow x$. **Hint two**: if you want to establish a proposition of the form $\neg p$, you typically want to begin by assuming the opposite of what you want to prove for a proof by contradiction; that is, the first line would read "1. p by assumption for a proof by contradiction." Then, your goal is to derive a contradiction, that is, a claim of the form q & $\neg q$, which would permit you to complete the proof.

Exercise 2.25 Teacups Imagine that there are 1000 cups of tea lined up in front of you. The cups are identical except for one difference: the cup to the far left (c_1) contains one grain of sugar, the second from the left (c_2) contains two grains of sugar, the third from the left (c_3) contains three grains of sugar, and so on. Since you cannot tell the difference between any two adjacent cups, you are indifferent between c_n and c_{n+1} for all n between 1 and 999 inclusive. Assuming that your preference relation is rational, what is your preference between the cup to the far left (c_1) and the one to the far right (c_{1000})?

Your findings from Exercise 2.21 are likely to come in handy when answering these questions.

2.5 Preference orderings

The preference relation is often referred to as a **preference ordering**. This is so because a rational preference relation allows us to order all alternatives in a list, with the best at the top and the worst at the bottom. Figure 2.3 shows an example of a preference ordering.

A rational preference ordering is simple. Completeness ensures that each person will have exactly one list, because completeness entails that each element can be compared with all other elements. Transitivity ensures that the list will be linear, because transitivity entails that the strict preference relation will never have cycles, as when $x \succ y$, $y \succ z$, and $z \succ x$. Here are two helpful exercises about cycling preferences.

Exercise 2.26 Cycling preferences Using the definitions and propositions discussed so far, show that it is impossible for a rational strict preference relation to cycle. To do so, suppose (for the sake of the argument) that $x \succ y \ \& \ y \succ z \ \& \ z \succ x$ and show that this leads to a contradiction.

Exercise 2.27 Cycling preferences, cont. By contrast, it is possible for the weak preference relation to cycle. This is to say that there may well be an x, y, and z such that $x \succcurlyeq y \ \& \ y \succcurlyeq z \ \& \ z \succcurlyeq x$. If this is so, what do we know about the agent's preferences over x, y, and z? Prove it.

In cases of indifference, the preference ordering will have ties. As you may have noticed, Figure 2.3 describes a preference ordering in which two items are equally good. Assuming that the universe is {Heavenly Bliss, Coke, Pepsi, Eternal Suffering}, this preference ordering is perfectly rational.

In economics, preference orderings are frequently represented using **indifference curves**, also called **indifference maps**. See Figure 2.4 for an example of a set of indifference curves. You can think of these as analogous to contour lines on a topographic map. By convention, each bundle on one of these curves is as good as every other bundle on the same curve. When two bundles are on different curves, one of the two bundles is strictly preferred to the other. Insofar as people prefer more of each good to less, bundles on curves to the top right will be strictly preferred to bundles on curves to the bottom left.

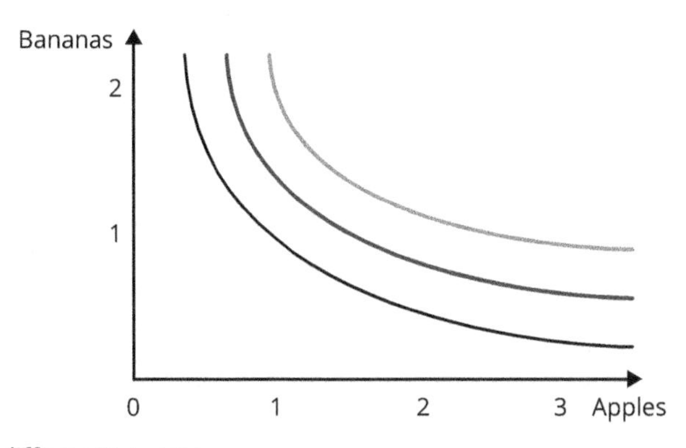

Figure 2.4 Indifference curves

Exercise 2.28 Indifference curves Represent the following sets of indifference curves graphically:
(**a**) Suppose that an apple for you is always as good as two bananas.
(**b**) Suppose that one apple is always as good, as far as you are concerned, as a banana.
(**c**) Suppose that you do not care for tea without milk or for milk without tea. However, every time you have two units of tea and one unit of milk, you can make yourself a cup of tea with milk. You love tea with milk, and the more the better, as far as you are concerned.

2.6 Choice under certainty

To make a **choice under certainty** is to face a menu. A **menu** is a set of options such that you have to choose exactly one option from the set. This is to say that the menu has two properties. First, the items in the menu are **mutually exclusive**; that is, you can choose at most one of them at any given time. Second, the items in the menu are **exhaustive**; that is, you have to choose at least one of them.

Example 2.29 The menu If a restaurant offers two appetizers (soup and salad) and two entrées (chicken and beef) and you must choose one appetizer and one entrée, what is your set of alternatives?

Since there are four possible combinations, your set of alternatives is {soup-and-chicken, soup-and-beef, salad-and-chicken, salad-and-beef}.

Exercise 2.30 The menu, cont. If you can also choose to eat an appetizer only, or an entrée only, or nothing at all, what would the new menu be?

There is no assumption that a menu is small, or even finite, though we frequently assume that it is.

In economics, the menu is often referred to as the **budget set**. This is simply that part of the set of alternatives that you can afford given your budget, that is, your resources at hand. Suppose that you can afford at most three apples (if you buy no bananas) or two bananas (if you buy no apples). This would be the case, for instance, if you had \$6 in your pocket and bananas cost \$3 and apples \$2. If so, your budget set – or your menu – is represented by the shaded area in Figure 2.5. Assuming that fruit is infinitely divisible, the menu is infinitely large. The line separating the items in your budget from the items outside of it is called the **budget line**.

Exercise 2.31 Budget sets Suppose that your budget is \$12. Use a graph to answer the following questions:
(**a**) What is the budget set when apples cost \$3 and bananas cost \$4?
(**b**) What is the budget set when apples cost \$6 and bananas cost \$2?
(**c**) What is the budget set when apples always cost \$2, the first banana costs \$4, and every subsequent banana costs \$2?

So what does it mean for a person **to be rational**? To be rational, or **to make rational choices**, means (i) that you have a rational preference ordering, and (ii) that whenever you are faced with a menu, you choose the most preferred item, or (in the case of ties) one of the most preferred items. The second condition can also be expressed as follows: (ii') that … you choose an item such that no other item in the menu is strictly preferred to it. Or like this: (ii") that … you do not choose an item that is strictly less preferred to another item in the menu. *This is all we mean when we say that somebody is rational in the context of choice under certainty.* If you have the preferences of Figure 2.3 and are facing a menu offering Coke, Pepsi, and Eternal Suffering, the rational choice is to pick either the Coke or the Pepsi option. When there is no unique best choice, as in this case, the theory says that you have to choose one of the best options; it does not specify which one.

The rational decision can be determined if we know the agent's indifference curves and budget set. If you superimpose the former (from Figure 2.4) onto the latter (from Figure 2.5), you get a picture like Figure 2.6. The consumer will choose the bundle marked X, because it is the most highly preferred bundle in the budget set. As you can tell, there is no more highly preferred bundle in the budget set.

It is important to note what the theory of rationality does *not* say. The theory does not say why people prefer certain things to others, or why they choose so as to satisfy their preferences. It does not say that people prefer apples to bananas because they

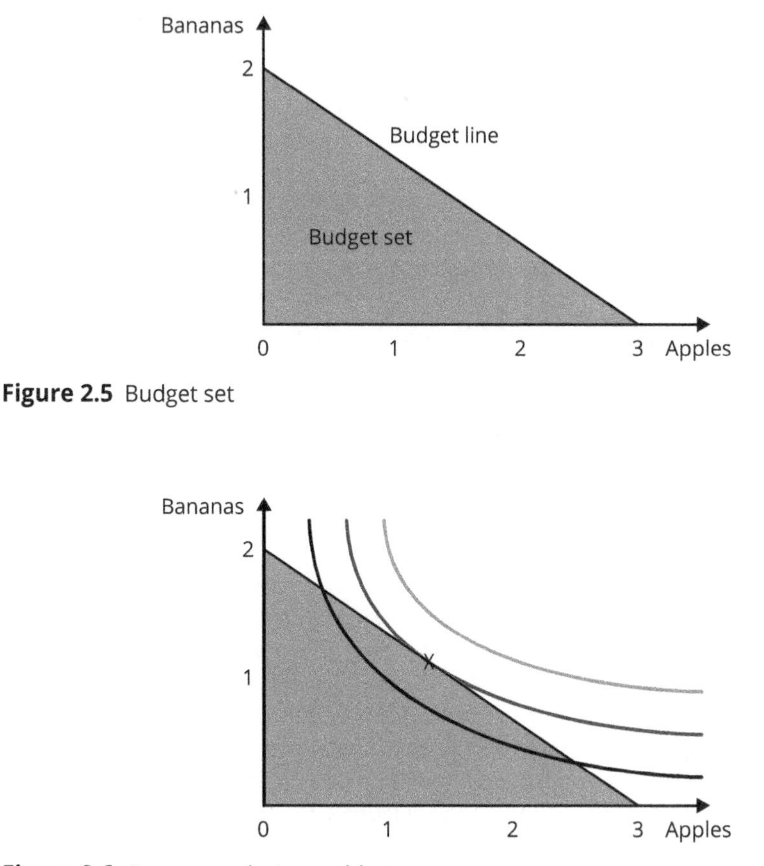

Figure 2.5 Budget set

Figure 2.6 Consumer choice problem

think that they will be happier, feel better, or be more satisfied if they get apples than if they get bananas (although that may, in fact, be the case). This theory says nothing about feelings, emotions, moods, or any other subjectively experienced state. As far as this theory is concerned, the fact that you prefer a cool drink on a hot day to being roasted over an open fire is just a brute fact; it is not a fact that needs to be grounded in an account of what feels good or bad, pleasant or unpleasant, rewarding or aversive. Similarly, the theory does not say why people choose the most preferred item on the menu; as far as this theory is concerned, they just do.

Moreover, the theory does not say that people are selfish, in the sense that they care only about themselves; or that they are materialistic, in the sense that they care only about material goods; or that they are greedy, in the sense that they care only about money. The definition of rationality implies that a rational person is **self-interested**, in the sense that her choices reflect her own preference ordering rather than somebody else's. But this is not the same as being selfish: the rational individual may, for example, prefer dying for a just cause over getting rich by defrauding others. The theory in itself specifies only some formal properties of the preference relation; it does not say anything about the things people prefer. The theory is silent about whether or not they pursue respectable and moral ends. Rational people may be weird, evil, sadistic, selfish, and morally repugnant, or saintly, inspiring, thoughtful, selfless, and morally admirable; they can act out of compulsion, habit, feeling, or as a result of machine-like computation. This conception of rationality has a long and distinguished history. The Scottish eighteenth-century philosopher and economist David Hume wrote:

> 'Tis not contrary to reason to prefer the destruction of the whole world to the scratching of my finger. 'Tis not contrary to reason for me to choose my total ruin, to prevent the least uneasiness of an Indian or person wholly unknown to me. 'Tis as little contrary to reason to prefer even my own acknowledge'd lesser good to my greater, and have a more ardent affection for the former than the latter.

Rational people cannot have preferences that are intransitive or incomplete, and they cannot make choices that fail to reflect those preferences.

2.7 Utility

The notion of **utility**, which is central to modern economics, has generated a great deal of confusion. It is worth going slowly here. Suppose that you want to use numbers to express how much a person prefers something, then how would you do it? One solution is obvious. Remember that a rational person's preferences allow us to arrange all alternatives in order of preference. Consider, for example, the preference ordering in Figure 2.3. The preference ordering has three "steps." In order to represent these preferences by numbers, we assign one number to each step, in such a way that higher steps are associated with higher numbers. See Figure 2.7 for an example.

A **utility function** associates a number with each member of the set of alternatives. In this case, we have associated the number 3 with Heavenly Bliss (HB). That number is called the utility of HB and is denoted $u(\text{HB})$. In this case, $u(\text{HB}) = 3$. The number

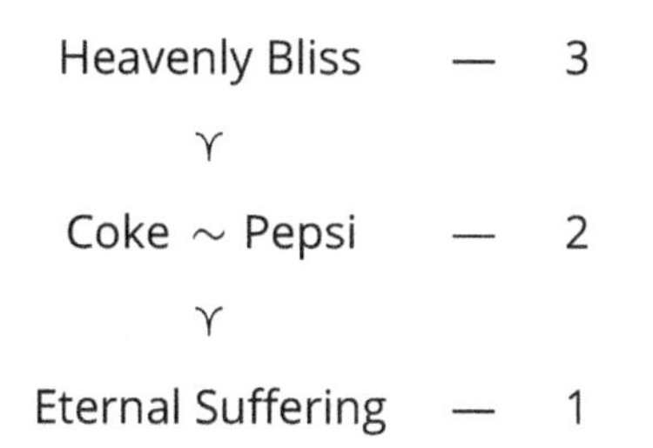

Figure 2.7 Preference ordering with utility function

associated with Eternal Suffering (ES) is called the utility of ES and is written u(ES). In this case, $u(\text{ES}) = 1$. If we use C to denote Coke and P to denote Pepsi, then $u(\text{C}) = u(\text{P}) = 2$. Because we designed the utility function so that higher utilities correspond to more preferred items, we say that the utility function $u(\cdot)$ **represents** the preference relation $\succsim$.

As the example suggests, two conditions must hold in order for something to be a utility function. First, it must be a function (or a mapping) from the set of alternatives into the set of real numbers. This means that every alternative gets assigned exactly one number. If Figure 2.7 had empty spaces in the right-hand column, or if the figure had several numbers in the same cell, we would not have a proper utility function. While the utility function needs to assign some number to every alternative, it is acceptable (as the example shows) to assign the same number to several alternatives. Second, for something to be a utility function, it must assign larger numbers to more preferred alternatives; that is, if x is at least as good as y, the number assigned to x must be greater than or equal to the number assigned to y. To put it more formally:

Definition 2.32 Definition of $u(\cdot)$ *A function $u(\cdot)$ from the set of alternatives into the set of real numbers is a utility function representing the preference relation $\succsim$ just in case $x \succsim y \Leftrightarrow u(x) \geq u(y)$ (for all x and y).*

A function $u(\cdot)$ that satisfies this condition can be said to be an **index** or a **measure** of the preference relation $\succsim$. Historically, the word "utility" has been used to refer to many different things, including the pleasure, happiness, and satisfaction of receiving, owning, or consuming something. Though most people (including economics professors) find it hard to stop speaking in this way, as though utility is somehow floating around "in your head," this usage is archaic. Utility is nothing but an index or measure of preference.

Given a rational preference relation, you may ask whether it is always possible to find a utility function that represents it. When the set of alternatives is finite, the answer is yes. The question is answered by means of a so-called **representation theorem**.

Proposition 2.33 Representation theorem *If the set of alternatives is finite then $\succsim$ is a rational preference relation just in case there exists a utility function representing $\succsim$.*

Proof.
Omitted. □

When the set of alternatives is infinite, representing preference relations gets more complicated. It remains true that if a utility function represents a preference relation, then the preference relation is rational. However, even if the preference relation is rational, it is not always possible to find a utility function that represents it.

As you may suspect, a utility function will associate strictly higher numbers with strictly preferred alternatives, and equal numbers with equally preferred alternatives. That is, the following proposition is true:

Proposition 2.34 Properties of $u(\cdot)$ *Given a utility function $u(\cdot)$ representing the preference relation $\succcurlyeq$, the following conditions hold:*

(i) $x \succ y \Leftrightarrow u(x) > u(y)$
(ii) $x \sim y \Leftrightarrow u(x) = u(y)$

Proof.

(i) First, assume that $x \succ y$, so that $x \geq y$ and $\neg y \geq x$. Using Definition 2.32 twice, we can infer that $u(x) \geq u(y)$ and that not $u(y) \geq u(x)$. Simple math tells us that $u(x) \succ u(y)$. Second, assume that $u(x) \succ u(y)$, which implies that $u(x) \geq u(y)$ and that not $u(y) \geq u(x)$. Using Definition 2.32 twice, we can infer that $x \succcurlyeq y$ and $\neg y \succcurlyeq x$, which in turn implies that $x \succ y$.

(ii) See Exercise 2.35. □

Recall (from the text box on page 20) that if you want to prove something of the form A ⇔ B, your proof must have two parts.

Exercise 2.35 Prove Proposition 2.34(ii).

It is easy to confirm that the proposition is true of the utility function from Figure 2.7.

One important point to note is that utility functions are not unique. The sequence of numbers ⟨1, 2, 3⟩ in Figure 2.7 could have been chosen very differently. The sequence ⟨0, 1, 323⟩ would have done as well, as would ⟨–1000, –2, 0⟩ and ⟨$-\pi$, e, 1077⟩. All these are utility functions, in that they associate higher numbers with more preferred options. As these examples show, it is important not to ascribe any significance to absolute numbers. To know that the utility I derive from listening to Justin Bieber is 2 tells you *absolutely nothing* about my preferences. But if you know that the utility I derive from listening to Rihanna is 4, you know something, namely, that I strictly prefer Rihanna to Justin Bieber. It is equally important not to ascribe any significance to ratios of utilities. Even if the utility of Rihanna is twice the utility of Justin Bieber, this does not mean that I like Rihanna "twice as much." The same preferences could be represented by the numbers 0 and 42, in which case the ratio would not even be well defined. In brief, for every given preference relation, there are many utility

A final word about proofs

While the proofs discussed in this chapter may at first blush seem intimidating, notice that the basic principles are fairly simple. So far, we have introduced only two axioms, namely, the transitivity of the weak preference relation (Axiom 2.5 on page 14) and the completeness of the weak preference relation (Axiom 2.6 on page 14); three definitions, namely, the definition of indifference (Definition 2.11 on page 16), the definition of strict preference (Definition 2.15 on page 18), and the definition of utility (Definition 2.32 on page 25); and two hints (see text box on page 20). In order to complete a proof, there are only seven things that you need to know.

functions representing it. Utility as used in this chapter is often called **ordinal utility**, because all it does is allow you to order things.

How do utilities relate to indifference curves? A utility function in effect assigns one number to each indifference curve, as in Figure 2.8. This way, two bundles that fall on the same curve will be associated with the same utility, as they should be. Two bundles that fall on different curves will be associated with different utilities, again as they should be. Of course, higher numbers will correspond to curves that are more strongly preferred. For a person who likes apples and bananas, $u_1 < u_2 < u_3$.

How does utility relate to behavior? Remember that you choose rationally insofar as you choose the most preferred item (or one of the most preferred items) on the menu. The most preferred item on the menu will also be the item with the highest utility. So to choose the most preferred item is to choose the item with the highest utility. Now, **to maximize utility** is to choose the item with the highest utility. Thus, you choose rationally insofar as you maximize utility. Hence, *to maximize utility is to choose rationally*. Notice that you can maximize utility in this sense without necessarily going through any particular calculations; that is, you do not need to be able to solve mathematical maximization problems in order to maximize utility. Similarly, you can maximize utility without maximizing feelings of pleasure, satisfaction, contentment, happiness, or whatever; utility (like preference) still has nothing to do with subjectively experienced states of any kind. This is a source of endless confusion.

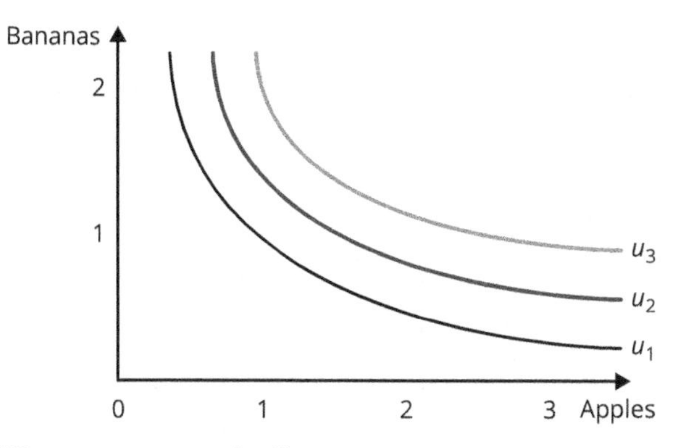

Figure 2.8 Indifference curves and utility

Consumer-choice theory

The theory of rational choice under certainty is for practical purposes often supplemented with various other, **auxiliary**, assumptions. The following three assumptions tend to appear under the heading of **consumer-choice theory** in microeconomics textbooks, and are worth knowing because they are so common. It is important to notice, however, that these auxiliary assumptions are not strictly speaking part of the theory of rational choice; they are useful add-ons.

Non-satiation Non-satiation says that no matter what you have, there is always a nearby bundle that you would rather have. Consider any bundle x in the set of alternatives (as in Figure 2.1) and draw a circle around it. Non-satiation says that no matter how small the circle around x, there is always another bundle within the circle that is strictly preferred to x. Diners who exhibit non-satiation, for example, will never be perfectly satisfied with their meal, no matter how good: they could always use a little more salad, or a little more champagne. Non-satiation is a convenient assumption, because it guarantees that the solution to the consumer choice problem (Figure 2.6) lies on the budget line. How do you know? Suppose, for a proof by contradiction, that the most highly preferred bundle in your budget set, call it x, is not in fact on the budget line. By non-satiation every circle drawn around x contains some bundle that you strictly prefer to x. If that circle is small enough, the preferred bundle will also be inside the budget set – but then x cannot in fact be the most highly preferred bundle in your budget set. QED.

Convexity Convexity in preferences captures a preference for variety or combination. Start with any two points on one indifference curve and draw a straight line between them. Convexity requires that points on the chord (excluding the end points) are preferred to the end points. Whenever this condition is satisfied, indifference curves will bulge toward the origin (the point where the axes intersect) in the manner of Figures 2.4 and 2.8. A person with convex preferences will always prefer one unit of gin & tonic to either one unit of gin or one unit of tonic. Convexity excludes snake-shaped indifference curves and guarantees that the consumer choice problem in Figure 2.6 will have a unique solution.

Continuity Continuity in preferences says that a person has similar preferences for similar bundles. Suppose that x is weakly preferred to y, and that there is another bundle x'' which is similar to x. Continuity requires that when x'' becomes ever more similar to x, in the limit, x'' must also be weakly preferred to y. This assumption guarantees that there are no "jumps," where a person has radically different preferences over very similar bundles. Continuity excludes so-called **lexicographic** preferences. You have lexicographic preferences whenever you always prefer a bundle with the largest amount of a and only in cases of ties prefer the one with more b. Lexicographic preferences involve jumps, since they would make you strictly prefer $\langle a+\varepsilon, b\rangle$ to $\langle a, b\rangle$, no matter how small $\varepsilon > 0$ is. The continuity assumption is especially useful in the context of utility theory, where it guarantees that representation theorems (analogous to Proposition 2.33) go through even when the set of alternatives is infinitely large.

2.8 Discussion

The first thing to notice is how much mileage you can get out of a small number of relatively weak assumptions. Recall that we have made only two fundamental assumptions: that preferences are rational and that people choose so as to satisfy their preferences. As long as these two assumptions are true, and the set of alternatives is not too large, we can define the concept of utility and make sense of the idea of utility maximization. That is the whole theory. The second thing to notice is what the theory does not say. The theory does not say that people are selfish, materialistic, or greedy; it says nothing about why people prefer one thing over another; it does not presuppose that people solve mathematical maximization problems in their heads; and it makes no reference to things like pleasure, satisfaction, and happiness. The fact that the theory is relatively noncommittal helps explain why so many economists are comfortable using it: after all, the theory is compatible with a great deal of behavior.

Though brief, this discussion sheds light on the nature of economics as some economists see it. Nobel laureate Gary Becker defines the economic approach to behavior in terms of three features: "The combined assumptions of maximizing behavior, market equilibrium, and stable preferences, used relentlessly and unflinchingly, form the heart of the economic approach as I see it." Because this part focuses on individual choice, I have little to say about market equilibrium here. However, what Becker has in mind when he talks about maximizing behavior and stable preferences should be eminently clear from what has already been said. In this analysis, preferences are **stable** in the sense that they are not permitted to change over time.

Exercise 2.36 Misguided criticism Many criticisms of standard economics are quite mistaken. Explain where the following critics go wrong.

(**a**) An otherwise illuminating article about behavioral economics in *Harvard Magazine* asserts that "the standard model of the human actor – Economic Man – that classical and neoclassical economics have used as a foundation for decades, if not centuries … is an intelligent, analytic, selfish creature."

(**b**) A common line of criticism of standard economics begins with some claim of the form "the most fundamental idea in economics is that money makes people happy" and proceeds to argue that the idea is false.

Is this a plausible theory of human behavior under conditions of certainty? To answer this question we need to separate the descriptive from the normative question. The first question is whether the theory is descriptively adequate, that is, whether people's choices *do as a matter of fact* reflect a rational preference ordering. This is the same as asking whether people maximize utility. Though both transitivity and completeness may seem obviously true of people's preferences, there are many cases in which they do not seem to hold: a person's preference relation over prospective spouses, for example, is unlikely to be complete. The second question is whether the theory is normatively correct, that is, whether people's choices *should* reflect a rational preference ordering. This is the same as asking whether people *should* maximize utility. Though transitivity and completeness may seem rationally required, it can be argued that they are neither necessary nor sufficient for being rational.

Next, we explore what happens when the theory is confronted with data.

Additional exercises

Exercise 2.37 For each of the relations and properties in Table 2.1, use a check mark to identify whether or not the relation has the property.

Table 2.1 Properties of weak preference, indifference, and strong preference

	Property	Definition	$\succcurlyeq$	$\sim$	$\succ$
(a)	Transitivity	xRy & $yRz \rightarrow xRz$ (for all x, y, z)			
(b)	Completeness	$xRy \vee yRx$ (for all x, y)			
(c)	Reflexivity	xRx (for all x)			
(d)	Irreflexivity	$\neg xRx$ (for all x)			
(e)	Symmetry	$xRy \rightarrow yRx$ (for all x, y)			
(f)	Antisymmetry	$xRy \rightarrow \neg yRx$ (for all x, y)			

Exercise 2.38 More properties of the preference relation Here are two relations: "is married to" and "is not married to." Supposing the universe is the set of all living human beings, which of these is...

(a) reflexive
(b) irreflexive
(c) symmetric
(d) asymmetric
(e) antisymmetric

Exercise 2.39 As part of your answer to the following questions, make sure to specify what the universe is.

(a) Give an example of a relation that is complete but not transitive.
(b) Give an example of a relation that is transitive but not complete.

Exercise 2.40 Irrationality Explain (in words) why each of the characters below is irrational according to the theory you have learned in this chapter.

(a) In the drama *Sophie's Choice*, the title character finds herself in a Nazi concentration camp and must choose which one of her children is to be put to death. She is not indifferent and cannot form a weak preference either way.
(b) An economics professor finds that he prefers a \$10 bottle of wine to an \$8 bottle, a \$12 bottle to a \$10 bottle, and so on; yet he does not prefer a \$200 bottle to an \$8 bottle.
(c) Buridan's ass is as hungry as it is thirsty and finds itself exactly midway between a stack of hay and a pail of water. Unable to decide which is better, the animal expires.

Exercise 2.41 Consumer-choice theory Do the following quotations agree or disagree with the assumptions of consumer-choice theory (see box on page 28)? **(a)** "Variety is the spice of life" – William Cowper. **(b)** "(I Can't Get No) Satisfaction" – The Rolling Stones. **(c)** "People will always choose more money over more sex" – Douglas Coupland. **(d)** "Greed, for lack of a better word, is good" – Gordon Gecko.

Exercise 2.42 Consumer-choice theory and indifference curves For each of the sets of indifference curves in Figure 2.9, what assumption rules it out?

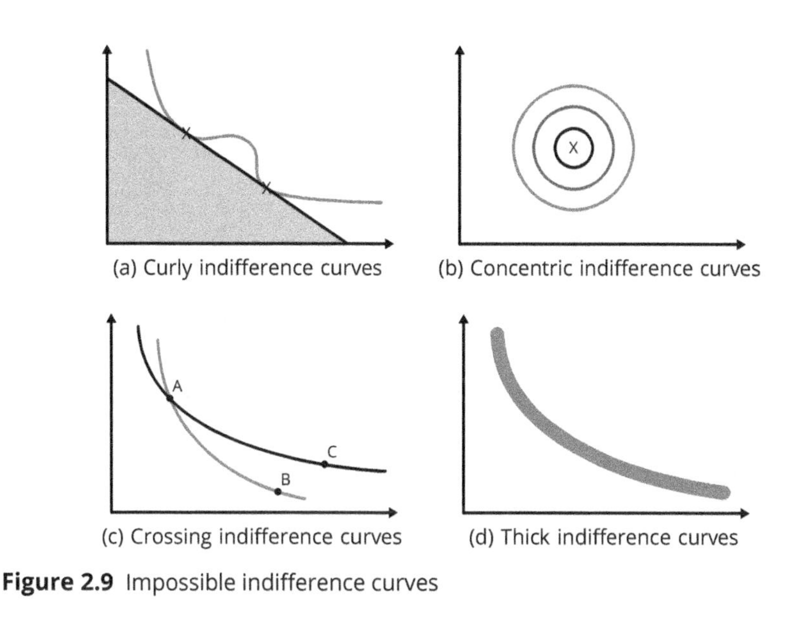

Figure 2.9 Impossible indifference curves

Further reading

A nontechnical introduction to decision theory is Allingham (2002). More technical accounts can be found in Mas-Colell et al. (1995, Chapters 1–2). The paragraph from David Hume comes from Hume (2000 [1739–40], p. 267). The Becker quotation is from Becker (1976, p. 5). The *Harvard Magazine* article is Lambert (2006), and the critics talking about happiness Dutt and Radcliff (2009, p. 8). The quotations in Exercise 2.42(a) and (c) are from Cowper (1785) and Coupland (2008).

3 DECISION-MAKING UNDER CERTAINTY

Learning objectives

After studying this chapter you will:

- Be able to identify common behavior patterns that violate the theory of rational choice under certainty
- Know some building blocks of the (descriptive) behavioral theories, including the biases-and-heuristics program and prospect theory
- Apply the (normative) theory of rationality in the real world – but also appreciate how demanding the theory is

3.1 Introduction

The previous chapter showed how an extensive theory of choice under certainty can be built upon the foundation of a modest number of assumptions. Though the assumptions may seem weak, their implications can be challenged on both descriptive and normative grounds. In this chapter, we confront the theory with data. We explore some of the phenomena that behavioral economists argue are inconsistent with the theory of choice under certainty, as we know it. We focus on a couple of different phenomena, beginning with the failure to consider opportunity costs. Moreover, we will begin discussing what behavioral economists do when they discover phenomena that appear inconsistent with standard theory. In particular, we will discuss some of the building blocks of prominent behavioral alternatives, including prospect theory and the heuristics-and-biases program.

3.2 Opportunity costs

Imagine that you invest a small amount of money in real estate during a period when it strikes you as a safe and profitable investment. After you make the investment, the markets become unstable and you watch nervously as prices rise and fall. Finally, you sell your assets and realize that you have made a profit. "Wow," you say to yourself, "that turned out to be a great investment!" But when you boast to your friends, somebody points out that you could have earned even more money by investing in the stock market. At some level, you knew this. But you still feel that investing in real

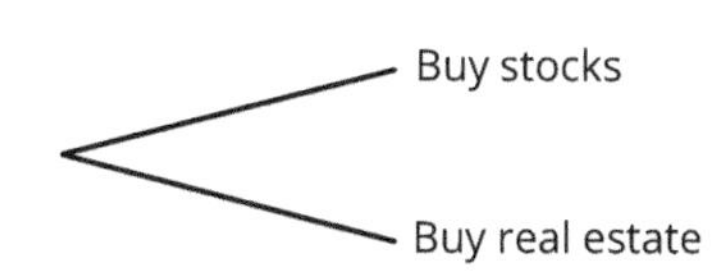

Figure 3.1 Simple decision tree

estate was a good choice: at least you did not lose any money. This is a case where you may have been acting irrationally because you failed to consider **opportunity costs.**

In order to analyze this kind of situation, let us stand back for a moment. An agent's decision problem can be represented using a **decision tree**: a graphical device showing what actions are available to the agent. Given that you only have two available actions – buying stocks and buying real estate – your decision problem can be represented as a decision tree (see Figure 3.1). Because this chapter is about choice under certainty, I will pretend that there is no uncertainty about the consequences that follow from each of these choices. (We will abandon this pretense in our discussion of choice under risk and uncertainty in Part 3.)

Suppose that you are tempted to buy real estate. What is the cost of doing so? There would be an out-of-pocket or **explicit cost**: the seller of the property would want some money to give it up. The real cost, however, is what you forgo when you buy the real estate. The opportunity cost – or **implicit cost** – of an alternative is the value of what you would have to forgo if you choose it. In dollar terms, suppose that stocks will gain \$1000 over the next year and that real estate will gain \$900. If so, the opportunity cost of buying real estate is \$1000 and the opportunity cost of buying stocks is \$900. If you buy real estate, then, your economic profit will be \$900 – \$1000 = –\$100. If you buy stock, your economic profit would be \$1000 – \$900 = \$100. If there are more than two options, the opportunity cost is the value of the *most valuable* alternative option. Suppose that you can choose between stocks, real estate, and bonds, and that bonds will gain \$150 over the next year. The opportunity cost of buying stocks would remain \$900, and the economic profit would still be \$100.

Exercise 3.1 Investment problem

(a) Draw a decision tree illustrating this decision problem.
(b) What is the opportunity cost of buying real estate?
(c) What is the opportunity cost of buying bonds?

Decision trees make it clear that you cannot choose one alternative without forgoing another: whenever you choose to go down one branch of the tree, there is always another branch that you choose not to go down. When you vacation in Hawaii, you cannot at the same time vacation in Colorado; when you use your life savings to buy a Ferrari, you cannot at the same time use your life savings to buy a Porsche; when you spend an hour reading sociology, you cannot spend the same hour reading anthropology; when you are in a monogamous relationship with *this* person, you cannot at the same time be in a monogamous relationship with *that* person; and so on. Consequently, there is an opportunity cost associated with every available option in every decision problem.

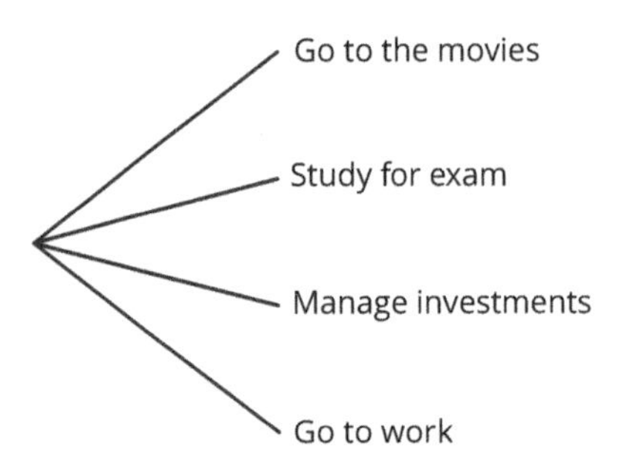

Figure 3.2 Everyday decision tree

For another example, imagine that you are considering going to the movies. On an ordinary evening, the decision that you are facing might look like Figure 3.2. Remember that the opportunity cost of going to the movies is the value of the most valuable option that you would forgo if you went to the movies; that is, the opportunity cost of going to the movies is the greatest utility you could get by going down one of the other branches of the decision tree, which is the utility of the most valuable alternative use for some \$20 and two hours of your time.

As a matter of notation, we write $a_1, a_2, \ldots, a_n$ to denote the n different acts available to you; $u(a_1), u(a_2), \ldots, u(a_n)$ to denote the utilities of those acts; and $c(a_1), c(a_2), \ldots, c(a_n)$ to denote the opportunity costs of those acts. The opportunity cost $c(a_i)$ of act a_i can then be defined as follows:

Definition 3.2 Opportunity cost

$$c(a_i) = \max\left\{u(a_1),\ u(a_2),\ \ldots,\ u(a_{i-1}),\ u(a_{i+1}),\ \ldots,\ u(a_n)\right\}$$

This is just to say that the opportunity cost of act a_i equals the maximum utility of the other acts.

Figure 3.3 represents a decision problem in which utilities and opportunity costs of four acts have been identified. The number on the left is the utility; the number in parentheses is the opportunity cost. You can compute the profit (in utility terms) by subtracting the latter from the former.

Exercise 3.3 Opportunity costs This exercise refers to Figure 3.3. Suppose that a fifth act (call it a_5) becomes available. Assume that a_5 has a utility of 9.

(**a**) What would the tree look like now?

(**b**) What would happen to the opportunity costs of the different alternatives?

There is a tight connection between opportunity costs, utilities, and the rational thing to do. As it happens, you are rational – that is, you maximize utility – just in case you take opportunity costs properly into account.

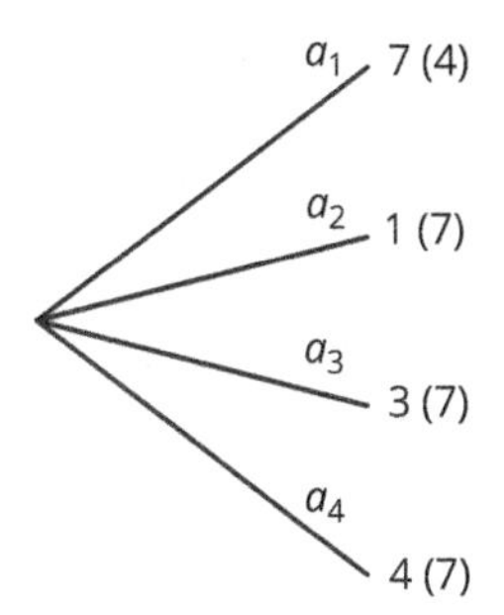

Figure 3.3 Decision tree with utilities and opportunity costs (in parentheses)

Proposition 3.4 *a_i is a rational choice* $\Leftrightarrow u(a_i) \geq c(a_i)$

Proof.
We need to prove the claim both ways. The first part goes as follows: assume that a_i is the rational choice. Given our definition of rationality, this means that $u(a_i) \geq u(a_j)$ for all $j \neq i$. If so,

$$u(a_i) \geq \max\{u(a_1), u(a_2), \ldots, u(a_{i-1}), u(a_{i+1}), \ldots, u(a_n)\}$$

But because of the way we defined opportunity costs in Definition 3.2, this means that $u(a_i) \geq c(a_i)$. The second part is the same thing over again, except reversed. □

This proposition establishes formally what we already knew to be the case, namely, that it is irrational to invest in real estate whenever there is a more valuable alternative available, even if investing in real estate would generate a profit. Notice that the condition holds even if there is more than one optimal element. Whenever this happens, the utility of the optimal alternative will equal its opportunity cost.

Example 3.5 Gangnam Style By mid-2014, the goofy music video "Gangnam Style" became the most watched YouTube clip of all time. According to a review by *The Economist*, in the amount of time people spent watching the clip, they could instead have built three of the largest type of aircraft carrier, four Great Pyramids of Giza, six Burj Khalifas (the world's tallest building in Dubai), or 20 Empire State Buildings. As this example shows, opportunity costs can be huge.

Exercise 3.6 Opportunity costs What is the opportunity cost of (**a**) staying in an unfulfilling relationship, (**b**) pursuing a course of study that does not excite you, and (**c**) sleeping until noon?

The concept of opportunity cost has considerable explanatory power.

Exercise 3.7 Opportunity costs, cont. Using the language of opportunity cost, explain why highly paid people are less likely than poor people to mow their own lawns, clean their own houses, maintain their own cars, and so on.

In practice, however, people frequently overlook or underestimate opportunity costs. In the context of investment decisions, many people are pleased with their performance if their investments increase in value over time, whether or not there is another investment that would have generated a larger profit. Overlooking opportunity costs can make you behave in self-destructive or otherwise suboptimal ways, as Exercise 3.6 shows. A person who ignores the opportunity cost of a bad relationship, for example, could miss out on a lot. If you were surprised by the opportunity costs of watching the "Gangnam Style" video, chances are you underestimated them. Here is another example.

Example 3.8 Ignoring opportunity costs Imagine that after visiting your parents in Kansas a few times, you earn a voucher that can be exchanged for a free airplane ticket anywhere in the country. You decide to go to Las Vegas. You would not actually have bought a ticket to Vegas, but because it was free you figured you might as well. Now you would like to visit your parents in Kansas, and wish you did not have to pay so much money for the ticket.

In this case, you may be acting irrationally because you did not consider the opportunity cost of using the ticket to go to Vegas. Insofar as you would have preferred to use the voucher to visit your parents in Kansas, you failed to consider what you could have used it for instead. Though the ticket was purchased using a voucher rather than cash, the decision to use it to go to Vegas is associated with a substantial opportunity cost. As this example illustrates, people are particularly likely to ignore the opportunity cost of spending a **windfall**, that is, an unexpectedly large or unforeseen profit. Yet the best way to spend a dollar is not a function of how it ended up in your pocket.

What follows is a classic example. There are two different versions of this question, one with the original numbers and one with the numbers in square brackets.

Example 3.9 Jacket/calculator problem Imagine that you are about to purchase a jacket for $125 [$15], and a calculator for $15 [$125]. The calculator salesman informs you that the calculator you wish to buy is on sale for $10 [$120] at the other branch of the store, located 20 minutes' drive away. Would you make the trip to the other store?

In the original study, 68 percent of respondents were willing to make the drive to save $5 on the $15 calculator. Yet only 29 percent were willing to make the drive to save $5 on the $125 calculator.

Here, many respondents seem to have failed to take opportunity costs properly into account. Standard theory requires you to make your decision based on the opportunity cost of the 20-minute drive. The opportunity cost does not depend on how you saved the $5. If people took the opportunity cost properly into account, therefore, they would drive in the one scenario just in case they would drive in the other. But that is not what we observe.

Exercise 3.10 Room service Suppose that you are staying at an expensive hotel that offers in-room dining for $60. Ordinarily you would not pay that kind of money for a meal, but you say to yourself: "I already paid $220 for this room, so what's $60 more?" Explain in what way your decision to pay $60 for dinner ignores the opportunity cost of doing so.

In general, thinking about outcomes in terms of ratios is a good way to go wrong. Asking questions such as "What's another $10,000 when I'm already spending $100,000 on a new house?" is one way to ignore the opportunity costs of spending $10,000 that way. (We will return to the topic in Section 7.2.)

Exercise 3.11 Whatever it takes Politicians sometimes promise to do "whatever it takes" to eliminate poverty, defeat terrorism, etc. Use the concept of opportunity cost to explain why that is not the best idea.

There are systematic data on how much money is lost by people who ignore opportunity costs. Here is one example.

Example 3.12 One day at a time The field study mentioned in Section 1.3 found that New York City cab drivers who set their own hours ignore opportunity costs in a big way. The study found that cab drivers operate with a daily income target, meaning that they will work until they have earned that amount and then quit. The result is that they will work fewer hours on more profitable (e.g., rainy) days, and more hours on less profitable (e.g., sunny) days. This is exactly the opposite of what they ought to be doing, since the opportunity cost of taking time off is *higher* on profitable days. The authors of the study estimate that holding the total number of hours constant, drivers could increase their earnings by as much as 5 percent by working the same number of hours every day, and as much as 10 percent by working more on more profitable days.

It is easy to think of other scenarios where people might ignore opportunity costs. Think about unpaid work. Even if you can save money by mowing your own lawn, this does not mean that it is rational to do so: if your hourly wage exceeds the hourly charge to have a lawn-care company mow your lawn, and you have no particular desire to mow the lawn rather than spending another hour at work, then the rational choice is to stay at work and pay somebody to mow your lawn. Or think about investments in public safety. Even if recruiting more police officers might save lives, this does not necessarily mean that it is rational to do so. If there is an alternative investment that would save even more lives – street lights, for example – or in other ways generate more value, it would be irrational to recruit more police officers. In general, even if some business move, political reform, or any other initiative can be shown to have hugely beneficial consequences, this does not automatically make it rational: everything hinges on what the opportunity cost is.

Exercise 3.13 Advertising campaigns Your latest efforts to boost revenue led to an advertising campaign that turned out to be hugely successful: an investment of $1000 led to a $5000 boost in revenue. Does this mean that the investment in the advertising campaign was rational?

Exercise 3.14 War or terror Imagine that a new terrorist group is threatening the lives of innocent civilians, and that all things equal, it would be a good thing if the group vanished. Does this necessarily make it a good idea to launch military action to destroy it?

Why do people ignore opportunity costs? The first thing to note is how very difficult it would be to live up to the requirement to take opportunity costs into proper account. The requirement does not say that you have to consciously consider all the different alternatives available to you. But it does say that you must never choose an alternative whose opportunity cost is higher than its utility. This is an extremely demanding condition. Consider what would be required in order to rationally choose whom to marry. You must have complete and transitive preferences over all alternatives, and you must make sure that your choice of spouse is not inferior to any other choice. And the set of alternatives in this case might include half of humankind or more, though for many of us the budget set would be rather smaller. Hence, we should not be surprised that people sometimes overlook opportunity costs. (Section 3.5 discusses another reason why people may fail to take opportunity costs properly into account.)

It is irrational to fail to take opportunity costs properly into account, given the account of rationality that we developed in the previous chapter and use in this one. Certainly, in many cases this is right: if you fail to consider the opportunity cost of using your free ticket to go to Vegas, for example, it might be acceptable to call you irrational. But the fact that considering all possible alternatives is so very demanding – when getting married, for instance – means that there can be legitimate disagreement about whether failing to consider opportunity costs under those conditions is irrational or not. And if it can be rational to ignore opportunity costs, the theory that we have studied here is normatively incorrect. Still, an awareness of opportunity costs can be helpful, as Exercise 3.6 shows. Articulating what the opportunity costs of a given action are can help you see other and better opportunities open to you.

It is important not to exaggerate people's inability to take opportunity costs properly into account. Although the opportunity cost of going to college can be huge, since it includes the amount that can be gained from working instead of studying, people still do go to college. Does this mean they are irrational? Not necessarily. The opportunity cost of *not* going to college can be even greater, since the forgone alternative in this case includes the higher lifetime earnings, more enjoyable work conditions, and so on, that a college degree can confer. If so, going to college can be perfectly rational in spite of the sizeable opportunity cost of doing so. (The decision problem here – as in the case of investments in real estate and the stock market – is complicated by the fact that it involves choice over time, which is discussed further in Part 4 of this book.)

Problem 3.15 The opportunity cost of an economics education *What is (or would be) the opportunity cost, for you, of taking a course in behavioral economics? Upon reflection, is it (or would it be) worth it?*

The notion of opportunity cost has other applications. According to psychologist Barry Schwartz, the fact that we face opportunity costs helps explain why many of us are so unhappy in spite of the extraordinary freedom that we enjoy. In this analysis, it is the very fact that we are free, in the sense of having many options available to us, that prevents us from being happy:

> [The] more alternatives there are from which to choose, the greater our experience of the opportunity costs will be. And the greater our experience of the opportunity costs, the less satisfaction we will derive from our chosen alternative... [A] greater variety of choices actually makes us feel worse.

Schwartz calls it "the paradox of choice." Notice, though, that this would be a case where we pay too much, rather than too little, attention to opportunity costs.

There are behaviors that need not result from a lack of attention to opportunity costs, although it may look that way. When a person stays in a bad relationship because he or she would rather deal with "the devil you know than the devil you don't," the person is facing not just opportunity costs but also risk or uncertainty – which is something we will return to in Part 3.

3.3 Sunk costs

Suppose that you are the manager of the research and development (R&D) department of a big corporation and that you have to decide whether to spend $1M to complete a project that you know will fail. In one scenario, your corporation has already invested $9M in the project, and you have to decide whether to invest the additional $1M required to complete it. In another scenario, you have not yet invested anything in the project, and you have to decide whether to invest the $1M required to complete it. What would you do? You might be willing to invest in the first scenario but unwilling to do so in the second. Assuming that the two decision problems are otherwise identical, this would be irrational. And it would be irrational because you would, as people say, be "throwing good money after bad." One way to see this is to think of it in terms of opportunity costs: when you make a decision based on past investments, you in effect ignore or underweight opportunity costs – which is irrational (see the previous section).

Most decisions are not as simple as the decision trees we saw in Section 3.2. Some decision problems have several stages. Assuming that you want a soft drink, for example, you may first have to decide where to get one. Thus, you may have to choose, first, whether to go to Piggly Wiggly or Publix, and second, whether to get Coke or Pepsi. While the fact that this decision problem has two stages complicates matters somewhat, decision trees allow you to represent problems of this type too, as Figure 3.4 shows.

As always, the theory does not say whether you should choose Coke or Pepsi. However, it does say a few things about how your choices in one part of the tree should relate to your choices in other parts of the tree. Let us assume, for simplicity, that your universe consists of these two alternatives only, and that you are not indifferent between them. If you choose Coke rather than Pepsi at node #2, you also have to choose Coke rather than Pepsi at node #3; if you choose Pepsi rather than Coke at node #2, you also have to choose Pepsi rather than Coke at node #3. As we established

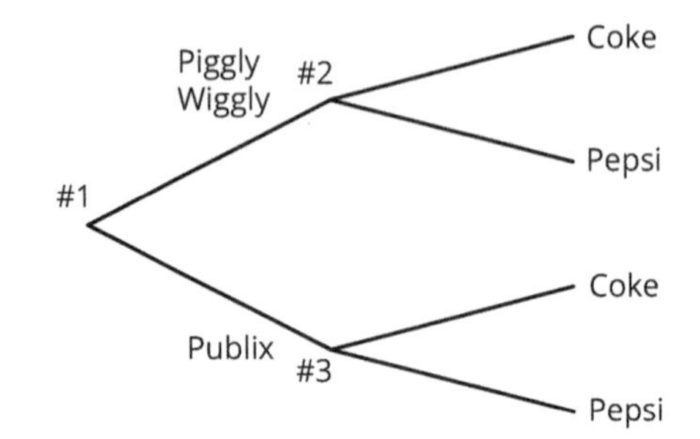

Figure 3.4 Multi-stage drinking problem

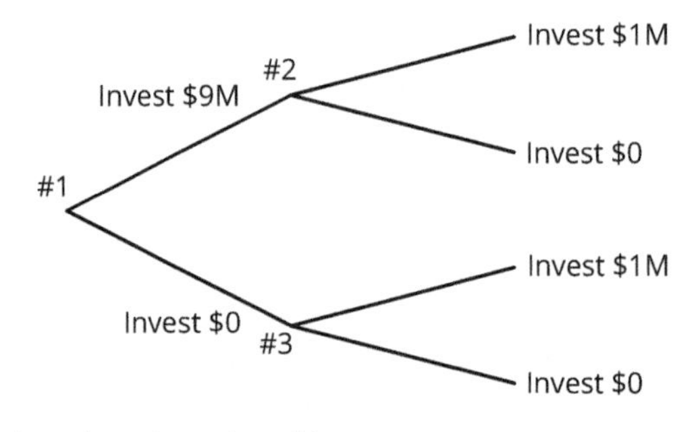

Figure 3.5 Multi-stage investment problem

in Exercise 2.21(b) on page 19, if you strictly prefer x to y, you must not also prefer y to x. And at #2 you are indeed facing the exact same options as you are at #3. (If you are indifferent between Coke and Pepsi, you can rationally choose either.)

Would anybody fail to act in the same way at nodes #2 and #3? There is a wide class of cases where people do. These are cases in which there are **sunk costs**: costs beyond recovery at the time when the decision is made. Consider the R&D scenario outlined earlier in this section. It can be represented as in Figure 3.5. Faced with these two problems, many people say they would invest at node #2 but not invest at node #3. Yet, at node #2, the $9M is a sunk cost: it cannot be recovered. Whether or not making the further $1M investment is worth it should not depend on whether you find yourself at node #2 or #3. Failing to ignore sunk costs is referred to as **honoring sunk costs**, or as committing the **sunk-cost fallacy**. The sunk-cost fallacy is sometimes called the **Concorde fallacy**, since French and British governments continued to fund the supersonic passenger jet long after it became clear that it would not be commercially viable, supposedly because they had already invested so much money and prestige in the project. As these examples illustrate, the sunk-cost fallacy can be costly.

Example 3.16 The basketball game Imagine that you paid $80 for a ticket for a college basketball game to be played about an hour's drive away. The ticket cannot be sold. On the day of the game, there is a freak snowstorm that makes driving hazardous. Would you go to the game?

Now, imagine that the ticket instead was given to you for free. Would you be more or less likely to go to the game?

Many people would be more likely to go to the game if they paid actual money for the ticket. Yet the cost of the ticket is a sunk cost at the time when the decision is made. Whether or not you paid for the ticket should not affect whether you go to the game or not.

The sunk-cost fallacy is evident in a wide range of everyday decisions. For one thing, it can make people hold on to failed investments. If you refuse to sell an underperforming asset on the basis that you would lose the difference between what you paid for it and what you would get for it if you sold it now, you are honoring the sunk cost of the original investment; the rational thing to do is to hold on to the asset if you think it is the best investment available to you right now and to sell if it is not. The sunk-cost fallacy can also make people stay in failed relationships. If a friend refuses to ditch her current boyfriend, even though she realizes that he is an utter loser, on the basis that leaving him would mean that she would have wasted some of the best years of her life, she would be honoring the sunk cost of the time and effort that she has already committed to him. As these examples show, honoring sunk costs can be expensive not just in terms of money, but also in terms of time, effort, and heartache.

Exercise 3.17 Sunk costs Draw decision trees for people who (**a**) hold on to failed investments and (**b**) stay in bad relationships, in such a way that it becomes clear that the people in question are committing the sunk-cost fallacy.

Note that the rational decision is determined by what is going on only to the right of the node where you find yourself. What happens in other parts of the tree – in particular, to the left of your node – is completely irrelevant. In this sense, rational choices are completely forward-looking. For as the ancient Greek philosopher Aristotle noted some years back: "We do not decide to do what is already past; no one decides, for instance, to have sacked Troy." Things that happened in the past matter only insofar as they affect future outcomes.

Example 3.18 Reading habits According to a survey administered by Goodreads.com, 38.1 percent of readers say they will finish a book they started reading no matter what – even if they hate it. This behavior can be understood as an effort to honor the sunk cost involved in buying and starting to read the book.

Exercise 3.19 The fridge When cleaning out your fridge, you find some food obviously rotting in the back. You have no trouble throwing away the cheap mass-produced cheese somebody gave you, but you just cannot bring yourself to toss the amazing cheese you brought back from Paris. So you put it back in the back of the fridge, where it will sit for another couple of months before it smells so badly that you have no choice but to get rid of it. How do sunk costs figure in these decisions?

Exercise 3.20 Course selection Students at an expensive liberal arts college may take courses at a nearby public university at no additional charge. One of their professors tells them that it would make no sense to do so, since they would be losing the money they paid for tuition at the pricier college. Given that a student has already paid tuition at the liberal arts college, but judges that the course offerings at the public university are better for her, where should she sign up? Explain.

The sunk-cost fallacy can start a vicious circle sometimes referred to as an **escalation situation**. Once a project – whether an R&D effort, a marriage, a financial investment, or whatever – is beginning to go downhill, the sunk-cost fallacy encourages people irrationally to make additional investments in the project. Once the additional investment has been made, unless it turns the project around, people find themselves with an even greater sunk cost, which is even harder to ignore, thereby encouraging even greater investments.

Example 3.21 The F-35 The F-35 is the US military's next-generation fighter jet. Plagued with technical difficulties, the project was $160 billion over budget already by early 2014. Critics argue that the project needs to be cancelled. But the US Department of Defense (DOD) still wants its planes. "I don't see any scenario where we are walking back away from this program. We're going to buy a lot of these airplanes," the officer in charge of the program said to the *Fiscal Times*. Why? "DOD is so far down the F-35 rabbit hole, both in terms of technology and cost – $400 billion for 2,400 planes – that it has no choice but to continue with the program." You are unlikely to ever find a cleaner example of the sunk-cost fallacy.

The sunk-cost fallacy and escalation behavior are often invoked when explaining why the US spent so many years fighting a losing war in Vietnam. According to this analysis, once soldiers were committed and started dying, it became impossible to withdraw for fear that the dead would have "died in vain"; thus, more soldiers were committed, more soldiers died, and it became even harder to withdraw. Interestingly, the scenario was outlined as early as 1965 by George Ball, then Undersecretary of State. In a memo to President Johnson, Ball wrote:

> The decision you face now is crucial. Once large numbers of US troops are committed to direct combat, they will begin to take heavy casualties in a war they are ill-equipped to fight in a noncooperative if not downright hostile countryside. Once we suffer large casualties, we will have started a well-nigh irreversible process. Our involvement will be so great that we cannot – without national humiliation – stop short of achieving our complete objectives. Of the two possibilities I think humiliation will be more likely than the achievement of our objectives – even after we have paid terrible costs.

Some wars are justified, and some wars are not. But the justification can never take the form of simply pointing out how much money and how many lives have already been sacrificed; if the war is justified, it must be for other, forward-looking reasons. One important insight is this: before you embark on a risky project, you may want to ask yourself whether it will be possible for you to call it off in case it starts going downhill. If not, this is a reason not to embark on the project in the first place.

An awareness of our tendency to honor sunk costs can be used to influence people for good and evil. The loser boyfriend from earlier in this section can appeal to sunk costs when trying to keep his girlfriend. Less obnoxiously, you can use knowledge of the sunk-cost fallacy to market your products. One of the reasons why outlet malls are located so far away from where people live is that executives want shoppers to think of the sunk cost of a long drive to the mall as an investment that will be lost if they do not shop enough. More upliftingly, it turns out that you can make money by teaching people about rational-choice theory.

Example 3.22 How to sell tires The following story was related by Cory, a former student of behavioral economics:

> I co-manage a local tire/automotive service retailer. Today, one of my good customers came into my store. He had purchased a set of tires from a separate online tire seller and came to get them installed. Before we started working on his car, he asked what other options would have been available to him. I proceeded to tell him about a brand new tire that I stocked that was overall much better than the one he bought – better traction features, better mileage, etc. I politely asked if he would like to buy them and have me send the others back. His response was: "No, that's okay, I've already bought these. I better just stick with them." I told him how easy the return process would be, how much longer the new tire would last, how it would save him money in the long run, etc., but his response remained the same: "Already paid for these; better stick with them." Finally, I told him that he was honoring sunk costs.
>
> So of course he asked what I meant, and I explained the concept at length. He was simply fascinated by this random lecture that he was receiving from somebody he thought to be just a "tire guy." I concluded the conversation by humorously declaring: "If you decide to stick with your original purchase, you are violating the theory of rational decision-making."
>
> He then looked at me with a ponderous stare. "You know what? I learned a lot from you. I think I *will* buy your tires and send the others back. Thanks for helping me!" So he bought the tires from me, had them installed, and I made a nice commission. I thought to myself: "Wow, I have been finished with behavioral economics for only one day, and already it is paying dividends!"

Knowledge of a tendency to honor sunk costs can also help us resist other people's manipulative behavior. If you are the loser's girlfriend, it might help to remind yourself that the wasted years in your past are no reason to stay with him, and if you are the shopper at the outlet mall, you can remind yourself that the long drive out there is no reason to buy stuff you do not want.

Example 3.23 Bullet trains In 2011, California Governor Jerry Brown pushed an effort to build a 520-mile high-speed railroad between the cities of Los Angeles and San Francisco. However, since it is very expensive to build railroads where people live, he hoped to start construction in a relatively remote and unpopulated part of the state. Why would he do this? According to critics, the *New York Times* reported at the time, Brown calculated "that future legislatures would not be able to abandon the project before it reached major population centers. 'What they are hoping is that this will be to high-speed rail what Vietnam was to foreign policy: that once you're in there, you have to get in deeper,' said Richard White, a professor of history at Stanford University." There is a good chance politicians hoped that the sunk cost of the initial investment would start an escalation situation from which future administrations could not escape.

There are cases that may superficially look like the sunk-cost fallacy, but that really are not. In order for you to commit the sunk-cost fallacy in a decision problem like that in Figure 3.5, the options available to you at nodes #2 and #3 need to be identical. If they are not, you could rationally choose any combination of options. For example, your boss might demote you for failing to invest at node #2 but not do so at node #3. If that is the case, it may well be rational for you to invest at #2 but not at #3, even if the investment would lead to large losses for the company. Similarly, if calling off a misguided military adventure would have unfortunate consequences – perhaps because the national humiliation would be unbearable, the military would look weak, and/or the next election would be lost – a president could rationally continue to fight the war. (Notice that we are concerned with rationality here, not morality.)

Either way, it is important not to accuse people of committing the sunk-cost fallacy if their behavior is better captured by standard theory.

Problem 3.24 Revenge *When wronged, many people feel a strong urge to take revenge. Assuming revenge is costly, would not a revenge simply be a matter of honoring the sunk cost of whatever injury they have already sustained? Or are there conditions under which taking revenge can be rational?*

3.4 Menu dependence and the decoy effect

Spend a moment considering which of the subscription offers in Table 3.1 you would prefer. You will notice that there is something strange about the three options. Why would anybody choose the print subscription, given that you can get an online and a print subscription for the very same price? Given that nobody in their right mind would choose option 2, why was it included? It turns out that there is a good reason for *The Economist* to present potential customers with all three options. When

Table 3.1 *The Economist* subscription offers

	Economist.com offers	Price
Option 1	Web subscription	$59
Option 2	Print subscription	$125
Option 3	Print + web subscription	$125

researchers presented MBA students with options 1 and 3 only, 68 percent chose option 1 and 32 percent chose option 3. When the authors presented MBA students with options 1, 2, and 3, 0 percent chose option 2. But only 16 percent chose option 1 whereas 84 percent chose option 3. Thus, it appears that the inclusion of an option that nobody in their right mind would choose can affect people's preferences over the remaining options.

Recall from Section 2.6 that the rational choice depends on your budget set, that is, your menu. When your menu expands, more options become available to you, and one of them may turn out to be more preferred than the one you would have chosen from the smaller menu. However, the theory from the previous chapter does impose constraints on what happens when your menu expands. Suppose you go to a burger restaurant and are told that you can choose between a hamburger and a cheeseburger. Imagine that you strictly prefer a hamburger and say so. Suppose, furthermore, that the server corrects herself and points out that there are snails on the menu as well. In this case, you can legitimately say that you will stick with the hamburger, in case hamburgers rank higher than snails in your preference ordering (as in columns (A) and (B) in Figure 3.6). Or you can switch to snails, in case snails rank higher than hamburgers (as in column (C) in the figure). It would be odd, however, if you said: "Oh, I see. In that case I'll have the cheeseburger." Why? No rational preference ordering permits you to change your mind in this way. Either you prefer cheeseburgers to hamburgers, in which case you should have ordered the cheeseburger from the start, or you do not, in which case you should not choose it whether or not there are snails on the menu. (This chain of reasoning assumes that we are talking about strict preference; if you are indifferent, you can choose anything.)

Formally speaking, the theory implies something we call the **expansion condition**.

Proposition 3.25 Expansion condition *If x is chosen from the menu $\{x, y\}$, assuming that you are not indifferent between x and y, you must not choose y from the menu $\{x, y, z\}$.*

Proof.
If you choose x when y is available, given that you are not indifferent between the two, you must strictly prefer x to y. If you choose y when x is available, given that you are not indifferent between the two, you must strictly prefer y to x. But we know from Proposition 2.16 (ii) on page 18 that the strict preference relation is anti-symmetric, so this is impossible. □

Plainly put, Proposition 3.25 simply says that the introduction of an inferior product should not change your choice. The choice of y from the expanded menu would signal that you changed your preferences between the first and second decision. The theory, however, does not permit you to change your preferences as a result of an

(A)	(B)	(C)
Hamburger	Hamburger	Snails
$\succ$	$\succ$	$\succ$
Cheeseburger	Snails	Hamburger
$\succ$	$\succ$	$\succ$
Snails	Cheeseburger	Cheeseburger

Figure 3.6 Preference orderings over food

expanding menu or for any other reason. As we know from Section 2.8, the theory assumes that preferences are stable and do not change over time. The plausibility of the expansion condition can also be seen by reflecting on the nature of indifference curves and budget sets. If some option is optimal, given a set of indifference curves and a budget line, this fact cannot be changed by adding another (clearly inferior) option to the budget set. Thus, the introduction of another alternative inside the shaded area of Figure 2.6 on page 23 would not make the alternative marked X suboptimal.

Yet there is evidence that people's preferences do change when the menu expands. We talk about this as a case of **menu dependence**. Suppose that you market a product, which we call the **target**. The problem is that another company markets a similar product, which we call the **competitor**. The consumer can afford each of these; both the target and the competitor are on the budget line. The problem is that the consumer prefers the competitor. Thus, the decision problem facing the consumer looks like Figure 3.7(a). The two products here (the target and the competitor) can be understood as before, as commodity bundles consisting of so many units of apples and so many units of bananas. They can also be understood as goods that differ along two dimensions: say, as cars that differ along the dimensions of speed and safety. Either way, it should be clear that the consumer in this figure will choose the competitor.

It turns out, however, that you can manipulate the consumer's choices by introducing a product that is in every respect inferior to the target. We say that one product x **dominates** another y just in case x is better than y in every possible respect. In terms of Figure 3.7(a), the target dominates every product in the boxes marked B and C. The competitor dominates every product in boxes A and B. Given a menu and a good x, we say that a product y is **asymmetrically dominated** by x just in case y is dominated by x but not by any other member of the menu. Suppose, now, that the menu includes a third item, which we call the **decoy**, and which is asymmetrically dominated by the target. This means that the decoy is located in the box marked C in Figure 3.7(a). In spite of the fact that few consumers would bother buying such a good – since it is in every respect worse than the target – its introduction can change people's choices. There is evidence that the introduction of the decoy changes people's indifference curves in the manner illustrated by Figure 3.7(b).

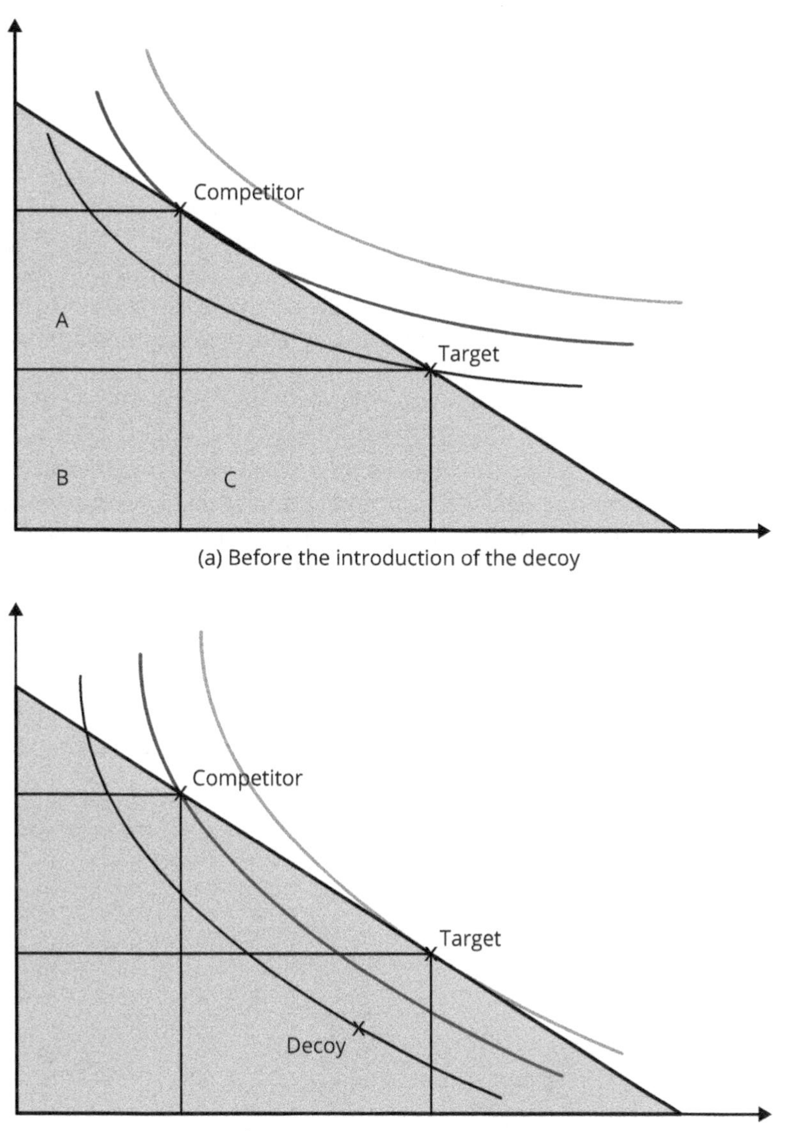

(b) After the introduction of the decoy

Figure 3.7 Decoy effect

Notice that the indifference curves appear to have rotated clockwise around the competitor, as it were, toward the decoy. As a result, the target is now on a higher indifference curve than the competitor. Thus, in spite of the fact that the rational consumer would not dream of buying the decoy, the introduction of it still succeeds in changing people's indifference curves and therefore their choices. Because the presence of the dominated option appears to increase the attractiveness to the consumer of the dominating alternative, the decoy effect is sometimes referred to as the **attraction effect**. Notice that such shifts violate the expansion condition (Proposition 3.25) and consequently are irrational.

Example 3.26 Speed vs. safety Imagine that you are in charge of selling a car that is very fast but not very safe. Let us call it a "Bugatti." (When people complained about the poorly designed brakes of his cars, Mr Bugatti himself is alleged to have said: "My cars are for driving, not for braking.") The problem is that your customers too often choose a car that is less fast but more safe. Let us call it a "Volvo." To increase sales of your car, you decide to start selling a decoy. What features should it have?

If you put speed on the x-axis and safety on the y-axis, you get a graph exactly like Figure 3.7, in which the Bugatti is the target and the Volvo the competitor. The decoy must go in the box marked C. The fact that the decoy is below both target and competitor means that it is less safe than both the Bugatti and the Volvo. The fact that the decoy is left of the target but right of the competitor means that it needs to be less fast than the Bugatti but faster than the Volvo. Notice that a vehicle that is inferior to both cars in both respects – let us call it a "golf cart" – would not do the trick.

Exercise 3.27 Decoy effect For this question, refer to Figure 3.8.

(**a**) If you are in charge of marketing the target product, in what area would you want to put the decoy?

(**b**) Assuming the decoy works as anticipated, does the figure show the indifference curves the way they look *before* or *after* the introduction of the decoy?

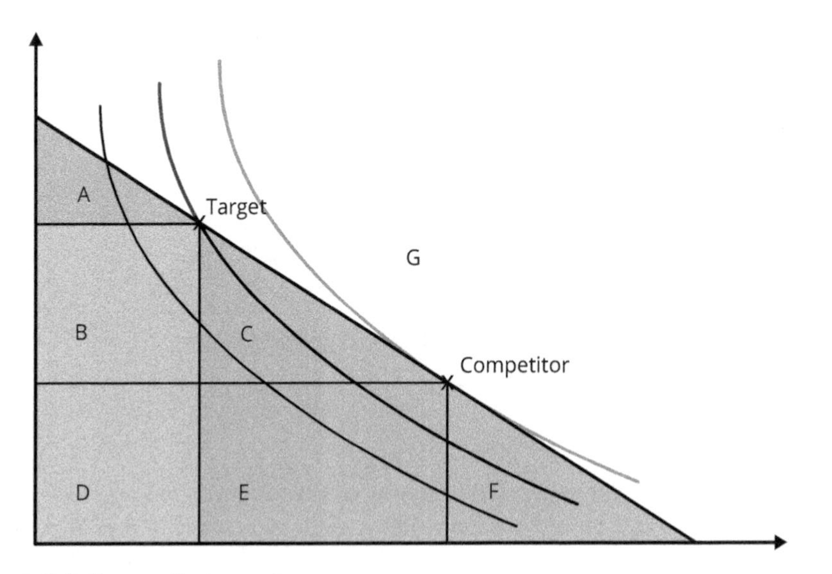

Figure 3.8 Decoy effect exercise

Exercise 3.28 Real-estate sales Suppose that you are a real-estate agent showing two properties to potential customers. The one is in good shape but far from the clients' office; the other is only in decent shape but close to the office.

(**a**) If you want the customers to choose the former, what third property should you show?

(**b**) If you want the customers to choose the latter, what third property should you show?

Exercise 3.29 Third-party candidate Politician A has decided to run for political office on a platform promising lower taxes and cuts in public services. His opponent B promises higher taxes and more funding for public services. All things equal, the voters prefer low taxes and generous public services. Now, Politician B is in the lead. Politician A decides to channel some of his campaign funds to a third-party candidate, C, who will act as a decoy. What sort of platform must the third-party politician endorse to act as a decoy for A?

The following example is designed to show just how useful the study of behavioral economics can be. Remember that with great power comes great responsibility.

Exercise 3.30 Wingmen and wingwomen To improve your chances on the dating scene, you have decided to recruit a wingman or wingwoman.

(**a**) How, in general terms, should you choose your wingman or wingwoman?

(**b**) Imagine that your attractiveness and intelligence are both rated 9 on a 10-point scale. You have two competitors: one whose attractiveness is 10 and intelligence 8, and another whose attractiveness is 8 and intelligence 10. In what range would you want your wingman or wingwoman's attractiveness and intelligence to fall?

(**c**) If somebody asks you to be his or her wingman or wingwoman, what does this analysis suggest he or she thinks about your attractiveness and intelligence?

How does this help us explain the subscription offers in Table 3.1? Each option can be represented as a bundle of three different goods: online access, paper subscription, and attractive price. Thus, option 1 can be represented as $\langle 1, 0, 1 \rangle$ because it includes online access, does not include a paper subscription, but has an attractive price. Similarly, option 2 can be represented as $\langle 0, 1, 0 \rangle$ and option 3 as $\langle 1, 1, 0 \rangle$. From this way of representing the options, it is quite clear that option 2 is (weakly) asymmetrically dominated by option 3. If the analysis offered in this section is correct, the introduction of the (inferior) option 2 might still drive customers to option 3, which is what *The Economist* wanted and expected.

The decoy effect is only one form of menu dependence. Another effect that has received a great deal of attention, especially in marketing literature, is the **compromise effect**: people's tendency to choose an alternative that represents a compromise or middle option in the menu. The phenomenon is sometimes described as resulting from **extremeness aversion**: a tendency to avoid options at the extremes of the relevant dimension. A high-end brand might try to drive business to their expensive products by introducing a super-expensive product; although the super-expensive product might never sell, it could make the expensive product stand out as an attractive compromise between the cheap and the super-expensive one. Low-end brands might try to do the same by introducing super-cheap products. This may be how we got diamond-studded swimsuits on the one end of the spectrum, and swimsuits made out of materials that degrade in the presence of sunlight, salt, and chlorine on the other. Various forms of menu dependence are sometimes described as **context effects**, because people's decisions appear to be responsive to the context in which the decisions are made.

Exercise 3.31 Speed vs. safety, cont. Suppose that you want to harness the power of the compromise effect to sell more Bugattis (see Exercise 3.26). What sort of vehicle would you need to introduce to do so?

Exercise 3.32 Third-party candidates, cont. Suppose that politician A in Exercise 3.29 above wanted to win elections instead by harnessing the power of the compromise effect. What sort of platform must the third-party politician endorse to have this effect?

How do we best explain the decoy effect and other cases of menu dependence? Perhaps consumers look for a *reason* to pick one option over another, or to reject one of the options, in order to feel better about their decision. The introduction of a decoy gives the consumer a reason to reject the competitor and to choose the target, in that the target no longer scores lowest along either one of the two dimensions. The introduction of an extreme option gives people a reason to choose the option in the middle. Such considerations would suggest that the search for reasons for action – or **reason-based choice** – may actually be responsible for making us behave irrationally. This is interesting, because having reasons for actions is otherwise frequently seen as the hallmark of rationality.

Either way, menu dependence can explain a wide variety of marketing practices, including why the number of options available to customers keeps increasing (see the discussion of the paradox of choice in Section 3.2). And it is obviously relevant to anybody who wants to sell things. Given the decoy and compromise effects, it might make sense to introduce a product you do not expect anyone to buy. For that reason, you cannot assess a product's contribution to corporate profits by simply looking at the sales of that product in isolation.

Notice that there are cases that look like menu dependence but which may be better described in some other way. Suppose you enter a restaurant in a part of town you do not know very well and are offered the choice between fish, veal, and nothing at all. You choose the fish. Suppose, next, that the waiter returns to tell you that the menu also includes crack-cocaine. It would be perfectly possible, at this point, to decide you would rather not have anything at all without violating rational-choice theory. Why? When the waiter comes back, he does not just expand your menu, he also tells you something about the establishment – and also that you do not want to eat any of their food. Notice that this is not even a case of choice under certainty. The decoy scenarios are different: in this analysis, it is not the case that you learn something about the target by learning that the decoy is also on the menu.

3.5 Loss aversion and the endowment effect

The theory that we studied in the previous chapter makes preferences independent of your endowment, meaning what you have at the time when you make the decision. Consider your preferences for coffee mugs. Obviously, the theory does not tell you how much you should value a mug. However, it does say a few things about how you must order mugs and other things. Assuming you prefer more money to less, you prefer \$1 to \$0, \$2 to \$1, and so on. Because of completeness, if you are rational, there must be a dollar amount p (not necessarily in whole dollars and cents) such that you are indifferent between p and the mug. If p is \$1, your preference ordering can be represented as Figure 3.9. If you have the mug, and somebody asks you what it would take to give it up, your answer would be "No less than \$1," which is to say that your willingness-to-accept (WTA) equals \$1. If you do not have a mug, and somebody asks you what you would pay in order to get one, your answer would be "No more than \$1," which is to say that your willingness-to-pay (WTP) equals \$1 too. Your preference between the mug and the dollar bill does not depend on whether or not you already have a mug, and your WTA equals your WTP. (Your preference for a second mug might depend, however, on whether you already have a mug, as Figure 3.10 shows.)

The independence of your preferences from your endowment is reflected in a fact about utility functions. Suppose your utility function over mugs looks like Figure 3.10.

$$
\begin{array}{c}
\$3 \\
\curlyvee \\
\$2 \\
\curlyvee \\
\$1 \sim \text{mug} \\
\curlyvee \\
\$0
\end{array}
$$

Figure 3.9 Preference ordering with mug

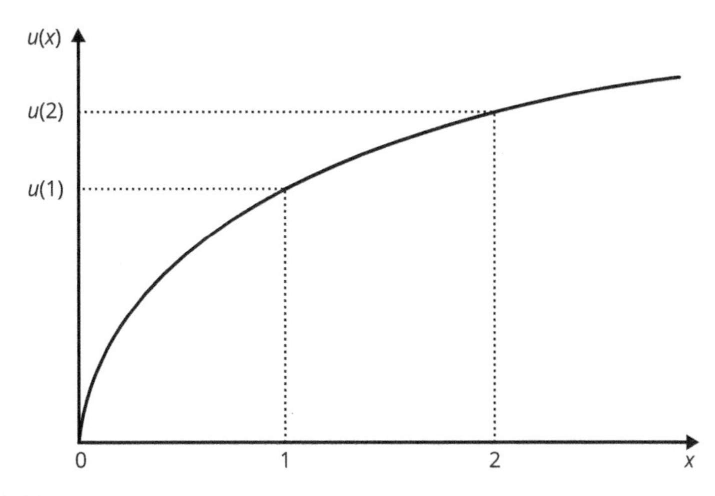

Figure 3.10 Utility over mugs

The numbers on the x-axis represent how many mugs you own, and the numbers on the y-axis represent how much utility they give you. Thus, owning one mug gives you $u(1)$ units of utility. When you move from zero to one mug (rightwards along the x-axis), your utility increases from $u(0) = 0$ to $u(1)$. Similarly, when you move from one to zero mugs (leftwards along the x-axis) your utility decreases from $u(1)$ to zero. Because the increase in utility resulting from receiving the mug equals the decrease in utility resulting from losing it, the utility of the first mug is independent of your endowment, that is, whether or not you have the mug. We can show this numerically. Suppose $u(x) = 3\sqrt{x}$, so that $u(1) = 3$ and $u(0) = 0$. If so, the amount of utility received from acquiring your first mug (3) equals the amount of utility lost when giving it up (3). Again, a second mug would increase your utility by a smaller amount: $u(2) - u(1) = 3\sqrt{2} - 3\sqrt{1} \approx 1.24$.

However, people do not in general behave this way. Frequently, people require a lot more to give up a cup when they already have one than they would be willing to pay when they do not. In one study using Cornell University coffee mugs, the median owner asked for \$5.25 when selling a mug, while the median buyer was only willing to pay \$2.25 to \$2.75 to purchase one. This phenomenon is referred to as the **endowment effect**, because people's preferences appear to depend on their endowment, or what they already possess. Since the manner in which people assess various options might depend on a reference point – in this case, their current endowment – phenomena like these are sometimes referred to as **reference-point phenomena**.

The endowment effect and reference-point phenomena are instances of **framing effects**, which occur when people's preferences depend on how the options are *framed*. There are many kinds of framing effects. In 2007, the Associated Press reported that Irishman David Clarke was likely to lose his license after being caught driving 180 km/h (112 mph) in a 100 km/h (62 mph) zone. However, the judge reduced the charge, "saying the speed [in km/h] seemed 'very excessive,' but did not look 'as bad' when converted into miles per hour." The judge's assessment appears to depend on whether Clarke's speeding was described in terms of km/h or mph. Similarly, people traveling to countries with a different currency sometimes fall prey to what is called **money illusion**. Even if you know that one British pound equals about one and a half US dollars, paying two pounds for a drink might strike you as better than paying three dollars.

The endowment effect and other reference-point phenomena are typically explained as the result of **loss aversion**: the apparent fact that people dislike losses more than they like commensurate gains. When the Brad Pitt character in *Moneyball* says "I hate losing more than I even wanna win," that is loss aversion. It is reflected in the fact that many people are more upset when they lose something than pleased when they find the same thing. Consider, for example, how upset you would be if you realized that you had lost a \$10 bill, as compared with how pleased you would be if you found one. Adam Smith noted this very phenomenon when he wrote "Pain … is, in almost all cases, a more pungent sensation than the opposite and correspondent pleasure." Using the language of framing, we will say that how much you value a \$10 bill depends on whether it is framed as a (potential) loss, as in the first case, or as a (potential) gain, as in the second case, and that *losses loom larger than gains*.

Example 3.33 WTA vs. WTP In the presence of loss aversion, your willingness-to-accept (WTA) does not in general equal your willingness-to-pay (WTP). When eliciting your WTA, you are asked to imagine that you have some good and to state

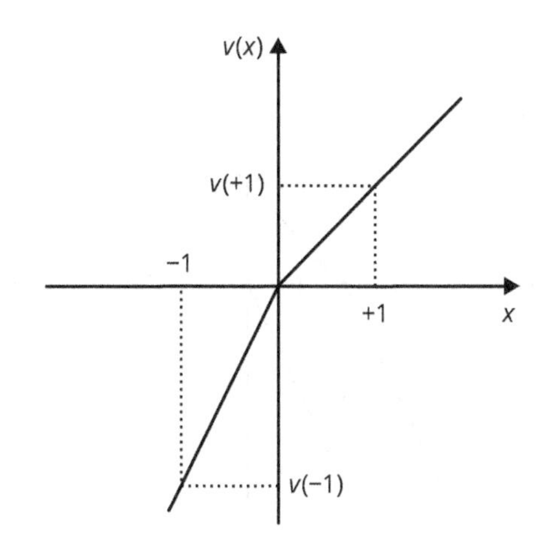

Figure 3.11 Value function

what dollar amount you would be willing to accept in order to give the good up, meaning that the good will be evaluated in the loss frame. When eliciting your WTP, you are asked to imagine that you do not have some good and to state what dollar amount you would be willing to pay in order to acquire the good, meaning that the good will be evaluated in the gain frame. Given that losses loom larger than gains, we should expect your WTA to exceed your WTP.

As the example shows, loss aversion has radical implications for the practice of cost–benefit analysis (to which we will return in Section 6.3). It is quite common to assess the value of goods by eliciting people's willingness-to-accept (WTA) or willingness-to-pay (WTP). The elicitation of WTAs and WTPs is particularly common in the case of public goods, like nature preserves, and other goods that are not traded on an open market. As we saw earlier in this section, standard theory entails that WTAs and WTPs should be more or less the same. Loss aversion entails that people value something that they have more than something that they do not have, meaning that we should expect their WTA to exceed their WTP and such analyses to generate distorted results.

Behavioral economists capture loss aversion by means of a **value function** $v(\cdot)$, which represents how an agent evaluates a change. The value function is an essential part of **prospect theory**, which is one of the most prominent theories to emerge from behavioral economics and to which we will return frequently below (e.g., in Sections 7.2, 7.3 and 7.6). The value function has two critical features. First, unlike the utility function, which ranges over *total* endowments, the value function ranges over *changes* relative to some reference point. Second, the value function has a kink at the reference point – in this case, the current endowment – in such a way that the curve is steeper to the left of the origin. See Figure 3.11 for an illustration of a typical value function. Notice that in this picture the vertical distance between the origin and $v(-1)$ is much greater than the vertical distance between the origin and $v(+1)$. Mathematically,

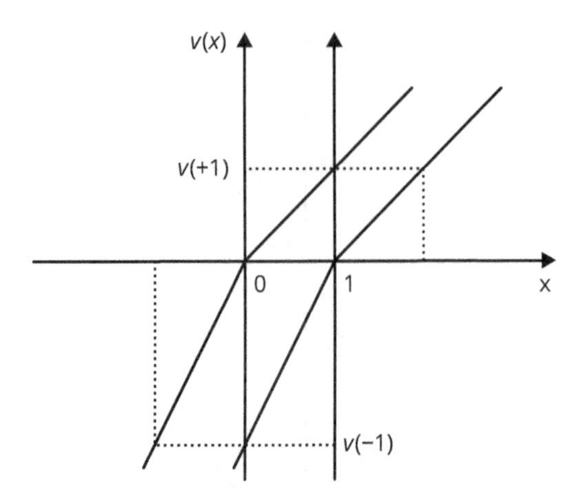

Figure 3.12 Value function and the car problem

$|v(-1)| > |v(+1)|$. Again, this captures the fact that losses loom larger than gains, that is, that people dislike losses more than they like the commensurate gains.

Another implication of loss aversion is that if you gain something and then lose it, you may feel worse off even though you find yourself where you started, which makes little sense from a traditional economic perspective. Suppose your parents promise to buy you a car if you graduate with honors, but after you do they reveal that they were lying. Figure 3.12 shows how you might represent this change. By just looking at the picture, it is evident that the gain of a car, $v(+1)$, is smaller than the (absolute value of the) loss of a car, $v(-1)$. Thus, the net effect is negative: you will be worse off (in terms of value) after you find out that your parents tricked you into thinking that you would get a new car, than you would have been if they had not, even though you find yourself with the same number of cars (zero).

Example 3.34 Toy Yoda In 2001, a Florida waitress working for an establishment that proudly describes itself as "tacky" and "unrefined" sued her employer for breach of contract, alleging that she had been promised a new car for winning a beer sales contest. The manager, the waitress said, had promised her a "new Toyota" but when she claimed her prize, she found out that she had won a "new Toy Yoda" – meaning a Star Wars figure. She was understandably upset. Loss aversion explains why: being subjected to that kind of prank leaves a person worse off than she was before, even if her total endowment remains unchanged.

It is possible to analyze loss aversion formally, by defining a value function $v(\cdot)$ that is steeper for losses than for gains. For example, we can define a value function along the following lines:

$$v(x) = \begin{cases} x/2 & \text{for gains } (x \geq 0) \\ 2x & \text{for losses } (x < 0) \end{cases}$$

It is easy to confirm that such a specification will generate graphs like Figure 3.11. The next examples and exercises illustrate how the value function can be used in practice.

Example 3.35 Losses and gains Use the value function above to answer the following two questions:
(**a**) What is the increase in value you would experience if you found \$10?
(**b**) What is the decrease in value you would experience if you lost \$10?
(**c**) In absolute terms, which is the greater number?

Here are the answers:
(**a**) When you go from your current endowment to your current endowment plus \$10, in terms of deviations from the existing endowment, you go from 0 to +10. In terms of value, this amounts to going from $v(\pm 0) = 0$ to $v(+10) = 10/2 = 5$. The change is $v(+10) - v(0) = +5 - 0 = +5$, meaning a gain of 5.
(**b**) When you go from your current endowment to your current endowment minus \$10, in terms of deviations from the existing endowment, you go from ±0 to −\$10. In terms of value, this amounts to going from $v(\pm 0) = 0$ to $v(-10) = -20$. The change is $v(-10) - v(\pm 0) = -20 - 0 = -20$, meaning a loss of 20.
(**c**) The absolute value of the loss (20) is greater than the absolute value of the gain (5).

Notice two things about these calculations. First, to compute the change in value, you want to compute the value of your endowment *after* the change and subtract the value of your endowment *before* the change. And second, to compute the value of an endowment, you must first express the endowment as a *deviation* from your reference point and not in terms of absolute wealth or anything of the sort.

Exercise 3.36 Toy Yoda, cont. Suppose that the same value function captures the waitress's value function over cars in Example 3.34. What is the total change in value she experiences after she gains a car and then loses it? Assume that she incorporates it into her endowment and therefore adjusts her reference point immediately upon receiving the promise of a car.

Exercise 3.37 Having and losing In fact, it is possible to be worse off in value terms even if you are better off in dollar terms. Working with the same value function, compute the net effect in value terms of gaining \$6 and subsequently losing \$4. Assume that you incorporate the \$6 into your endowment immediately.

If you do *not* incorporate the \$6 gain into your endowment, and therefore do not change your reference point before losing the \$4, the net effect in value terms would be $v(+6 - 4) = v(+2) = 1$. This phenomenon illustrates that it matters how complex outcomes are **bundled** – a topic to which we will return in Section 7.3 and again in Section 9.5.

How you frame various outcomes and what reference point you choose, whether you do so consciously or not, can have a huge effect on how you feel about the outcome. The next example and exercises illustrate various applications of this idea.

Exercise 3.38 The bond market An ill-fated investment of yours in the junk bond market just decreased from $1 to $0. Your value function is $v(x) = x/2$ for gains and $v(x) = 2x$ for losses.

(**a**) Suppose, first, that your reference point is $0. In value terms, how large is the loss you just experienced?

(**b**) Suppose, instead, that your reference point is $1. In value terms, how large is the loss you just experienced?

(**c**) Which framing makes you feel worse?

Exercise 3.39 The stock market Alicia, Benice, and Charlie own stock in the same company. When they bought the stock, it was worth $10. It later rose to $17, but then dropped to $12 before they sold it. The three are loss averse and have the same value function: $v(x) = x/2$ for gains and $v(x) = 2x$ for losses.

(**a**) Alicia uses the selling price ($12) as her reference point. If you ask her, how much would she say that she lost in terms of value when the price dropped from $17 to $12?

(**b**) Benice uses the peak price ($17) as her reference point. If you ask her, how much would she say that she lost in terms of value when the price dropped from $17 to $12?

(**c**) Charlie uses the buying price ($10) as her reference point. If you ask her, how much would she say that she lost in terms of value when the price dropped from $17 to $12?

(**d**) Who was more disappointed when the price dropped?

The key is to see that Alicia and Charlie evaluate the change as a forgone gain, whereas Benice evaluates the change as an actual loss. Given that losses loom larger than gains, Benice suffers more.

Exercise 3.40 Thievery A thief steals $100 from a victim. Let us suppose that the thief and victim have the same value function over money: $v(x) = x/2$ for gains and $v(x) = 2x$ for losses.

(**a**) How much, in value terms, does the thief gain as a result of his robbery?

(**b**) How much, in value terms, does the victim lose as a result of the robbery?

(**c**) Assuming that it makes sense to compare the thief's gain and the victim's loss, and ignoring any other consequences of the crime, what is the total effect of the robbery in value terms?

(**d**) Does this suggest that crime is a force for good or bad?

Exercise 3.41 The tax cut Suppose Alex and Bob are loss averse, so that their value function is $v(x) = x/2$ for gains and $v(x) = 2x$ for losses. Because of an upcoming election, politician R promises a tax cut which would give each citizen an additional $2 in his or her pocket every day. Politician D opposes the tax cut. Ultimately D wins the election. Neither Alex nor Bob receives the additional $2 per day.

(**a**) Alex thought D would win the election and never thought of the additional $2 as part of her endowment. She thinks of the $2 as a forgone gain. What would she say D's election cost her in terms of value?

(**b**) Bob was sure that R would win the election, and started thinking of the $2 as part of his endowment. He thinks of the $2 as an actual loss. What would he say D's election cost him in terms of value?

(**c**) Who is likely to be more disappointed, Alex or Bob?

Loss aversion can explain a wide range of phenomena. For example, it can explain why many companies have 30-day no-questions-asked return policies. Although costly in other ways, such policies may serve to convince a customer who otherwise would not make the purchase to take the product home and try it out. Once taken home, however, the product becomes part of the customer's endowment and loss aversion kicks in, meaning that the customer is unlikely to return the product. Loss aversion serves to explain why credit-card companies permit merchants to offer "cash bonuses" but prevent them from imposing "credit-card surcharges." Clients find it easier to forgo a cash bonus than to suffer the loss of a surcharge, so they are more likely to use a credit card in the presence of the former than of the latter. Loss aversion helps explain why politicians argue about whether cancelling tax cuts amounts to raising taxes. Voters find the forgone gain associated with a cancelled tax cut easier to stomach than they do the loss associated with a tax increase. Consequently, politicians favoring higher taxes will talk about "cancelled tax cuts" whereas politicians opposing higher taxes will talk about "tax increases." Loss aversion can also explain why so many negotiations end in stalemate, even in the presence of potential, mutually beneficial agreements. Suppose two partners are negotiating the division of a pie and that both partners think they are owed two-thirds of the pie. Any division that strikes an outside observer as fair (including a 50–50 split) will feel like a loss to both partners, and an agreement might be hard to come by. Loss aversion can also explain why the volume of real-estate sales decreases in an economic downturn. Sellers may find it so hard to sell their house at a loss, relative to the price at which the property was purchased, that they would rather not sell it at all. This kind of behavior even prevents people from upgrading from a smaller to a larger property, which is economically rational to do during a downturn.

Exercise 3.42 Latte discounts Many coffee shops give customers a small discount if they bring their own reusable mug. The coffee shops could equivalently lower all prices and add a small penalty to the bill of customers who do not bring their own mugs. Yet, few coffee shops go for the latter solution. Why?

In addition, loss aversion helps account for some of the phenomena studied earlier in this chapter, including the fact that people fail to take opportunity costs properly into account. If people treat out-of-pocket costs as losses and opportunity costs as forgone gains, loss aversion entails that out-of-pocket costs will loom larger than opportunity costs. Loss aversion may also help explain why people are so prone to honoring sunk costs. Since a sunk cost is often experienced as a loss, loss aversion entails that such costs will loom large, which in turn might drive people to honor sunk costs.

So far, we have largely assumed that the reference point is determined by a person's current endowment. This is not always the case. A person's reference point can be determined by her aspirations and expectations, among other things. The fact that reference points can be fixed by aspirations and expectations explains how people who get a five percent raise can feel cheated if they expected a ten percent raise, but be absolutely elated if they did not expect a raise at all.

Exercise 3.43 Bonuses Draw a graph that illustrates how a person who gets a five percent raise can be elated if she did not expect a raise at all, but feel cheated if she expected a ten percent raise.

Exercise 3.44 Exam scores Assume that Alysha and Billy have the following value function over exam scores: $v(x) = x/2$ for gains and $v(x) = 2x$ for losses. Both of them use the expected exam score as their reference point.

(**a**) Alysha expects to score 75 out of 100 on her upcoming midterm exam. She does better than expected. What is her gain, in terms of value, if her final score is 93?

(**b**) Billy also expects to score 75 out of 100 on his upcoming midterm exam. He does worse than expected. What is his loss, in terms of value, if his final score is 67?

(**c**) Insofar as you use your expectation as a reference point, what does the theory seem to say about maximizing value in your life: should you perform well or poorly in exams? Should you set high or low expectations for yourself?

A spectacular example of the outsize role that expectations can play is baseball player Barry Bonds:

> When Pittsburgh Pirate outfielder Barry Bonds's salary was raised from $850,000 in 1990 to $2.3 million in 1991, instead of the $3.25 million he had requested, Bonds sulked, "There is nothing Barry Bonds can do to satisfy Pittsburgh. I'm so sad all the time."

The Stoic philosopher Seneca diagnosed the problem about 2000 years ago. He wrote: "Excessive prosperity does indeed create greed in men, and never are desires so well controlled that they vanish once satisfied." Behavioral economists use the term **aspiration treadmill** to refer to a process where increasing endowments lead to rising aspirations. The result is that people like Barry Bonds are nowhere near as happy with their $2.3 million salary when they have it as they thought they would be before they started making that kind of money. Thus, the aspiration treadmill is often invoked to explain the fact that the marginal happiness of money is sharply diminishing (see Section 12.4).

The aspiration treadmill also has consequences for how we experience losses. A model like that represented in Figure 3.10 suggests that a rich person who loses money would experience a much smaller loss in utility than a poor person would. And yet, casual observation suggests many affluent people dislike, e.g., paying taxes no less than poor people do. Seneca had a comment on this too:

> For you are in error if you suppose that rich men put up with losses more cheerfully: the largest bodies feel the pain of a wound no less than the smallest ... You may be sure that rich and poor men are in the same position, that their suffering is no different; for money sticks fast to both groups, and cannot be torn away without their feeling it.

Some people succeed in jumping off the aspiration treadmill. A 2006 story in the *New York Times* described an internet entrepreneur who had just traded his Porsche Boxster for a Toyota Prius: "'I don't want to live the life of a Boxster, because when you get a Boxster you wish you had a 911,' he said, referring to a much more expensive Porsche. 'And you know what people who have 911s wish they had? They wish they had a Ferrari.'"

A person's reference point can also be determined by other people's achievements or endowments. Such **social comparisons** help explain the fact that whether a person is happy with his or her salary will depend in part on the salaries of neighbors, friends, and relatives. In this respect, income is similar to speed. Is 70 mph (110 km/h) fast or slow? If everybody else on the highway is driving 50 mph, 70 mph feels fast. But if everybody else is driving 80 mph, 70 mph feels slow. Thus, your sense for how fast you are traveling depends not just on your absolute speed, but on your relative speed. Similarly, your sense for how much money you earn depends not just on your absolute income, but on your relative income.

Exercise 3.45 Salary comparisons Insofar as you use other people's salaries as a reference point, what does the theory seem to say about maximizing value in your life?

Example 3.46 Salary comparisons, cont. At one university library, there is a single book that is so popular that it is chained to the checkout counter. You would hope that it is some important medical reference work. But it is the book that lists the salaries of all university employees. Presumably, no one would use the book to look up their own salary: that information is more easily accessible on one's monthly pay stub. In all likelihood, the book is so popular because people like to look up their colleagues' salaries.

As the last example illustrates, it is hard to deny that people engage in social comparisons. Social comparisons can also explain why bronze-medal winners can be more satisfied with their performance than silver-medal winners. Assuming that a bronze-medal winner compares himself or herself with the athletes who did not win a medal, the bronze medal represents achievement. But assuming that a silver-medal winner compares his or her performance with the gold-medal winner's, the silver medal represents defeat.

Loss aversion has other, and perhaps even more radical, implications for microeconomics. Because standard theory presupposes that preferences are independent of endowments, it implies that indifference curves are independent of endowments. Thus, if your indifference curves look like those in Figure 3.13(a), they do so independently of whether you happen to possess bundle x or y. This indifference curve is **reversible**, in the sense that it describes your preference over (actually, indifference between) x and y independently of your endowment.

By contrast, loss aversion entails that your indifference curves will not be independent of your current endowment. Suppose, for instance, that your value function is $v(x) = x$ for gains and $v(x) = 2x$ for losses, and that this is true for both apples and bananas. If you begin with bundle $y = \langle 3, 1 \rangle$ and lose an apple, you will require two additional bananas to make up for that loss. Hence, the indifference curve going through y will also go through $\langle 2, 3 \rangle$. If you start out with bundle y and lose a banana, you will require two additional apples to make up for that loss. Hence, the very same

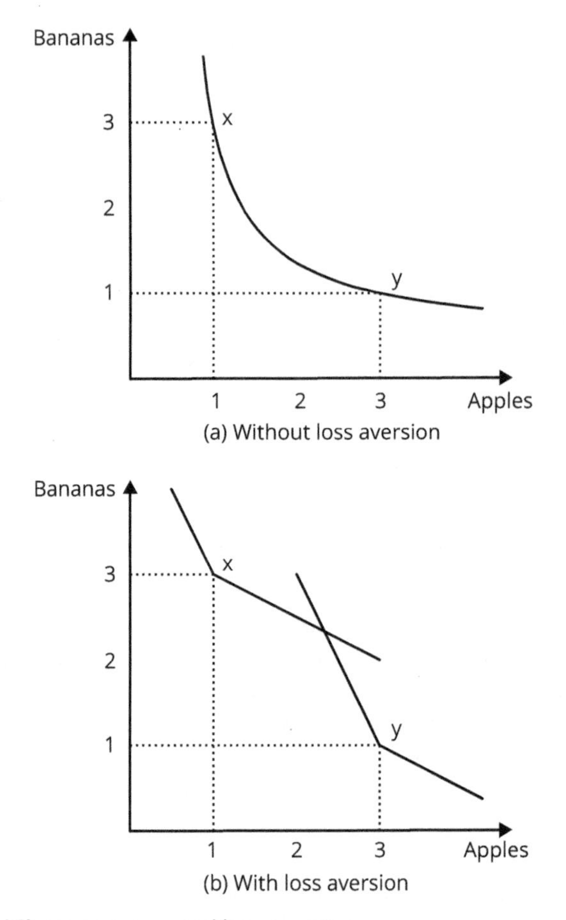

Figure 3.13 Indifference curves and loss aversion

indifference curve will also go through ⟨5, 0⟩. The result is similar if you begin with bundle $x = \langle 1, 3 \rangle$. If you are loss averse, then, your indifference curves will look like those in Figure 3.13(b). Notice that there are two indifference curves here – really, two sets of indifference curves – one for initial endowment x and one for initial endowment y.

Exercise 3.47 Value functions Suppose that you are loss averse, and that your value function is $v(x) = x$ for gains and $v(x) = 3x$ for losses.

(**a**) Represent the value function graphically, in the manner of Figure 3.11.

(**b**) Represent your indifference curves graphically, in the manner of Figure 3.13(b), assuming that your initial endowment is ⟨3, 4⟩.

A feature of having kinks in your indifference curves – as in Figure 3.13(b) – is that if you begin with x and are offered to trade it for y, you will reject the offer. At the same time, if you begin with y and are offered to trade it for x, you will reject this offer too. This phenomenon is sometimes referred to as **status quo bias**, because you exhibit a tendency to prefer the existing state of affairs under any circumstances. In the study of Cornell coffee mugs, a median number of 2 out of 22 mugs changed hands when participants were allowed to trade mugs for money; in the absence of status quo bias, one would expect the number of trades to equal about 11. This result is important among other things because it appears at odds with what has come to be known as the **Coase Theorem**. Loosely speaking, the theorem says that in the absence of transaction costs, bargaining will lead to an efficient allocation. But to the extent that people exhibit status quo bias, they may fail to reach an efficient bargaining solution even when transaction costs are zero.

Exercise 3.48 Health care Broadly speaking, in Europe, health care is provided by the government and paid for by taxes. That is, individuals are taxed by the government, which then uses the tax money to provide health-care services for the citizens. In the US, again broadly speaking, health care is largely purchased privately. That is, individuals pay lower taxes, and (if they so choose) use their money to purchase health-care services.

(a) Use the notion of status quo bias to provide a detailed explanation of the following paradox: most Americans prefer their system to a European-style one, whereas most Europeans prefer their system to an American-style one.
(b) Illustrate your answer with a graph showing the indifference curves of a typical American and a typical European.
(c) Imagine that the US were to adopt a European-style health-care system. How should we expect Americans to feel about their new health-care system then, and how easy would it be for the opposition party to switch back?

Exercise 3.49 Affordable Care Act Writing in the *Washington Examiner* in July of 2013, Byron York explained why Democrats were so eager to implement the Affordable Care Act, a.k.a. Obamacare, which aims to provide health insurance for otherwise uninsured Americans:

> Obamacare is designed to increase the number of Americans who depend on the government to pay for health insurance ... In all, the government will be transferring hundreds of billions of dollars to Americans for health coverage. The White House knows that once those payments begin, repealing Obamacare will no longer be an abstract question of removing legislation not yet in effect. Instead, it will be a very real matter of taking money away from people. It's very, very hard to do that.

What principle does this insight embody?

Exercise 3.50 Milton and Rose Friedman Nobel laureate Milton Friedman and Rose Friedman famously wrote: "Nothing is so permanent as a temporary government program."

(a) Use the concept of loss aversion to explain why supposedly temporary government programs have a tendency to last longer than originally intended.
(b) Some government programs come with a sunset provision, which states that the law will be cancelled after a specific date. Explain how such provisions aim to solve the problem you identified under (a).

For better or worse, loss aversion and status quo bias make it rational for those who want to see expanded government programs to push ahead even with imperfect proposals, since it is easier to fix a flawed program later than it is to enact it in the first place. Loss aversion and status quo bias also make it rational for people who oppose them to aggressively resist *any* expansion, since it is very difficult to go back once a program is in place.

Status quo bias can be a real obstacle not only to sensible policies, but also to personal growth.

Example 3.51 Clutter While many people across the world continue to struggle to meet fundamental needs, more and more people in the developed world have the opposite problem: too much stuff. Although American homes have grown dramatically over the last 50 years or so, the amount of stuff people own has increased even faster. Consequently, the self-storage industry brags that it is one of the fastest-growing sectors of commercial real estate, bringing in almost $40 billion annually in the US alone. There is an entire body of self-help literature advising people to get rid of clutter in exchange for simplicity, happiness, and peace of mind, signifying both that people want to do so and that they are unable to. Why is decluttering so hard?

Loss aversion makes the gain of simplicity, happiness, and peace of mind seem small relative to the loss of all that stuff – prompting people to forgo the former in order to avoid the latter. The result is that people are biased in favor of the status quo, even when they acknowledge that it is inferior. The good news is that it may be possible to hack your mind by framing the decisions differently. If it is hard to get rid of that collection of VHS tapes you inherited, do not ask yourself, "Can I throw this away?" Instead, ask yourself, "If I didn't already own this, would I buy it?" If the answer is no, you know what to do. Or, if it is hard to get rid of half of your CD collection, do not ask yourself, "Which CDs should I throw away?" Instead, tell yourself that you are going to get rid of the entire collection and then ask yourself, "Which CDs should I keep?" Chances are it will be an easier process when the decision to declutter is not framed as a loss.

Status quo bias can explain why many people oppose human genetic enhancement. Many of us would be unwilling to give up our natural, pristine "unenhanced" state in exchange for an increase in intelligence quotient (IQ); yet, if we were the beneficiaries of genetic enhancement, it is hard to imagine that we would willingly accept a decrease in IQ in exchange for the natural, pristine "unenhanced" state. Also, status quo bias explains why many people oppose free trade when they do not have it but are in favor of it when they do. Many people who reject free-trade agreements with other countries (with which they do not already trade freely) would protest loudly if somebody proposed eliminating free-trade agreements already in existence, for example, with adjoining states and regions. In all these cases, people are unmoved by the prospect of gaining a benefit that they do not already have, but deeply averse to losing some benefit that they already have.

Loss aversion must not be confused with diminishing marginal utility. If people would be willing to pay less for a mug that they do not own than they would accept in return for the mug that they do own, this may reflect diminishing marginal utility for mugs. Going back to Figure 3.10, notice that the utility derived from the second mug (that is, the marginal utility of the second mug) is much lower than the utility derived from the first mug. It is important not to attribute loss aversion to agents whose behavior is better explained in terms of diminishing marginal utility.

3.6 Anchoring and adjustment

Imagine that you subject people to the following two experiments. If your research participants protest that they do not know the answers to the questions, tell them to offer their best guesses.

Example 3.52 Africa and the UN Spin a wheel of fortune to come up with a number between 0 and 100, and invite your participants to answer the following two questions:

(a) Is the percentage of African nations in the United Nations (UN) greater than or less than the number?

(b) What is the actual percentage of African nations in the UN?

You probably would not expect the answer to (b) to reflect the random number generated by the wheel of fortune. Yet evidence suggests that you would find a correlation between the two. In one study, when the starting point was 10, the median answer to (b) was 25; when the starting point was 65, the median answer was 45. (The correct answer is 28 percent.)

Example 3.53 Multiplication Give people 5 seconds to come up with an answer to either one of the following multiplication problems:

(a) Compute: $1 * 2 * 3 * 4 * 5 * 6 * 7 * 8$

(b) Compute: $8 * 7 * 6 * 5 * 4 * 3 * 2 * 1$

Given that (a) and (b) are mathematically equivalent, you might expect your research participants to come up with more or less the same answer independently of which question they were asked. Yet, when a group of high school students were asked the first question, the median answer was 512; when they were asked the second question, the median answer was 2250. (The correct answer is 40,320.)

These phenomena are frequently explained by reference to **anchoring and adjustment**, which is a cognitive process that can be used when forming judgments. As the name suggests, anchoring and adjustment is a two-stage process: first, you pick an initial estimate called an **anchor**, and second, you adjust the initial estimate up or down (as you see fit) in order to come up with a final answer. When deciding what a used car with a $15k price tag is worth, for example, you might start by asking yourself whether it is worth $15k and then adjust your estimate as required. If you think $15k is too much, you adjust your estimate of its worth downward; if you think $15k is too little, you adjust it upward.

According to one prominent account of human judgment and decision-making – the **heuristics-and-biases program** – we make judgments not by actually computing probabilities and utilities but by following **heuristics**. A heuristic is a rule of thumb or mental shortcut that can be used when forming judgments. Heuristics are assumed to be functional, in the sense that they reduce the time and effort required to solve everyday problems and produce approximately correct answers under a wide range of conditions. But they are not assumed to be perfect: under certain circumstances, they can fail in predictable fashion. Because the consistent application of a heuristic can lead to answers that are systematically and predictably wrong, we say that it can lead to **bias**. Thus, an account according to which we follow heuristics can help explain both why we oftentimes are able to make quick and perfectly appropriate judgments, and why we sometimes go wrong.

Anchoring and adjustment is one of the heuristics identified by the heuristics-and-biases program. Like all heuristics, anchoring and adjustment is thought to be functional but can at the same time lead to bias under certain conditions. Evidence suggests that the adjustment is often insufficient. This means that the final judgment will to some extent be a function of the anchor, which may be perfectly arbitrary. If the anchor is very different from the true answer, anchoring and insufficient adjustment can generate highly inaccurate answers.

Consider Example 3.52. People's answer to the question about the percentage of African nations in the UN can be explained by saying that they take the random number as an anchor and adjust the answer up or down as they see fit. If the random number is 65, they begin with 65, then (assuming this number strikes them as too high) adjust downward. If, instead, the random number is 10, then (assuming this strikes them as too low) they adjust upward. Insofar as the adjustment is insufficient, we should expect the final estimate to be higher if the random number is 65 than if it is 10.

Consider Example 3.53. Under time pressure, students presumably perform a few steps of the multiplication problem (as time allows) and adjust upward to compensate for the missing steps. Insofar as the adjustment is insufficient, you would expect the answers to be too low. Moreover, because people who answer (a) will get a lower number after a few steps than people who answer (b), you would expect the former to offer a lower estimate than the latter. As you can tell, this is exactly what happened.

Anchoring and adjustment might affect a wide range of judgments. Recall the following famous story.

Exercise 3.54 Invention of chess According to legend, the inventor of chess was asked by the emperor what he (the inventor) wanted in return for his invention. The inventor responded: "One grain of rice for the first square on the chess-board, two grains for the second square, four grains for the third square, and so on." The emperor was happy to oblige. Only there are 64 squares on the chess board, so on the 64th day, the inventor could demand $2^{64-1} \approx 10^{19} = 10{,}000{,}000{,}000{,}000{,}000{,}000$ grains of rice, a figure much greater than what the emperor had expected and could afford. Use the idea of anchoring and adjustment to explain how the emperor could underestimate the number so dramatically.

So far we have talked about anchoring and adjustment as something that affects beliefs, but there is evidence that it also affects preferences. In one study, experimenters showed MBA students various products, and asked them, first, whether they would be willing to buy the product for a price equal to the last two digits of their social security number, and second, to state their WTP. When people in the lowest quintile (the lowest 20 percent of the distribution with respect to social security numbers) were willing to pay $8.64 on the mean for a cordless trackball, people in the highest quintile were willing to pay $26.18. When people in the lowest quintile were willing to pay $11.73 for a bottle of fine wine, people in the highest quintile were willing to pay $37.55. Thus, people in the highest quintile were sometimes willing to pay more than three times as much as people in the lowest quintile. These results are easily explained by anchoring and (insufficient) adjustment, if we assume that the study participants used the last two digits of their social security number as an anchor.

From the discussion at the beginning of Section 3.5, it should be clear why this behavior pattern is irrational. The behavior pattern can be said to violate **procedure invariance**: the proposition that a stated preference should not differ depending on the method used to elicit it.

Anchoring and adjustment can explain a whole range of phenomena. It can explain, for instance, why it is so common to lure customers by lines such as these: "Used to be $50! Now only $24.99!" or "A $500 value for only $399" or "Suggested retail price: $14.99. Now, only $9.99." Sellers might hope that potential customers will form a judgment about the dollar value of the product by using the first amount as an anchor. The seller might realize that the customer would not be willing to pay $24.99 for the product if asked a direct question; however, the seller may be hoping that the customer will start with the $50 figure and insufficiently adjust downward, and therefore end up with a final WTP exceeding $24.99. That is, the seller hopes that people will use the greater number as an anchor. Anchoring and adjustment might also explain why realtors often publish an asking price that is higher than what they expect to receive for the property: by publishing a higher number, they might hope to influence what potential buyers would be willing to pay.

Exercise 3.55 Toasters Suppose that you wanted to sell toasters for $160, which will strike customers like a lot. How does research on anchoring and adjustment suggest that you do it?

You can also use anchoring and adjustment to increase the quantity that customers buy. Promotions of the form "3 for $2," "Limit of 6 per person," and "Buy 10 get 2 for free" work this way. To the extent that customers use the suggested quantity as an anchor and adjust insufficiently downward when deciding what quantity to purchase, such promotions can dramatically increase sales.

Anchoring and adjustment might also play a role in other kinds of decision. A German study of experienced judges and prosecutors found that sentencing decisions reflected irrelevant information provided to them by the researchers. Before handing down their decisions in a realistic but fictional case of sexual assault, the legal professionals were asked to imagine that a journalist called to ask if the sentence would be higher or lower than x years. The call should have no influence on the decision, but the researchers found that it did: when x was 1, the recommended sentence was 25 months; when x was 3, the recommended sentence was 33 months. Most amazingly, the difference between conditions remained significant even when the participants in the study were told that the anchor was generated randomly – *and when it was generated randomly right before their eyes by the rolling of dice.*

It is important not to exaggerate people's susceptibility to anchoring-and-adjustment-related bias. It may be that people respond to suggested retail prices (and the like) because they take a high suggested retail price to signal high quality. If so, their behavior may not be due to anchoring and adjustment at all. They would not even be making a choice under certainty. That said, this line of argument cannot explain how roulette wheels and social security numbers can affect behavior, on the assumption that no one would take such numbers as a mark of quality.

Either way, the heuristics-and-biases program has been enormously influential, and we will continue to discuss it below (for example in Sections 5.2 and 5.6). Because it carries so much explanatory power, we will return to anchoring and adjustment repeatedly (see, for example, Exercise 4.30 on page 83).

3.7 Discussion

This chapter has reviewed a number of different phenomena that appear to pose a problem for the theory that we learned in Chapter 2. Most of these phenomena are presented as challenges to the descriptive adequacy of rational-choice theory. Thus, behavioral economists think of the manner in which people ignore opportunity costs but honor sunk costs, exhibit menu dependence, overweight losses relative to gains, and permit arbitrary anchors to unduly affect their behavior as inconsistent with the view that people actually behave in accordance with rational-choice theory. Though far from universal, the deviations appear to be substantial, systematic, and predictable. Examples have illustrated that these phenomena can be costly indeed, not just in terms of time, effort, and money, but also in terms of human lives. Some of the phenomena that we have studied can also be construed as challenges to the normative adequacy of the theory. When it comes to opportunity costs, for example, we noted how extraordinarily demanding the theory can be. It has been argued that this makes the theory unsuitable as a normative theory. Obviously, this chapter does not pretend to have presented a complete list of phenomena that are at odds with standard theory of choice under certainty.

We have also reviewed some basic building blocks of theories that behavioral economists have proposed to account for phenomena that cannot be captured within standard economics. We have also studied the value function of prospect theory, which is one of the most prominent theories to emerge from behavioral economics. And we have come across the notion of a heuristic, which is essential to the enormously influential heuristics-and-biases program. We will return to prospect theory and the heuristics-and-biases program repeatedly below. Studying these theories will give you a better idea of what behavioral economists do, beyond finding fault with standard theory.

We have come across the idea of reason-based choice, which says that choices are made on the basis of a search for reasons (which need not involve probabilities and utilities). Although it is often proposed that choosing on the basis of reasons is the hallmark of rationality, as we saw, reason-based choice can generate irrational choice patterns.

Exercise 3.56 Reasons to accept vs. reasons to reject In a 1993 study, participants were asked to adjudicate an only-child sole-custody case following a messy divorce. Parent A was described as average in terms of income, health, work–life balance, rapport with child, and social stability. Parent B was described as having big strengths but also big weaknesses: a very close relationship with the child, but also lots of work-related travel, and so on. When participants were asked to whom custody should be awarded, 64 percent favored B. But when they were asked to whom custody should be denied, 55 percent also said B. Use the idea of reason-based choice to explain this pattern.

In the process, we have come across a variety of ways in which knowledge of behavioral economics permits you to influence the behavior of others. By appealing to sunk costs, introducing asymmetrically dominated alternatives, altering the frame in which options are presented, or introducing arbitrary anchors, you can affect other people's evaluation of various options. Since standard economic models frequently treat the options available to consumers as ordered *n*-tuples of price and product characteristics, they leave out many variables of interest to marketers, medical doctors, public health officials, and the like. By bringing things such as framing and reference points into the picture, behavioral economics offers a wider range of levers that can be used to influence people's behavior (for good or for evil). But knowledge of behavioral economics can also help you resist other people's efforts to manipulate your behavior. By being aware of the manner in which sunk costs, inferior alternatives, framing, and heuristics affect your behavior, you may be less likely to fall for other people's tricks.

We will return to these themes in later chapters, after we have studied the standard theory of judgment under risk and uncertainty.

Additional exercises

Exercise 3.57 AEA The following question was famously asked of 200 professional economists at the 2005 meeting of the American Economic Association (AEA). By looking up the answer in the answer key, you can compare your performance with theirs:

> You won a free ticket to see an Eric Clapton concert (which has no resale value). Bob Dylan is performing on the same night and is your next-best alternative activity. Tickets to see Dylan cost $40. On any given day, you would be willing to pay up to $50 to see Dylan. Assume there are no other costs of seeing either performer. Based on this information, what is the opportunity cost of seeing Eric Clapton? (a) $0, (b) $10, (c) $40, or (d) $50.

Exercise 3.58 Does money buy happiness? Research by economists Betsey Stevenson and Justin Wolfers finds that the marginal happiness of money is positive (though sharply diminishing) at all levels of income. This means that all things equal, more money would make the average person happier (though less and less so when she becomes richer).

Writing in *The Atlantic*, Derek Thompson concludes: "Money Buys Happiness and You Can Never Have Too Much." But it would be a mistake to infer that the average person would be happier if she made more money. What is overlooked here?

Exercise 3.59 Hot yoga "You're back; you must have liked it!," says the hot-yoga studio receptionist the second time you use your one-week pass. What fallacy might the receptionist be unfamiliar with?

Exercise 3.60 Mr Humphryes The following quotation is from a 2009 news story titled "Tension builds around courthouses' reopening." The controversy concerns whether to reopen a satellite courthouse in a building that the county owns or one in a building that the county leases. What fallacy does Mr Humphryes commit?

> The county owns the Centerpoint Building and leases Gardendale and Forestdale buildings. [Commissioner Bobby] Humphryes believes it would make economic sense to open one of the buildings [the] county leases. "I think it's senseless to shut down a building we have leases on when we let the others remain idle, we don't have leases on and don't owe money on," Humphryes said.

Exercise 3.61 Pear Corporation The Pear computer company is introducing a new line of tablet computers. The Macro has huge storage capacity but is not very affordable. The Micro has limited storage capacity but is very affordable.

(**a**) Market research suggests that a typical consumer tends to be indifferent between the Micro and the Macro. Draw a graph with storage capacity on the x-axis and affordability on the y-axis. Use a solid line to represent a typical consumer's indifference curves going through the Micro and the Macro.

(**b**) Pear wants to steer customers toward the more expensive tablet computer. They decide to use a decoy, which they will call the Dud, to accomplish this goal. Use an "X" to mark the location of the Dud in the graph.

(**c**) Use dashed lines to show what the typical consumer's indifference curves would look like if the introduction of the Dud had the intended effect.

Exercise 3.62 Tim and Bill Tim and Bill are addicted to Apple products. They are looking at the new iPhone at a store that offers a no-questions-asked return policy. They are not sure the new features are worth it. Tim decides to take one home, thinking that he can always return it tomorrow. Bill decides against taking one home, thinking that he can always come back and pick one up tomorrow. They are both loss averse, with a value function over iPhones of $v(x) = x$ for gains and $v(x) = 3x$ for losses. Ignore transaction costs.

(**a**) After Tim has taken his phone home, he incorporates it into his endowment. How much of a loss, in value terms, would he incur by returning it tomorrow?

(**b**) Bill, who does not take his phone home, does not incorporate it into his endowment. How much of a forgone gain, in value terms, does the phone he does not own represent to him?

(**c**) Who is more likely to end up the owner of an iPhone, Tim or Bill?

Exercise 3.63 Larry and Janet Larry and Janet are loss averse: their value function is $v(x) = x/3$ for gains and $v(x) = 3x$ for losses. The two hold stock in the same company. They bought it yesterday, when the stock was worth $7. Today, unfortunately, it dropped to $4.

(a) Larry uses the original price ($7) as his reference point. If you ask him, how much value did he lose when the price dropped to $4?

(b) Janet uses the new price ($4) as her reference point. If you ask her, how much value did she lose when the price dropped to $4?

(c) Who is more disappointed, Larry or Janet?

Exercise 3.64 W. E. B. Du Bois The great African–American scholar and activist W. E. B. Du Bois said: "The most important thing to remember is this: To be ready at any moment to give up what you are for what you might become." Why is this so hard, and why do people have to be reminded?

Exercise 3.65 Seneca What phenomenon might Seneca have had in mind when he wrote the following? "Let us turn now to inheritances, the greatest cause of human distress; for should you compare all the other ills that make us suffer…with the evils that our money causes us, this portion will easily preponderate."

Exercise 3.66 Match each of the vignettes below with one of the following phenomena: *anchoring and adjustment, compromise effect, failure to consider opportunity costs, loss aversion*, and *sunk-cost fallacy*. If in doubt, pick the best fit.

(a) Adam has just arrived at the movie theater when he realizes that he has lost the $10 ticket he bought just before dinner. The theater staff informs him that there are more tickets available. "I'm not buying another ticket," he tells them. "This movie may be worth ten bucks but there's no way I'm going to pay $20 for it." He is angry with himself all night.

(b) Bruce is buying a new car, which he intends to keep for a long time. The car dealership has two new cars for sale. There is no difference except for the color: one is red and one is metallic blue. The red one used to be $15,995 and the metallic blue one $16,495; both are on sale for $14,995. Bruce has no particular preference for blue, and fears that the metallic finish will make him look unmanly. Still, thinking about what these cars used to cost, it seems to him that it would be worth paying more for the blue one, which makes it seem like a better deal. He tells the car salesman that he will take the blue one, though he secretly wonders how his friends will react.

(c) The owners of the local organic grocery store decide that they want to encourage their customers to use fewer plastic bags. They decide that the customers would be upset if the store started charging money for the bags – after all, people expect their plastic bags to be free. Instead, they gradually raise the prices of their goods by an average of 25 cents per customer, and give people who bring their own bags a 25-cent discount. Current customers have no problem with this arrangement.

(d) A philosophy department is hiring a new professor. There are two candidates. Dr A does esthetics. Dr E does ethics. Everybody agrees that Dr E is the most accomplished philosopher. However, Prof. P maintains that Dr A should be hired anyway, because the Department used to have a spectacularly talented professor in esthetics.

(e) Erica is a highly paid neurosurgeon with no shortage of work. Every second Friday afternoon she leaves work early to go home and mow the lawn. She takes no pleasure in mowing the lawn, but she cannot bring herself to pay the lawn company $75 to do it for her.

(f) Frank is looking at strollers for his first child. He is undecided between the basic SE model and the slightly more upscale CE model. Suddenly, he realizes that there exists a third option, the extra-special XS model. He settles on the CE model.

Problem 3.67 *Drawing on your own experience, make up stories like those in Exercise 3.66 to illustrate the various ideas that you have read about in this chapter.*

Further reading

The jacket/calculator problem appears in Tversky and Kahneman (1981, p. 457). The idea that choices can lead to unhappiness is explored in Schwartz (2004, pp. 122–3). The classic analysis of sunk costs is Arkes and Blumer (1985), but see also Hastie and Dawes (2010, pp. 34–42). Reading habits are discussed in Goodreads.com (2013), and the F-35 in Francis (2014). The line from Aristotle is from the *Nicomachean Ethics* (1999 [*c* 350 BCE], p. 87); escalation behavior is analyzed in Staw and Ross (1989), the source of the George Ball quotation (p. 216); and the California railroad is discussed in Nagourney (2011). Decoy effects are discussed in Huber et al. (1982) and Ariely (2008); the latter is the source of the example involving subscription offers (pp. 1–6). The classic text on reason-based choice is Shafir et al. (1993). A great review of loss aversion, the endowment effect, and status quo bias is Kahneman et al. (1991); the fate of the speeding Irishman was reported by the Associated Press (2007); the Adam Smith quotation is from Smith (2002 [1759], p. 151); the fate of the Florida waitress was chronicled by the *St. Petersburg Times* (2001); and Barry Bonds was quoted in Myers (1992, p. 57). The passages from Seneca appear in Seneca (2007 [*c* 49], pp. 124–5); the internet entrepreneur is cited in Hafner (2006). Byron York (2013) discusses taking money away from people; Friedman and Friedman (1984) criticize "temporary" government programs. The classic discussion of anchoring and adjustment is Tversky and Kahneman (1974), which is also the source of the examples and data cited early in the section; the follow-up experiments are described in Ariely et al. (2003). The German study of judges and prosecutors is Englich et al. (2006). The study of professional economists is discussed in Frank (2005), and the marginal happiness of money in Thompson (2013). Mr Humphryes is cited by FOX6 WBRC (2009) and Du Bois in Hampton (2012, p. 185).

PART 2

JUDGMENT UNDER RISK AND UNCERTAINTY

4 PROBABILITY JUDGMENT

Learning objectives

After studying this chapter you will:

- Know the theory of probability
- Be able to prove important principles of probability, including Bayes's theorem, on the basis of fundamental rules of probability
- Understand how the theory of probability can be used as a theory of rational belief

4.1 Introduction

Though the theory of choice explored in Part 1 is helpful for a range of purposes, most real-life decisions are not choices under certainty. When you decide whether to start a company, buy stocks, propose to the love of your life, or have a medical procedure, you will not typically know at the time of making the decision what the outcome of each available act would be. In order to capture what people do, and what they should do, in such situations, we need another theory. Part 2 explores theories of judgment: how people form and change beliefs. In Part 3, we will return to the topic of decision-making.

In this chapter, we explore the theory of probability. There is wide – but far from complete – agreement that this is the correct normative theory of probabilistic judgment, that is, that it correctly captures how we should make probabilistic judgments. Consequently, the theory of probability is widely used in statistics, engineering, finance, public health, and elsewhere. Moreover, the theory can be used as a descriptive theory about how people make judgments, and it can be used as part of a theory about how they make decisions.

Like the theory of rational choice under certainty, probability theory is axiomatic. Thus, we begin by learning a set of axioms – which will be called "rules" – and which you will have to take for granted. Most of the time, this is not hard: once you understand them, the rules may strike you as intuitively plausible. We will also adopt a series of definitions. Having done that, everything else can be derived. Thus, we will spend a great deal of time below proving increasingly interesting and powerful principles on the basis of axioms and definitions.

4.2 Fundamentals of probability theory

Here are two classic examples of probability judgment.

Example 4.1 Mrs Jones's children You are visiting your new neighbor, Mrs Jones. Mrs Jones tells you that she has two children, who are playing in their room. Assume that each time somebody has a child, the probability of having a girl is the same as the probability of having a boy (and that there are no other possibilities). Moreover, whether the mother had a boy or a girl the first time around does not affect the probabilities involved the second time around. Now, Mrs Jones tells you that at least one of the children is a girl. What is the probability that the other child is a girl too?

Example 4.2 The Linda problem Linda is 31 years old, single, outspoken, and very bright. She majored in philosophy. As a student, she was deeply concerned with issues of discrimination and social justice and also participated in anti-nuclear demonstrations.
(**a**) What is the probability that Linda is a bank teller?
(**b**) What is the probability that Linda is a bank teller and a feminist?

Answers to these questions will be given once we have developed the apparatus required to address them rigorously. For now, I will just note that one reason why this theory is interesting is that people's intuitive probability judgments – and therefore many of their decisions – tend to fail in predictable ways.

Before we start, we need to develop a conceptual apparatus that will permit us to speak more clearly about the subject matter. For example, we want to talk about the different things that can conceivably happen. When you flip a coin, for instance, you can get heads or you can get tails; when you roll a six-sided die, you can get any number from one to six.

Definition 4.3 Definition of outcome space *The **outcome space** is the set of all possible individual outcomes.*

We represent outcome spaces following standard conventions, using curly brackets and commas. To denote the outcome space associated with flipping a coin, we write: {Heads, Tails} or {H, T}. To denote the outcome space associated with rolling a six-sided die, we write: {1, 2, 3, 4, 5, 6}.

Oftentimes, we want to talk about what actually happened or about what may happen. If so, we are talking about actual outcomes, as in "the coin came up tails" and "I might roll snake eyes (two ones)."

Definition 4.4 Definition of outcome *An **outcome** is a subset of the outcome space.*

We write outcomes following the same conventions. Thus, some of the outcomes associated with one roll of a six-sided die include: {1} for one, {6} for six, {1, 2, 3} for a number less than or equal to three, and {2, 4, 6} for an even number. There is one exception: when the outcome only has one member, we may omit the curly brackets and write 6 instead of {6}. Notice that in all these cases, the outcomes are subsets of the outcome space.

Definition 4.5 Definition of probability *The* ***probability function*** *is a function* $Pr(\cdot)$ *that assigns a real number to each outcome. The* ***probability*** *of an outcome A is the number* $Pr(A)$ *assigned to A by the probability function* $Pr(\cdot)$.

Hence, the probability of rolling an even number when rolling a six-sided die is denoted Pr({2, 4, 6}). The probability of rolling a six is denoted Pr({6}), or relying on the convention introduced above, Pr(6). The probability of getting heads when flipping a coin is denoted Pr({H}) or Pr(H). The probability of an outcome, of course, represents (in some sense) the chance of that outcome happening. Sometimes people talk about odds instead of probabilities. Odds and probabilities are obviously related, but they are not identical. Refer to the text box on page 81 for more about odds.

The next propositions describe the properties of this probability function. They will be referred to as the **rules** or **axioms** of probability.

Axiom 4.6 The range of probabilities *The probability of any outcome A is a number between 0 and 1 inclusive; that is,* $0 \leq Pr(A) \leq 1$.

Thus, probabilities have to be numbers no smaller than zero and no greater than one. Equivalently, probabilities can be no lower than 0 percent and no greater than 100 percent. You might not know the probability that your internet startup company will survive its first year. But you do know this: the probability is no lower than 0 percent and no greater than 100 percent.

In general, it can be difficult to compute probabilities. People such as engineers and public health officials spend a lot of time trying to determine the probabilities of events such as nuclear disasters and global pandemics. There is one case in which computing probabilities is easy, however, and that is in the case when individual outcomes are equally probable, or **equiprobable**.

Axiom 4.7 The EQUIPROBABILITY rule *If there are n equally probable individual outcomes* $\{A_1, A_2, \ldots, A_n\}$*, then the probability of any one individual outcome* A_i *is* $1/n$*; that is,* $Pr(A_i) = 1/n$.

Suppose we are asked to compute the probability of getting a four when rolling a fair die. Because all outcomes are equally likely (that is what it means for the die to be fair) and because there are six outcomes, the probability of getting a four is 1/6.

So Pr(4) = 1/6. Similarly, the probability of getting heads when flipping a fair coin is 1/2. So Pr(H) = 1/2.

Exercise 4.8 Matching cards Suppose that you are drawing one card each from two thoroughly shuffled but otherwise normal decks of cards. What is the probability that you draw the same card from the two decks?

You could answer this question by analyzing all $52^2 = 2704$ different outcomes associated with drawing two cards from two decks. The easiest way to think about it, though, is to ask what it would take for the second card to match the first.

Exercise 4.9 The Large Hadron Collider According to some critics, the Large Hadron Collider has a 50 percent chance of destroying the world. In an interview with the *Daily Show*'s John Oliver on April 30, 2009, science teacher Walter Wagner argued: "If you have something that can happen, and something that won't necessarily happen, it's going to either happen or it's going to not happen, and so the best guess is one in two." Why is this not a correct application of the EQUIPROBABILITY rule?

As it happens, we have already developed enough of an apparatus to address Example 4.1. First, we need to identify the outcome space associated with having two children. Writing G for "girl" and B for "boy," and BG for "the oldest child is a boy and the youngest child is a girl," the outcome space is {GG, GB, BG, BB}. Once you learn that at least one of the children is a girl, you know for a fact that it is not the case that both children are boys. That is, you know that BB does not obtain. This means that the outcome space has been reduced to {GG, GB, BG}. In only one of three cases (GG) is the other child a girl also. Because these three outcomes are equally likely, you can apply the EQUIPROBABILITY rule to find that the probability that the other child is a girl equals Pr(GG) = 1/3.

Exercise 4.10 Mrs Jones's children, cont. Instead of telling you that at least one of the children is a girl, Mrs Jones tells you that her oldest child is a girl. Now, what is the probability that the other child is also a girl?

Exercise 4.11 Mr Peters's children Your other neighbor, Mr Peters, has three children. Having just moved to the neighborhood, you do not know whether the children are boys or girls. Let us assume that every time Mr Peters had a child, he was equally likely to have a boy and a girl (and that there are no other possibilities).

(**a**) What is the relevant outcome space?

(**b**) Imagine that you learn that at least one of the children is a girl. What is the new outcome space?

(**c**) Given that you know that at least one of the children is a girl, what is the probability that Mr Peters has three girls?

(**d**) Imagine that you learn that at least two of the children are girls. What is the new outcome space?

(**e**) Given that you know that at least two of the children are girls, what is the probability that Mr Peters has three girls?

Exercise 4.12 Three-card swindle Your friend Bill is showing you his new deck of cards. The deck consists of only three cards. The first card is white on both sides. The second card is red on both sides. The third card is white on one side and red on the other. Now Bill shuffles the deck well, occasionally turning individual cards over in the process. Perhaps he puts them all in a hat and shakes the hat for a long time. Then he puts the stacked deck on the table in such a way that you can see the visible face of the top card only.

(**a**) What is the outcome space? Write "W/R" to denote the outcome where the visible side of the top card is white and the other side is red, and so on.

(**b**) After shuffling, the visible side of the top card is white. What is the new outcome space?

(**c**) Given that the visible side of the top card is white, what is the probability that the other side of the top card is red?

This last exercise is called the "three-card swindle," because it can be used to fool people into giving up their money. If you bet ten dollars that the other side is white, you will find that many people are willing to accept the bet. This is so because they (mistakenly) believe that the probability is 50 percent. You might lose. Yet, because you have got the probabilities on your side, on average you will make money. It is not clear that this game deserves the name "swindle" since it involves no deception. Still, because this might be illegal where you live, you did not hear it from me.

Exercise 4.13 Four-card swindle Your other friend Bull has another deck of cards. This deck has four cards: one card is white on both sides; one card is black on both sides; one card is red on both sides; and one card is white on one side and red on the other. Imagine that you shuffle the deck well, including turning individual cards upside down every so often.

(**a**) What is the outcome space? Write "W/R" to denote the outcome where the visible side of the top card is white and the other side is red, and so on.

(**b**) Suppose that after shuffling, the visible side of the top card is black. What is the new outcome space?

(**c**) Given that the visible side of the top card is black, what is the probability that the other side of the card is black as well?

(**d**) Suppose that after shuffling, the visible side of the top card is red. What is the new outcome space?

(**e**) Given that the visible side of the top card is red, what is the probability that the other side of the card is white?

We end this section with one more exercise.

Exercise 4.14 The Monty Hall problem You are on a game show. The host gives you the choice of three doors, all of which are closed. Behind one door there is a car; behind the others are goats. Here is what will happen. First, you will point to a door. Next, the host, who knows what is behind each door and who is doing his best to make sure you do not get the car, will open one of the other two doors (which will have a goat). Finally, you can choose to open either one of the remaining two closed doors; that is, you can keep pointing to the same door, or you can switch. If you do not switch, what is the probability of finding the car?

4.3 Unconditional probability

The theory should also allow us to compute unknown probabilities on the basis of known probabilities. In this section we study four rules that do this.

Axiom 4.15 The OR rule *If two outcomes A and B are mutually exclusive (see below), then the probability of A OR B equals the probability of A plus the probability of B; that is, $Pr(A \vee B) = Pr(A) + Pr(B)$.*

Suppose that you want to know the probability of rolling a one or a two when you roll a fair six-sided die. The OR rule tells you that the answer is $\Pr(1 \vee 2) = \Pr(1) + \Pr(2) = 1/6 + 1/6 = 1/3$. Or suppose that you want to know the probability of flipping heads or tails when flipping a fair coin. The same rule tells you that $\Pr(H \vee T) = \Pr(H) + \Pr(T) = 1/2 + 1/2 = 1$.

Notice that the rule requires that the two outcomes be **mutually exclusive**. What does this mean? Two outcomes *A* and *B* are mutually exclusive just in case at most one of them can happen. In the previous two examples, this condition holds. When flipping a coin, H and T are mutually exclusive since at most one of them can occur any time you flip a coin. No coin can land heads and tails at the same time. Similarly, when you roll one die, one and two are mutually exclusive, since at most one of them can occur. Notice that the latter two outcomes are mutually exclusive even though neither one may occur.

Exercise 4.16 Mutual exclusivity Which pairs of outcomes are mutually exclusive? More than one answer may be correct.
(**a**) It is your birthday; you have a test.
(**b**) It rains; night falls.
(**c**) You get Bs in all of your classes; you get a 4.0 GPA.
(**d**) Your new computer is a Mac; your new computer is a PC.
(**e**) You are a remarkable student; you get a good job after graduation.

Exercise 4.17 What is the probability of drawing an ace when drawing one card from a regular (well-shuffled) deck of cards? If you intend to apply the OR rule, do not forget to check that the relevant outcomes are mutually exclusive.

The importance of checking whether two outcomes are mutually exclusive is best emphasized by giving an example. What is the probability of rolling a fair die and getting a number that is either strictly less than six or strictly greater than one? It is quite obvious that you could not fail to roll a number strictly less than six or strictly greater than one, so the probability must be 100 percent. If you tried to take the probability that you roll a number strictly less than six *plus* the probability that you roll a number strictly greater than one, you would end up with a number greater than one, which would be a violation of Axiom 4.6. So there is good reason for the OR rule to require that outcomes be mutually exclusive.

The answer to the question in the previous paragraph follows from the following straightforward rule:

Axiom 4.18 The EVERYTHING rule *The probability of the entire outcome space is equal to one.*

So, Pr({1, 2, 3, 4, 5, 6}) = 1 by the EVERYTHING rule. We could also have computed this number by using the OR rule, because Pr({1, 2, 3, 4, 5, 6}) = Pr(1 OR 2 OR 3 OR 4 OR 5 OR 6) = Pr(1) + Pr(2) + Pr(3) + Pr(4) + Pr(5) + Pr(6) = 1/6 * 6 = 1. The OR rule (Axiom 4.15) applies because the six individual outcomes are mutually exclusive; the EQUIPROBABILITY rule (Axiom 4.7) applies because all outcomes are equally probable. What the EVERYTHING rule tells us that we did not already know is that the probability of the entire outcome space equals one whether or not the outcomes are equiprobable. The next rule is easy too.

Axiom 4.19 The NOT rule *The probability that some outcome A will NOT occur is equal to one minus the probability that it does. That is,* $Pr(\neg A) = 1 - Pr(A)$.

For example, suppose that you want to know the probability of rolling anything but a six when rolling a six-sided die. $Pr(\neg 6) = 1 - Pr(6) = 1 - 1/6 = 5/6$. Given that the outcomes are mutually exclusive, we could have computed this using the OR rule too. (How?) In general, it is good to check that you get the same number when solving the same problem in different ways. If you do not, there is something wrong with your calculations.

Axiom 4.20 The AND rule *If two outcomes A and B are independent (see below), then the probability of A AND B equals the probability of A multiplied by the probability of B; that is,* $Pr(A \& B) = Pr(A) * Pr(B)$.

Suppose you flip a fair coin twice. What is the probability of getting two heads? Writing H_1 for heads on the first coin, and so on, by the AND rule, $Pr(H_1 \& H_2) = Pr(H_1) * Pr(H_2) = 1/2 * 1/2 = 1/4$. You could also solve this problem by looking at the outcome space $\{H_1H_2, H_1T_2, T_1H_2, T_1T_2\}$ and using the EQUIPROBABILITY rule. Similarly, it is easy to compute the probability of getting two sixes when rolling a fair die twice: $Pr(6_1 \& 6_2) = Pr(6_1) * Pr(6_2) = 1/6 * 1/6 = 1/36$.

Exercise 4.21 Are you more likely to get two sixes when rolling one fair die twice or when simultaneously rolling two fair dice?

Notice that the AND rule requires that the two outcomes be **independent**. What does this mean? Two outcomes A and B are independent just in case the fact that one occurs does not affect the probability that the other one does. This condition is satisfied when talking about a coin flipped twice. H_1 and H_2 are independent since the coin has no memory: whether or not the coin lands heads or tails the first time you flip it will not affect the probability of getting heads (or tails) the second time.

Exercise 4.22 Independence What pairs of outcomes are independent? More than one answer may be correct.
(**a**) You sleep late; you are late for class.
(**b**) You are a remarkable student; you get a good job after graduation.
(**c**) You write proper thank-you notes; you get invited back.
(**d**) The first time you flip a silver dollar you get heads; the second time you flip a silver dollar you get tails.
(**e**) General Electric stock goes up; General Motors stock goes up.

Exercise 4.23 Luck in love According to a well-known saying: "Lucky in cards, unlucky in love." Is this to say that luck in cards and luck in love are independent or not independent?

The importance of checking whether two outcomes are independent is best emphasized by giving an example. What is the probability of simultaneously getting a two and a three when you roll a fair die once? The answer is not $1/6 * 1/6 = 1/36$, of course, but zero. The outcomes are not independent, so you cannot use the AND rule. This example also tells you that when two outcomes are mutually exclusive they are not independent. (We will return to the topic of independence in Section 5.2.)

Exercise 4.24 When rolling two fair dice, what is the probability that the number of dots add up to 11? If you intend to use the OR rule, make sure the relevant outcomes are mutually exclusive. If you intend to use the AND rule, make sure the relevant outcomes are independent.

Exercise 4.25 Suppose you draw two cards from a well-shuffled deck of cards *with replacement*, meaning that you put the first card back into the deck (and shuffle the deck once more) before drawing the second card.
(**a**) What is the probability that you draw the ace of spades twice?
(**b**) What is the probability that you draw two aces? (Here, you can use your answer to Exercise 4.17.)

We are now in a position to address Example 4.2. Obviously, the theory of probability by itself will not tell you the probability that Linda is a bank teller. But it can tell you something else. Let F mean that Linda is a feminist and B that Linda is a bank teller.

Then the probability that she is both a feminist and a bank teller is $\Pr(B \& F) = \Pr(B) * \Pr(F)$. (In order to apply the AND rule here, I am assuming that the outcomes are independent; the general result, however, holds even if they are not.) Because $\Pr(F) \leq 1$ by Axiom 4.6, we know that $\Pr(B) * \Pr(F) \leq \Pr(B)$. For, if you multiply a positive number x with a fraction (between zero and one) you will end up with something less than x. So whatever the relevant probabilities involved are, it must be the case that $\Pr(B \& F) \leq \Pr(B)$; that is, the probability that Linda is a bank teller and a feminist has to be smaller than or equal to the probability that she is a bank teller. Many people will tell you that Linda is more likely to be a bank teller and a feminist than she is to be a bank teller. This mistake is referred to as the **conjunction fallacy**, about which you will hear more in Section 5.3.

We end this section with one more exercise.

Exercise 4.26 For the following questions, assume that you are rolling two fair dice:
(**a**) What is the probability of getting *two* sixes?
(**b**) What is the probability of getting *no* sixes?
(**c**) What is the probability of getting *exactly one* six?
(**d**) What is the probability of getting *at least one* six?

To compute the answer to (c), note that there are two ways to roll exactly one six. When answering (d), note that there are at least two ways to compute the answer. You can recognize that *rolling at least one six* is the same as *rolling two sixes or rolling exactly one six* and add up the answers to (a) and (c). Or you can recognize that *rolling at least one six* is the same as *not rolling no sixes* and compute the answer using the NOT rule.

Exercise 4.27 In computing the answer to Exercise 4.26(d), you may have been tempted to add the probability of rolling a six on the one die (1/6) to the probability of rolling a six on the other die (1/6) to get the answer 2/6 = 1/3. That, however, would be a mistake. Why?

If the answers to Exercise 4.26 are not completely obvious already, refer to Figure 4.1. Here, the numbers to the left represent what might happen when you roll the first die and the numbers on top represent what might happen when you roll the second die.

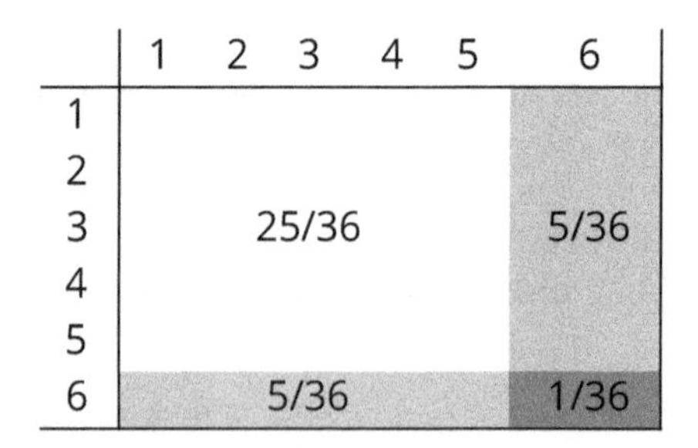

Figure 4.1 The two dice

Odds

Sometimes probabilities are expressed in terms of **odds** rather than probabilities. Imagine that you have an urn containing 2 black and 3 white balls, so that the probability of drawing a black ball is 2/5. One way to get this figure is to divide the number of favorable outcomes (outcomes in which the event of interest obtains) by the total number of outcomes. By contrast, you get the odds of drawing a black ball by dividing the number of favorable outcomes by the number of unfavorable outcomes, so that the odds of drawing a black ball are 2 to 3 or 2:3. Under the same assumptions, the odds of drawing a white ball are 3:2. If there is an equal number of black and white balls in the urn, the odds are 1 to 1 or 1:1. Such odds are also said to be **even**. When people talk about a 50–50 chance, they are obviously talking about even odds, since 50/50 = 1. How do odds relate to probabilities? If you have the probability p and want the odds o, you apply the following formula:

$$o = \frac{p}{1-p}$$

When p equals 2/5, it is easy to confirm that o equals 2/5 divided by 3/5 which is 2/3 or 2:3. If the probability is 1/2, the odds are 1/2 divided by 1/2 which is 1 or 1:1. If you have the odds o and want the probability p, you apply the inverse formula:

$$p = \frac{o}{o+1}$$

When o equals 2:3, you can quickly confirm that p equals 2/3 divided by 5/3 which is 2/5. If the odds are even, the probability is 1 divided by 1+1, which is 1/2. The use of odds instead of probabilities can come across as old-fashioned. But there are areas – for example, some games of chance and some areas of statistics – where odds are consistently used. It is good to know how to interpret them.

Thus, the table has 6 * 6 = 36 cells representing all the possible outcomes of rolling two dice. The dark gray area represents the possibility that both dice are sixes; because there is only one way to roll two sixes, this area contains but one cell and the answer to (a) is 1/36. The white area represents the possibility that both dice are non-sixes; because there are 5 * 5 ways to roll two non-sixes, this area contains 25 cells and the answer to (b) is 25/36. The light gray areas represent the possibility that one die is a six and the other one is a non-six; because there are 5 + 5 ways to attain this outcome, these areas contain ten cells and the answer to (c) is 10/36. You can compute the answer to (d) by counting the 5 + 5 + 1 = 11 cells in the two light gray and the one dark gray areas and get an answer of 11/36. But a smarter way is to realize that the gray areas cover everything that is not white, which allows you to get the answer by computing 1 – 25/36 = 11/36. Why this is smarter will be clear in Section 5.3. The figure also illustrates why you cannot compute the probability of getting at least one six by adding the probability of rolling a six on the first die to the probability of rolling a six on the second one. If you were to do that, you would add the number of cells in the bottom row to the number of cells in the right-most column – thereby double-counting the cell on the bottom right.

4.4 Conditional probability

In Exercise 4.25, you computed the probability of drawing two aces when drawing two cards with replacement. Suppose, instead, that you draw two cards *without replacement*, meaning that you put the first card aside after looking at it. What is the probability of drawing two aces without replacement? You know you cannot use Axiom 4.20, since the two outcomes we are interested in (drawing an ace the first time and drawing an ace the second time) are not independent. You can, however, approach the problem in the following way. First, you can ask what the probability is that the first card is an ace. Because there are 52 cards in the deck, and 4 of those are aces, you know that this probability is 4/52. Second, you can ask what the probability is that the second card is an ace, *given that the first card was an ace*. Because there are 51 cards left in the deck, and only 3 of them are aces, this probability is 3/51. Now you can multiply these numbers and get:

$$\frac{4}{52} * \frac{3}{51} = \frac{12}{2652} = \frac{1}{221}$$

This procedure can be used to calculate the probability of winning certain types of lotteries. According to the Consumer Federation of America, about one in five Americans believe that "the most practical way for them to accumulate several hundred thousand dollars is to win the lottery." The poor, least educated, and oldest are particularly likely to think of the lottery as a smart way to get rich. So it might be useful to ask just how likely or unlikely it is to win common lotteries.

Exercise 4.28 Lotto 6/49 Many states and countries operate lotteries in which the customer picks n of m numbers, in any order, where n is considerably smaller than m. In one version of this lottery, which I will call Lotto 6/49, players circle 6 numbers out of 49 using a ticket like that in Figure 4.2. The order in which numbers are circled

LOTTO 6/49						
1	2	3	4	5	6	7
8	9	10	11	12	13	14
15	16	17	18	19	20	21
22	23	24	25	26	27	28
29	30	31	32	33	34	35
36	37	38	39	40	41	42
43	44	45	46	47	48	49

Figure 4.2 Lotto 6/49 ticket

does not matter. You win the grand prize if all 6 are correct. What is the probability that you win the Lotto 6/49 any one time you play? Notice that this is similar to picking six consecutive aces out of a deck with 49 cards, if 6 of those cards are aces.

The fact that the probability of winning the lottery is low does not imply that it is necessarily irrational to buy these tickets. (We will return to this topic in Part 3.) Nevertheless, it may be fun to ask some questions about these lotteries.

Problem 4.29 Lotto 6/49, cont. *What does the probability of winning the Lotto 6/49 tell you about the wisdom of buying Lotto tickets? What does it tell you about people who buy these tickets?*

Exercise 4.30 Lotto 6/49, cont. Use the idea of anchoring and adjustment from Section 3.6 to explain why people believe that they have a good chance of winning these lotteries.

Considerations like these clarify why state lottery schemes are sometimes described as a tax on innumeracy.

The probability that something happens given that some other thing happens is called a **conditional probability**. We write the probability that A given C, or the probability of A conditional on C, as follows: $\Pr(A \mid C)$. Conditional probabilities are useful for a variety of purposes. It may be easier to compute conditional probabilities than unconditional probabilities. Knowing the conditional probabilities is oftentimes quite enough to solve the problem at hand.

Notice right away that $\Pr(A \mid C)$ is not the same thing as $\Pr(C \mid A)$. Though these two probabilities may be identical, they need not be. Suppose, for example, that S means that Joe is a smoker, while H means that Joe is human. If so, $\Pr(S \mid H)$ is the probability that Joe is a smoker given that Joe is human, which is a number strictly between zero and one. Meanwhile, $\Pr(H \mid S)$ is the probability that Joe is human given that Joe is a smoker, which is one (or at least close to one). Joe may not be a human being for quite as long if he is a smoker, but that is another matter.

Exercise 4.31 Conditional probabilities Suppose that H means "The patient has a headache" and T means "The patient has a brain tumor."

(**a**) How do you interpret the two conditional probabilities $\Pr(H \mid T)$ and $\Pr(T \mid H)$?
(**b**) Are the two numbers more or less the same?

It should be clear that the two conditional probabilities in general are different, and that it is important for both doctors and patients to keep them apart. (We will return to this topic in Sections 5.4 and 5.6.)

Suppose that you draw one card from a well-shuffled deck, and that you are interested in the probability of drawing the ace of spades given that you draw an ace. Given that you just drew an ace, there are four possibilities: the ace of spades, ace of clubs, ace of hearts, or ace of diamonds. Because only one of the four is the ace of spades, and because all four outcomes are equally likely, this probability is 1/4. You can get the same answer by dividing the probability that you draw the ace of spades by the probability that you draw an ace: 1/52 divided by 4/52, which is 1/4. This is no coincidence, as the formal definition of conditional probability will show.

Definition 4.32 Conditional probability *If A and B are two outcomes,* $Pr(A\,|\,B) = Pr(A\,\&\,B)/Pr(B)$.

As another example of conditional probability, recall the problem with the two aces. Let A denote "the second card is an ace" and let B denote "the first card is an ace." We know that the probability of drawing two aces without replacement is 1/221. This is $\Pr(A\ \&\ B)$. We also know that the probability that the first card is an ace is 1/13. This is $\Pr(B)$. So by definition:

$$\Pr(A|B) = \frac{\Pr(A\,\&\,B)}{\Pr(B)} = \frac{1/221}{1/13} = \frac{3}{51}$$

But we knew this: 3/51 is the probability that the second card is an ace given that the first card was: $\Pr(A\,|\,B)$. So the formula works.

Exercise 4.33 Ace of spades Use Definition 4.32 to compute the probability that you draw an ace of spades conditional on having drawn an ace when you draw one card from a well-shuffled deck. You can imagine a game show host who draws a card at random and announces that the card is an ace, and a contestant who has to guess what kind of ace it is. Given what you know about that card, what is the probability that it is the ace of spades?

Because you cannot divide numbers by zero, things get tricky when some probabilities are zero; here, I will ignore these complications.

One implication of the definition is particularly useful:

Proposition 4.34 The general AND rule $Pr(A\ \&\ B) = Pr(A\,|\,B) * Pr(B)$.

Proof.
Starting off with Definition 4.32, multiply each side of the equation by $\Pr(B)$. □

According to this proposition, the probability of drawing two aces equals the probability of drawing an ace the first time multiplied by the probability of drawing an ace the second time given that you drew an ace the first time. But again, we knew this. In fact, we implicitly relied on this rule when computing the answers to the first exercises in this section. Notice that this rule allows us to compute the probability of A AND B without requiring that the outcomes be independent. This is why it is called the general AND rule.

Exercise 4.35 The general AND rule Use the general AND rule to compute the probability that you will draw the ace of spades twice when drawing two cards from a deck *without* replacement.

The general AND rule permits us to establish the following result.

Proposition 4.36 $Pr(A \mid B) * Pr(B) = Pr(B \mid A) * Pr(A)$.

Proof.
By Proposition 4.34, $\Pr(A \& B) = \Pr(A \mid B) * \Pr(B)$ but also $\Pr(B \& A) = \Pr(B \mid A) * \Pr(A)$. Because by logic $\Pr(A \& B) = \Pr(B \& A)$, it must be the case that $\Pr(A \mid B) * \Pr(B) = \Pr(B \mid A) * \Pr(A)$. □

Suppose that you draw one card from a well-shuffled deck, and that **A** means that you draw an ace and that ♦ means that you draw a diamond. If so, it follows that $\Pr(\mathbf{A} \mid \blacklozenge) * \Pr(\blacklozenge) = \Pr(\blacklozenge \mid \mathbf{A}) * \Pr(\mathbf{A})$. You can check that this is true by plugging in the numbers: $1/13 * 13/52 = 1/4 * 4/52 = 1/52$.

This notion of conditional probability allows us to sharpen our definition of independence. We said that two outcomes A and B are independent if the probability of A does not depend on whether B occurred. Another way of saying this is to say that $\Pr(A \mid B) = \Pr(A)$. In fact, there are several ways of saying the same thing.

Proposition 4.37 Independence conditions *The following three claims are equivalent:*

(i) $Pr(A \mid B) = Pr(A)$

(ii) $Pr(B \mid A) = Pr(B)$

(iii) $Pr(A \& B) = Pr(A) * Pr(B)$

Proof.
See Exercise 4.38. □

Exercise 4.38 Independence conditions Prove that the three parts of Proposition 4.37 are equivalent. The most convenient way of doing so is to prove (**a**) that (i) implies (ii), (**b**) that (ii) implies (iii), and (**c**) that (iii) implies (i).

Notice that part (iii) is familiar: it is the principle that we know as the AND rule (Axiom 4.20). Thus, the original AND rule follows logically from the general AND rule and the assumption that the two outcomes in question are independent. This is pretty neat.

4.5 Total probability and Bayes's rule

Conditional probabilities can also be used to compute unconditional probabilities. Suppose that you are running a frisbee factory and that you want to know the probability that one of your frisbees is defective. You have two machines producing frisbees: a new one (B) producing 800 frisbees per day and an old one (¬B) producing 200 frisbees per day. Thus, the probability that a randomly selected frisbee from your factory was produced by machine B is $\Pr(B) = 800/(800 + 200) = 0.8$; the probability that it was produced by machine ¬B is $\Pr(\neg B) = 1 - \Pr(B) = 0.2$. Among the frisbees produced by the new machine, one percent are defective (D); among those produced by the old one, two percent are. The probability that a randomly selected frisbee produced by machine B is defective is $\Pr(D \mid B) = 0.01$; the probability that a randomly selected frisbee produced by machine ¬B is defective is $\Pr(D \mid \neg B) = 0.02$. It may be useful to draw a tree illustrating the four possibilities (see Figure 4.3).

What is the probability that a randomly selected frisbee from your factory is defective? There are two ways in which a defective frisbee can be produced: by machine B and by machine ¬B. So the probability that a frisbee is defective Pr(D) equals the following probability: that the frisbee is produced by machine B and turns out to be defective or that the frisbee is produced by machine ¬B and turns out to be defective; that is, $\Pr([D \& B] \vee [D \& \neg B])$. These outcomes are obviously mutually exclusive, so the probability equals $\Pr(D \& B) + \Pr(D \& \neg B)$. Applying the general AND rule twice, this equals $\Pr(D \mid B) * \Pr(B) + \Pr(D \mid \neg B) * \Pr(\neg B)$. But we have all these numbers, so:

$$\Pr(D) = \Pr(D|B) * \Pr(B) + \Pr(D|\neg B) * \Pr(\neg B) = 0.01 * 0.8 + 0.02 * 0.2 = 0.012$$

The probability that a randomly selected frisbee produced by your factory is defective is 1.2 percent. These calculations illustrate **the rule of total probability**.

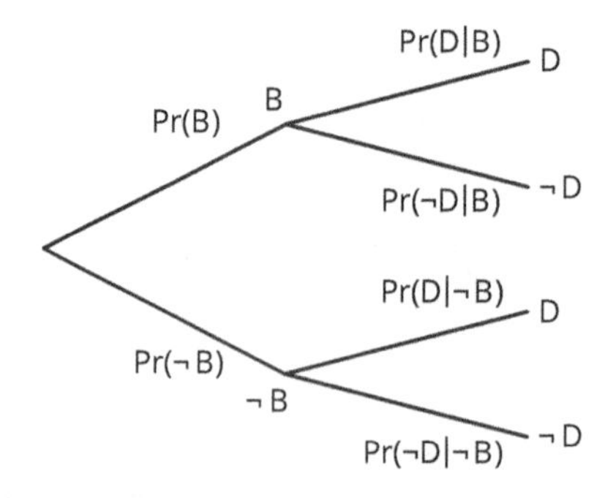

Figure 4.3 The frisbee factory

Proposition 4.39 The rule of total probability $\Pr(D) = \Pr(D \mid B) * \Pr(B) + \Pr(D \mid \neg B) * \Pr(\neg B)$.

Proof.
By logic, D is the same as $[D \& B] \vee [D \& \neg B]$. So $\Pr(D) = \Pr([D \& B] \vee [D \& \neg B])$. Because the two outcomes are mutually exclusive, this equals $\Pr(D \& B) + \Pr(D \& \neg B)$ by the OR rule (Axiom 4.15). Applying the general AND rule (Proposition 4.34) twice, we get $\Pr(D) = \Pr(D \mid B) * \Pr(B) + \Pr(D \mid \neg B) * \Pr(\neg B)$. □

Exercise 4.40 Cancer Use the rule of total probability to solve the following problem. You are a physician meeting with a patient who has just been diagnosed with cancer. You know there are two mutually exclusive types of cancer that the patient could have: type A and type B. The probability that he or she has A is 1/3 and the probability that he or she has B is 2/3. Type A is deadly: four patients out of five diagnosed with type A cancer die (D) within one year. Type B is less dangerous: only one patient out of five diagnosed with type B cancer dies (D) within one year.
(**a**) Draw a tree representing the four possible outcomes.
(**b**) Compute the probability that your patient dies within a year.

Exercise 4.41 Scuba diving certification You are scheduled to sit the test required to be a certified scuba diver and very much hope you will pass (P). The test can be easy (E) or not. The probability that it is easy is 60 percent. If it is easy, you estimate that the probability of passing is 90 percent; if it is hard, you estimate that the probability is 50 percent. What is the probability that you pass (P)?

There is another type of question that you may ask as well. Suppose you pick up one of the frisbees produced in your factory and find that it is defective. What is the probability that the defective frisbee was produced by the new machine? Here you are asking for the probability that a frisbee was B conditional on D, that is, $\Pr(B \mid D)$.

We know that there are two ways in which a defective frisbee can be produced. Either it comes from the new machine, which is to say that $D \& B$, or it comes from the old machine, which is to say that $D \& \neg B$. We also know the probabilities that these states will obtain for any given frisbee (not necessarily defective): $\Pr(D \& B) = \Pr(D \mid B) * \Pr(B) = 0.01 * 0.8 = 0.008$ and $\Pr(D \& \neg B) = \Pr(D \mid \neg B) * \Pr(\neg B) = 0.02 * 0.2 = 0.004$. We want the probability that a frisbee comes from the new machine given that it is defective, that is, $\Pr(B \mid D)$. By looking at the figures, you can tell that the first probability is twice as large as the second one. What this means is that in two cases out of three, a defective frisbee comes from the new machine. Formally, $\Pr(D \mid B) = 0.008/0.012 = 2/3$. This may be surprising, in light of the fact that the new machine has a lower rate of defective frisbees than the old one. But it is explained by the fact that the new machine also produces far more frisbees than the old one.

The calculations you have just performed are an illustration of **Bayes's rule**, or **Bayes's theorem**, which looks more complicated than it is.

Proposition 4.42 Bayes's rule

$$\Pr(B|D) = \frac{\Pr(D|B) * \Pr(B)}{\Pr(D)}$$
$$= \frac{\Pr(D|B) * \Pr(B)}{\Pr(D|B) * \Pr(B) + \Pr(D|\neg B) * \Pr(\neg B)}$$

Proof.
The rule has two forms. The first form can be obtained from Proposition 4.36 by dividing both sides of the equation by $\Pr(D)$. The second form can be obtained from the first by applying the rule of total probability (Proposition 4.39) to the denominator. □

Exercise 4.43 Cancer, cont. Suppose that your patient from Exercise 4.40 dies in less than one year, before you learn whether he or she has type A or type B cancer. Given that the patient died in less than a year, what is the probability he or she had type A cancer?

Exercise 4.44 Scuba diving certification, cont. You passed the scuba diving test! Your friend says: "Not to rain on your parade, but you obviously got the easy test." Given that you passed, what is the probability that you got the easy test?

Bayes's rule is an extraordinarily powerful principle. To show how useful it can be, consider the following problem. If it is not immediately obvious how to attack this problem, it is almost always useful to draw a tree identifying the probabilities.

Exercise 4.45 The dating game You are considering asking L out for a date, but you are a little worried that L may already have started dating somebody else. The probability that L is dating somebody else, you would say, is 1/4. If L is dating somebody else, he/she is unlikely to accept your offer to go on a date: in fact, you think the probability is only 1/6. If L is not dating somebody else, though, you think the probability is much better: about 2/3.

(a) What is the probability that L is dating somebody else but will accept your offer to go on a date anyway?
(b) What is the probability that L is *not* dating somebody else and will accept your offer to go on a date?
(c) What is the probability that L will accept your offer to go on a date?
(d) Suppose L accepts your offer to go on a date. What is the probability that L is dating somebody else, given that L agreed to go on a date?

There are more exercises on Bayes's rule in Sections 4.6 and 5.4. See also Exercise 5.34 on page 112.

4.6 Bayesian updating

Bayes's rule is often thought to capture how we should update our beliefs in light of new evidence. We update beliefs in light of new evidence all the time. In everyday life, we update our belief that a particular presidential candidate will win the election in light of evidence about how well he or she is doing. The evidence here may include poll results, our judgments about his or her performance in presidential debates, and so on. In science, we update our assessment about the plausibility of a hypothesis or theory in light of evidence, which may come from experiments, field studies, or other sources. Consider, for example, how a person's innocent belief that the Earth is flat might be updated in light of the fact that there are horizons, the fact that the Earth casts a circular shadow onto the Moon during a lunar eclipse, and the fact that one can travel around the world. Philosophers of science talk about the **confirmation** of scientific theories, so the theory of how this is done is called **confirmation theory**. Bayes's rule plays a critical role in confirmation theory.

To see how this works, think of the problem of belief updating as follows: what is at stake is whether a given hypothesis is true or false. If the hypothesis is true, there is some probability that the evidence obtains. If the hypothesis is false, there is some other probability that the evidence obtains. The question is how you should change your belief – that is, the probability that you assign to the possibility that the hypothesis is true – in light of the fact that the evidence obtains. Figure 4.4 helps to bring out the structure of the problem.

Let H stand for the **hypothesis** and E for the **evidence**. The probability of H, $\Pr(H)$, is called the **prior probability**: it is the probability that H is true before you learn whether E is true. The probability of H given E, $\Pr(H|E)$, is called the **posterior probability**: it is the probability that H obtains given that the evidence E is true. The question is what the posterior probability should be. This question is answered by a simple application of Bayes's rule. Substituting H for B and E for D in Proposition 4.42, we can write Bayes's rule as follows:

$$\Pr(H|E) = \frac{\Pr(E|H) * \Pr(H)}{\Pr(E|H) * \Pr(H) + \Pr(E|\neg H) * \Pr(\neg H)}$$

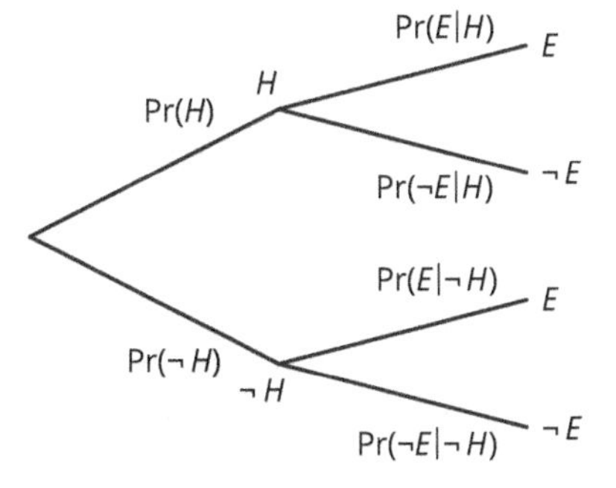

Figure 4.4 Bayesian updating

The result tells you how to update your belief in the hypothesis H in light of the evidence E. Specifically, Bayes's rule tells you that the probability you assign to H being true should go from $\Pr(H)$ to $\Pr(H|E)$. If you change your beliefs in accordance with Bayes's rule, we say that you engage in **Bayesian updating**.

Suppose that John and Wes are arguing about whether a coin brought to class by a student has two heads or whether it is fair. Imagine that there are no other possibilities. For whatever reason, the student will not let them inspect the coin, but she will allow them to observe the outcome of coin flips. Let H be the hypothesis that the coin has two heads, so that $\neg H$ means that the coin is fair. Let us consider John first. He thinks the coin is unlikely to have two heads: his prior probability, $\Pr(H)$, is only 0.01. Now suppose the student flips the coin, and that it comes up heads. Let E mean "The coin comes up heads." The problem is this: What probability should John assign to H given that E is true?

Given Bayes's rule, computing John's posterior probability $\Pr(H|E)$ is straightforward. We are given $\Pr(H) = 0.01$, and therefore know that $\Pr(\neg H) = 1 - \Pr(H) = 0.99$. From the description of the problem, we also know the conditional probabilities: $\Pr(E|H) = 1$ and $\Pr(E|\neg H) = 0.5$. All that remains is to plug the numbers into the theorem, as follows:

$$\Pr(H|E) = \frac{\Pr(E|H) * \Pr(H)}{\Pr(E|H) * \Pr(H) + \Pr(E|\neg H) * \Pr(\neg H)}$$

$$= \frac{1 * 0.01}{1 * 0.01 + 0.5 * 0.99} \approx 0.02$$

The fact that John's posterior probability $\Pr(H|E)$ differs from his prior probability $\Pr(H)$ means that he updated his belief in light of the evidence. The observation of heads increased his probability that the coin has two heads, as it should. Notice how the posterior probability reflects both the prior probability and the evidence E.

Now, if John gets access to ever more evidence about the coin, there is no reason why he should not update his belief again. Suppose that the student flips the coin a second time and gets heads again. We can figure out what John's probability should be after observing this second flip by simply treating his old posterior probability as the new prior probability and applying Bayes's rule once more:

$$\Pr(H|E) = \frac{1 * 0.02}{1 * 0.02 + 0.5 * 0.98} \approx 0.04$$

Notice that his posterior probability increases even more after he learns that the coin came up heads the second time.

Exercise 4.46 Bayesian updating Suppose Wes, before the student starts flipping the coin, assigns a probability of 50 percent to the hypothesis that it has two heads.
(**a**) What is his posterior probability after the first trial?
(**b**) After the second?

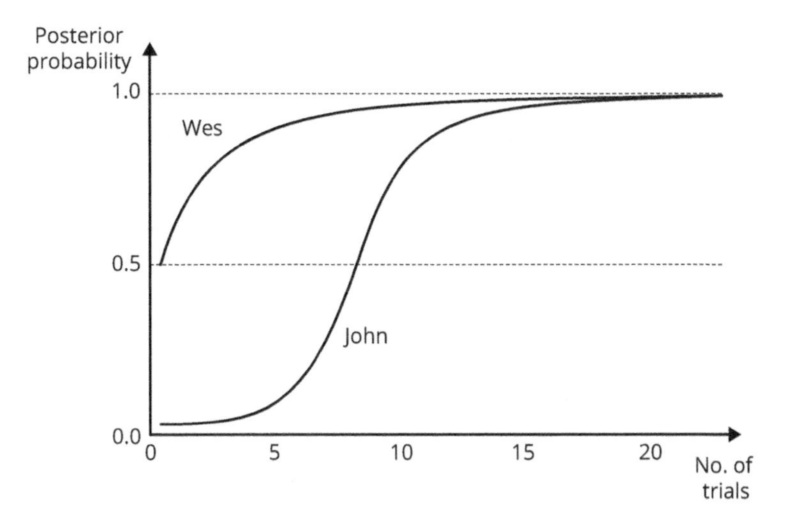

Figure 4.5 John's and Wes's probabilities after repeated trials

Figure 4.5 illustrates how John's and Wes's posterior probabilities develop as the evidence comes in. Notice that over time both increase the probability assigned to the hypothesis. Notice, also, that their respective probabilities get closer and closer. As a result, over time (after some 15–20 trials) they are in virtual agreement that the probability of the coin having two heads is almost 100 percent. We will return to questions of rational updating in the next chapter. Until then, one last exercise.

Exercise 4.47 Bayesian updating, cont. Suppose that, on the third trial, instead of flipping heads, the student flips tails. What would John's and Wes's posterior probability be? To solve this problem, let E mean "The coin comes up tails."

4.7 Discussion

In this chapter we have explored the theory of probability. The theory has something of the feel of magic to it. When you are facing these difficult problems where untutored intuitions are conflicting or way off the mark, all you have to do is to apply the incantations (rules) in the right way and in the right order and the answer pops right up! Anyway, probability theory is critical to a wide range of applications, among other things as the foundations of statistical inference. It is relevant here because it can be interpreted as a theory of judgment, that is, as a theory of how to revise beliefs in light of evidence. The section on Bayesian updating (Section 4.6) shows how the argument goes. As long as you are committed to the axioms and rules of the probability calculus, we established that you will (as a matter of mathematical necessity) update your probabilities consistently with Bayes's rule. Notice how neat this is. While the theory says nothing about what your prior probabilities are or should be, it does tell you exactly what your posterior probability will or should be after observing the evidence.

The theory also suggests that priors, over time, will make less of a difference than you might think (cf. Section 5.5).

How plausible is this theory? Again, we must separate the descriptive question from the normative question. Do people in fact update their beliefs in accordance with Bayes's rule? Should they? The axioms might seem weak and uncontroversial, both from a descriptive and from a normative standpoint. Yet the resulting theory is anything but weak. As in the case of the theory of choice under certainty, we have built a remarkably powerful theory on the basis of a fairly modest number of seemingly weak axioms. It is important to keep in mind that the theory is not as demanding as some people allege. It is not intended to describe the actual cognitive processes you go through when updating your beliefs: the theory does not say that you must apply Bayes's theorem in your head. But it does specify exactly how your posterior probability relates to your prior, given various conditional and unconditional probabilities.

In the next chapter, we will see how the theory fares when confronted with evidence.

Additional exercises

Exercise 4.48 SAT test When you take the SAT test, you may think that the correct answers to the various questions would be completely random. In fact, they are not. The authors of the test want the answers to *seem* random, and therefore they make sure that not all correct answers are, say, (d). Consider the following three outcomes. *A*: The correct answer to question 12 is (d). *B*: The correct answer to question 13 is (d). *C*: The correct answer to question 14 is (d). Are outcomes *A*, *B*, and *C* mutually exclusive, independent, both, or neither?

Exercise 4.49 Mr Langford Multiple lawsuits allege that area gambling establishments, on multiple occasions, doctored equipment so as to give Birmingham mayor Larry Langford tens of thousands of dollars in winnings. Langford, already in prison for scores of corruption-related charges, has not denied winning the money; he does deny that the machines were doctored.

We do not know exactly what the probability of winning the jackpot on a machine that has *not* been doctored might be, but we can make some intelligent guesses. Suppose that Langford on three occasions bet $1 and won $25,000. For each jackpot, in order to break even, a gambling establishment needs 24,999 people who bet $1 and do not win. So we might infer that the probability of winning when betting a dollar is somewhere in the neighborhood of 1/25,000. If the establishment wants to make a profit, which it does, the probability would have to be even lower, but let us ignore this fact.

What is the probability that Langford would win the jackpot three times in a row when playing three times on undoctored machines?

Note that the probability that Langford would win three times in a row given that the machines were not doctored is different from the conditional probability that the machines were not doctored given that Langford won three times in a row.

Exercise 4.50 Economists go to Vegas According to professional lore, economists are not welcome to organize large meetings in Las Vegas. The reason is a sort of sin of omission. What is it economists, unlike most normal people, allegedly do not do when in Vegas?

Exercise 4.51 Gender discrimination Imagine that an editorial board of 20 members is all male.

(**a**) What is the probability that this would happen by chance alone assuming that the board members are drawn from a pool of 1/2 men and 1/2 women?

(**b**) Perhaps the pool of qualified individuals is not entirely balanced in terms of gender. What is the answer if the pool consists of 2/3 men and 1/3 women?

(**c**) And what if it is 4/5 men and 1/5 women?

Exercise 4.52 Softball A softball player's batting average is defined as the ratio of hits to at bats. Suppose that a player has a 0.250 batting average and is very consistent, so that the probability of a hit is the same every time she is at bat. During today's game, this player will be at bat exactly three times.

(**a**) What is the probability that she ends up with three hits?

(**b**) What is the probability that she ends up with no hits?

(**c**) What is the probability that she ends up with exactly one hit?

(**d**) What is the probability that she ends up with at least one hit?

Exercise 4.53 Gov. Schwarzenegger After vetoing a bill from the California State Assembly in 2009, California Governor Arnold Schwarzenegger published a letter (see Figure 4.6). People immediately noticed that the first letter on each line together spelled out a vulgarity. When confronted with this fact, a spokesperson said: "It was just a weird coincidence."

(**a**) Assuming that a letter has eight lines, and that each of the 26 letters in the alphabet is equally likely to appear at the beginning of each line, what is the probability that this exact message would appear by chance?

(**b**) It is true that the Governor writes many letters each year, which means that the probability of any one letter spelling out this vulgarity is higher than your answer to (a) would suggest. Suppose that the Governor writes 100 eight-line letters each year. What is the probability that at least one of them will spell out the vulgarity?

Exercise 4.54 Max's bad day Max is about to take a multiple-choice test. The test has ten questions, and each has two possible answers: "true" and "false." Max does not have the faintest idea of what the right answer to any of the questions might be. He decides to pick answers randomly.

(**a**) What is the probability that Max will get all ten questions right?

(**b**) What is the probability that Max will get the first question wrong and the other questions right?

(**c**) What is the probability that Max will get the second question wrong and the other questions right?

(d) What is the probability that Max will get exactly nine questions right?
(e) Max really needs to get an A on this test. In order to get an A, he needs to get nine or more questions right. What is the probability that Max will get an A?

```
To the Members of the California State Assembly:

I am returning Assembly Bill 1176 without ...

For some time now I have lamented the fact tha ...
unnecessary bills come to me for consideration ...
care are major issues my Administration has br ...
kicks the can down the alley.

Yet another legislative year has come and gone ...
overwhelmingly deserve. In light of this, and ...
unnecessary to sign this measure at this time.

Sincerely,
Arnold Schwarzenegger
```

Figure 4.6 Governor Schwarzenegger's letter

Exercise 4.55 Pregnancy tests You are marketing a new line of pregnancy tests. The test is simple. You flip a fair coin. If the coin comes up heads, you report that the customer is pregnant. If the coin comes up tails, you report that the customer is not pregnant.
(a) Your first customer is a man. What is the probability that the test accurately predicts his pregnancy status?
(b) Your second customer is a woman. What is the probability that the test accurately predicts her pregnancy status?
(c) After you have administered the test ten times, what is the probability that you have not correctly predicted the pregnancy status of any of your customers?
(d) After you have administered the test ten times, what is the probability that you correctly predicted the pregnancy status of *at least one* of your customers?

Notice how high the probability of getting at least one customer right is. This suggests the following scheme for getting rich. Issue ten, or a hundred, or whatever, newsletters offering advice about how to pick stocks. No matter how unlikely each newsletter is to give good advice, if you issue enough of them it is extremely likely that at least one will give good advice. Then sell your services based on your wonderful track record, pointing to the successful newsletter as evidence. You would not be the first. We will return to this kind of problem in Section 5.3.

Problem 4.56 Pregnancy tests, cont. *The pregnancy test of Exercise 4.55 is needlessly complicated. Here is another test that is even simpler: just report that the customer is not pregnant. Roughly, what is the probability that you would get the pregnancy status of a randomly selected college student right when using the simplified test?*

Further reading

There are numerous introductions to probability theory. Earman and Salmon (1992) deals with probability theory in the context of the theory of confirmation and is the source of the stories about frisbees and coins (pp. 70–4); it also contains a discussion about the meaning of probability (pp. 74–89). The Consumer Federation of America (2006) discusses people's views about the most practicable way to get rich. The fate of Birmingham mayor Larry Langford is chronicled in Tomberlin (2009), and that of California governor Arnold Schwarzenegger in McKinley (2009).

5 JUDGMENT UNDER RISK AND UNCERTAINTY

Learning objectives

After studying this chapter you will:

- Be able to identify common belief patterns that violate probability theory
- Know all the building blocks of the biases-and-heuristics program
- Apply the theory of probability in the real world – but also appreciate that it may not be applicable in all contexts

5.1 Introduction

The previous chapter showed how a powerful theory of probabilistic judgment can be built on the foundation of a limited number of relatively uncontroversial axioms. Though the axioms might seem weak, the resulting theory is anything but. The theory is open to criticism, especially on descriptive grounds. In this section, we consider whether the theory can serve as a descriptively adequate theory, that is, whether it captures how people actually make probabilistic judgments, and we explore a series of phenomena that suggest that it does not. The discrepancy suggests that a descriptively adequate theory of judgment must differ from the theory that we just learned. We will also continue our study of the building blocks of behavioral theory. In particular, we will continue the discussion of the heuristics-and-biases program in Chapter 3, by reviewing more heuristics and discussing the biases that these heuristics can lead to. Thus, this chapter gives a better idea of how behavioral economists respond to discrepancies between observed behavior and standard theory.

5.2 The gambler's fallacy

The notion of independence that we encountered in Section 4.3 is absolutely critical in economics and finance. If you are managing investments, for example, you are supposed to diversify. It would be unwise to put all of your eggs in one basket, investing all your money in a single asset such as Google stock. But in order to diversify properly, it is not enough to invest in two or more assets: if you invest in stocks that will rise and fall together, you do not actually have a diversified portfolio. What you should do is to invest your money in assets that are sufficiently independent. In real-world investment management, a great deal of effort goes into exploring whether assets can be assumed to be independent or not.

The notion of independence is also very important in fields such as engineering. If you are in the process of designing a new nuclear power plant, you should include multiple safety systems that can prevent nuclear meltdown. But having five safety systems instead of one gives you additional safety only when a breakdown in one system is sufficiently independent from a breakdown in the other. If, for example, all safety systems are held together with one bolt, or plugged into the same outlet, a breakdown in the one system is not independent from a breakdown in the other, and your plant will not be as safe as it could be. In nuclear power plant design, and elsewhere, a great deal of effort goes into making sure that different systems (if they are all critical to the operation of the machine) are sufficiently independent from each other.

Exercise 5.1 The eggs and the basket Use the concept of independence to explain why you should not put all of your eggs in one basket.

Exercise 5.2 The alarms Knowing that alarm clocks sometimes fail to go off, some people set multiple alarms to make sure they will wake up in the morning.

(**a**) In order to work as intended, should the different alarms be dependent or independent?

(**b**) If you set multiple alarms on your smartphone, are they dependent or independent?

In principle, there are two ways in which you might make a mistake about independence: you may think that two outcomes are independent when in fact they are not, or you may think that two outcomes are not independent when in fact they are. People make both kinds of mistake. We will consider them in order.

Thinking that two outcomes are independent when in fact they are not happens, for instance, when people invest in stocks and bonds on the assumption that they are completely independent. In reality they are not. One of the important take-home lessons of the global financial crisis is that a vast range of assets – US stocks, Norwegian real estate, etc. – are probabilistically dependent because of the highly international nature of modern finance and the complicated ways in which mortgages and the like are packaged and sold. Thinking that two outcomes are independent when in fact they are not also happens when people build nuclear power plant safety systems that have parts in common or that depend on the sobriety of one manager or the reliability of one source of electric power.

Thinking that two outcomes are dependent when in fact they are not occurs, for example, when people think that they can predict the outcome of a roulette wheel based on its previous outcomes. People cannot: these things are set up in such a way as to make the outcomes completely independent. And they are set up that way because it is a good way for the casino to make sure customers are unable to predict the outcomes. Nevertheless, many people believe that they can predict the outcome of a roulette game. For evidence, the internet offers a great deal of advice about how to beat the casino when playing roulette: "monitor the roulette table," "develop a system," "try the system on a free table before operating it for financial gain," and so on. Do an internet search for "roulette tips," and you will find a long list of webpages encouraging you to think of various outcomes as probabilistically dependent, when they are not.

Exercise 5.3 Winners Purveyors of lottery tickets are fond of posting signs such as that in Figure 5.1. Of the two mistakes we have identified in this section, which one are they hoping you will make today?

$	*WINNING*	$
$	*$1,000,000*	$
$	*LOTTERY*	$
$	*TICKET*	$
$	*SOLD HERE*	$

Figure 5.1 Winning ticket sold here

Exercise 5.4 Threes "Bad things always happen in threes," people sometimes say, offering as evidence the fact that Janis Joplin, Jimi Hendrix, and Jim Morrison all died within a few months of each other in late 1970 and early 1971. What sort of mistake are these people making?

One specific case of thinking that two outcomes are dependent when in fact they are not is the **gambler's fallacy**: thinking that a departure from the average behavior of some system will be corrected in the short term. People who think they are "due for" a hurricane or a car accident or the like because they have not experienced one for a few years are committing the gambler's fallacy. Here, I am assuming that hurricanes and car accidents are uncorrelated from year to year. It is possible that thinking you are due for a car accident makes you more likely to have one; if so, a number of accident-free years might in fact make it more likely for you to have an accident.

The following exercises illustrate how easy it is to go wrong.

Exercise 5.5 Gambler's fallacy Carefully note the difference between the following two questions:

(**a**) You intend to flip a fair coin eight times. What is the probability that you end up with eight heads?

(**b**) You have just flipped a fair coin seven times and ended up with seven heads. What is the probability that when you flip the coin one last time you will get another heads, meaning that you would have flipped eight heads in a row?

The gambler's fallacy is sometimes explained in terms of **representativeness**. We came across heuristics in Section 3.6 on anchoring and adjustment. According to the heuristics-and-biases program, people form judgments by following heuristics, or rules of thumb, which by and large are functional but which sometimes lead us astray. The **representativeness heuristic** is such a heuristic. When you employ the representativeness heuristic, you estimate the probability that some outcome was the result of a given process by reference to the degree to which the outcome is representative of that process. If the outcome is highly representative of the process, the probability that the former was a result of the latter is estimated to be high; if the outcome is highly unrepresentative of the process, the probability is estimated to be low.

The representativeness heuristic can explain the gambler's fallacy if we assume that a sequence such as HHHHHHHH seems less representative of the process of flipping a fair coin eight times than a sequence such as HHHHHHHT, which seems less representative than a sequence such as HTTTHHTH. If you use the representativeness heuristic, you will conclude that the first sequence is less likely than the second, and that the second is less likely than the third. In reality, of course, the three are equally likely *ex ante*.

Exercise 5.6 Representativeness Which of the following two outcomes will strike people who use the representativeness heuristic as more likely: getting 4-3-6-2-1 or 6-6-6-6-6 when rolling five dice?

The representativeness heuristic might be perfectly functional in a wide variety of contexts. If it is used, for example, to infer that kind acts are more likely to be performed by kind people, and that mean acts more likely to be performed by mean people, the representativeness heuristic can protect us from adverse events. But because it can generate predictable and systematic patterns of mistakes, it can lead to bias, just as anchoring and adjustment can. For another example, consider the following case:

Exercise 5.7 Twins Let us assume that whenever one gets pregnant, there is a 1/100 chance of having twins, and that being pregnant with twins once will not affect the probability of being pregnant with twins later.

(**a**) You are not yet pregnant, but will get pregnant twice. What is the probability that you will be pregnant with twins twice?

(**b**) You have just had a set of twins, and will get pregnant one more time. What is the probability that you will end up pregnant with twins again, that is, that you will have been pregnant with twins twice?

Again, having two sets of twins might strike a person as extraordinarily unrepresentative of the process that generates children. Thus, people relying on the representativeness heuristic will think of the probability of having a second set of twins conditional on having one set already as considerably smaller than the probability of having a set of twins the first time around. But by assumption, of course, these probabilities are equal.

Exercise 5.8 Mr Langford, cont. Suppose that Langford from Exercise 4.49 on page 92 has just won two jackpots in a row and is about to play a third time. What is the probability that he will win a third time, so as to make it three jackpots in a row?

One way to explain the ubiquity of the gambler's fallacy is to say that people believe in the **law of small numbers**. That is, people exaggerate the degree to which small samples resemble the population from which they are drawn. In the case of the coins, the "population" consists of half heads and half tails. A believer in the law of small numbers would exaggerate the degree to which a small sample (such as a sequence of eight coin flips) will resemble the population and consist of half heads and half tails.

It is important to note, however, that there are games of chance in which outcomes are correlated. In Blackjack, for example, cards are drawn from a deck without being replaced, which means that the probability of drawing a given card will vary from draw to draw. In principle, then, you can beat the house by counting cards, which is why casinos reserve the right to throw you out if you do.

5.3 Conjunction and disjunction fallacies

We have already (in Section 4.3) come across the conjunction fallacy. "*A* AND *B*" is a conjunction; you commit the conjunction fallacy when you overestimate the probability of a conjunction. Consider the probability of winning the Lotto 6/49 (see Exercise 4.28 on page 82). When people learn the answer for the first time they are often shocked at how low it is. They are, in effect, overestimating the probability that the first number is right AND the second number is right AND ... and so on. Because they are overestimating the probability of a conjunction, they are committing the conjunction fallacy.

Example 5.9 Boeing aircraft A Boeing 747–400 has around 6 million parts. Suppose that each part is very reliable and only fails with probability 0.000,001. Assuming that failures are independent events, what is the probability that all parts work?

The probability that any one part works is 0.999,999, so the probability that all parts work is $(0.999{,}999)^{6{,}000{,}000} \approx 0.0025 = 0.25$ percent.

Given these numbers, the probability that all parts in a 747 work is only about a quarter of a percent! If this figure was lower than you expected, you may have committed the conjunction fallacy. Still, airplane crashes remain rare because planes are built with a great deal of redundancy, so that any one failure does not necessarily lead to a crash. That said, not all machines can be built in this way: some helicopters famously depend on a single rotor-retaining nut in such a way that if the nut fails, the whole machine will come crashing down. The term "Jesus nut" is sometimes used to denote a part whose failure would lead to a breakdown of the whole system. Presumably, the name is due to the only thing that can save you if the nut fails, though this assumes that a Jesus intervention is sufficiently independent of a nut failure.

The conjunction fallacy is particularly important in the context of **planning**. Complex projects are puzzles with many pieces, and typically each piece needs to be in place for the project to be successful. Even if the probability that any one piece will fall into place is high, the probability that all pieces of the puzzle will fall into place may be low. Planners who commit the conjunction fallacy will overestimate the probability of the conjunction – the proposition that the first piece is in place AND the second piece is in place AND the third piece is in place, and so on – meaning that they will overestimate the probability that the project will succeed.

The **planning fallacy** is the mistake of making plans based on predictions that are unreasonably similar to best-case scenarios. Many projects – senior theses, doctoral dissertations, dams, bridges, tunnels, railroads, highways, and wars – frequently take longer, and cost more, than planned. (Embarrassingly, that includes the book you are reading.) Here are two famous examples:

Example 5.10 The Sydney Opera House Many people consider the Sydney Opera House to be the champion of all planning disasters. According to original estimates in 1957, the opera house would be completed early in 1963 for $7 million. A scaled-down version of the opera house finally opened in 1973 at a cost of $102 million.

Example 5.11 Rail projects In more than 90 percent of rail projects undertaken worldwide between 1969 and 1998, planners overestimated the number of passengers who would use the systems by 106 percent. Cost overruns averaged 45 percent. Even more interestingly, estimates over the course of this period did not improve. That is, knowing about others' failed forecasts did not prompt planners to make theirs more realistic.

There is a related fallacy called the **disjunction fallacy**. "*A* OR *B*" is a disjunction; you commit the disjunction fallacy when you underestimate the probability of a disjunction. To illustrate, let us build upon Exercise 4.26(d) on page 80, in which you computed the probability of rolling at least one six when rolling two dice.

Example 5.12 Compute the probability of getting at least one six when rolling (**a**) one die, (**b**) two dice, (**c**) three dice, and (**d**) ten dice.

(**a**) The probability of rolling at least one six when rolling one die equals one minus the probability of rolling a non-six, which equals $1 - 5/6 \approx 16.6$ percent.

(**b**) The probability of rolling at least one six when rolling two dice equals one minus the probability of rolling no sixes in two trials, which equals $1 - (5/6)^2 \approx 30.6$ percent.

(**c**) The probability of rolling at least one six when rolling three dice equals one minus the probability of rolling no sixes in three trials, which equals $1 - (5/6)^3 \approx 42.1$ percent.

(**d**) Finally, the probability of rolling at least one six when rolling ten dice equals one minus the probability of rolling no sixes in ten trials, which equals $1 - (5/6)^{10} \approx$ 83.8 percent.

Notice how quickly the probability of rolling at least one six rises as the number of trials increases. The probability will never reach 100 percent, but it will approach it asymptotically, so that it will get closer and closer as the number of trials increases. If the resulting numbers here are greater than you expected, you may have committed the disjunction fallacy. For the probability of rolling at least one six in multiple trials equals the probability of rolling a six in the first trial, OR rolling a six in the second trial, OR ... And given the definition above, if you underestimate the probability of the disjunction, you are committing the disjunction fallacy.

Exercise 5.13 Hiking You plan to go on a hike in spite of the fact that a tornado watch is in effect. The national weather service tells you that for every hour in your area, there is a 30 percent chance that a tornado will strike. That is, there is a 30 percent chance that a tornado will strike your area between 10 am and 11 am, a 30 percent chance that a tornado will strike your area between 11 am and noon, and so on.

(**a**) What is the probability of a tornado striking your area at least once during a two-hour hike?
(**b**) What is the probability of a tornado striking your area at least once during a three-hour hike?
(**c**) What is the probability of a tornado striking your area at least once during a ten-hour hike?

Exercise 5.14 Flooding Imagine that you live in an area where floods occur on average every ten years. The probability of a flood in your area is constant from year to year. You are considering whether to live in your house for a few more years and save up some money, or whether to move before you lose everything you own in the next flood.

(**a**) What is the probability that there will be no floods in your area over the course of the next two years?
(**b**) What is the probability that there will be exactly one flood in your area over the course of the next two years?
(**c**) What is the probability that there will be at least one flood over the course of the next two years?
(**d**) What is the probability that there will be at least one flood over the course of the next ten years?

Exercise 5.15 Terrorism Compute the probability that at least one major terrorist attack occurs over the course of the next ten years, given that there are 365.25 days in an average year, if the probability of an attack on any given day is 0.0001.

That last exercise illustrates an infamous statement by the Irish Republican Army (IRA), which for decades fought a guerilla war for the independence of Northern Ireland from the United Kingdom and a united Ireland. In the aftermath of an unsuccessful attempt to kill British Prime Minister Margaret Thatcher in 1984 by planting a bomb in her hotel, the IRA released a statement that ended with the words: "You have to be lucky all the time. We only have to be lucky once."

The disjunction fallacy is particularly important in the context of **risk assessment**. When assessing the risk that some complex system will fail, it is often the case that the system as a whole – whether a car or an organism – critically depends on multiple elements in such a way that the failure of any one of these elements would lead to a breakdown of the system. Even if the probability that any one element will fail is low, the probability that at least one element will fail may be high. Assessors who commit the disjunction fallacy will underestimate the probability of the disjunction – the proposition that the first element fails or the second element fails or the third element fails, and so on – meaning that they will underestimate the probability of a system breakdown.

There is an obvious symmetry between the two fallacies discussed in this section. According to de Morgan's law, $A \& B$ is logically equivalent to $\neg[\neg A \vee \neg B]$

(see Section 2.4). So if you overestimate the probability $\Pr(A \& B)$, this is the same as saying that you overestimate $\Pr(\neg [\neg A \vee \neg B])$. But by the NOT rule, that is the same as saying that you overestimate $1 - \Pr(\neg A \vee \neg B)$, which is to say that you underestimate $\Pr(\neg A \vee \neg B)$. In the context of the Linda example, overestimating the probability that she is a feminist bank teller is (according to de Morgan's law) the same as underestimating the probability that she is a non-feminist or a non-bank teller. In sum, if you adhere to de Morgan's law, then you commit the conjunction fallacy if and only if you commit the disjunction fallacy.

Both the conjunction and disjunction fallacies can be explained in terms of anchoring and adjustment (see Section 3.6). People overestimate the probability of conjunctions – and therefore commit the conjunction fallacy – if they use the probability of any one conjunct as an anchor and adjust downwards insufficiently. They underestimate the probability of disjunctions – and therefore commit the disjunction fallacy – if they use the probability of any one disjunct as an anchor and adjust upwards insufficiently. Here are more exercises:

Exercise 5.16 What is the probability of drawing at least one ace when drawing cards from an ordinary deck, with replacement, when you draw: (**a**) 1 card, (**b**) 2 cards, (**c**) 10 cards, and (**d**) 52 cards?

Exercise 5.17 The birthday problem Suppose that there are 30 students in your behavioral-economics class. What is the probability that no two students have the same birthday? To make things easier, assume that every student was born the same non-leap year and that births are randomly distributed over the year.

Exercise 5.18 The preface paradox In the preface to your new book, you write that you are convinced that every sentence in your book is true. Yet you recognize that for each sentence there is a 1 percent chance that the sentence is false. (**a**) If your book has 100 sentences, what is the probability that at least one sentence is false? (**b**) What if your book has 1000 sentences?

Finally, an exercise about aviation safety.

Exercise 5.19 Private jet shopping Suppose you are fortunate (or delusional) enough to be shopping for a private jet. You have to decide whether to get a jet with one or two engines. Use p to denote the probability that an engine fails during any one flight. A "catastrophic engine failure" is an engine failure that makes the plane unable to fly.

(**a**) One of the jets you are looking to buy has only one engine. What is the probability of a catastrophic engine failure during any one flight with this plane?
(**b**) Another jet you are looking to buy has two engines, but is unable to fly with only one functioning engine. Assume that engine failures are independent events. What is the probability of a catastrophic engine failure during any one flight with this plane?
(**c**) Which jet strikes you as safer?
(**d**) What if the twin-engine jet can fly with only one functioning engine?

The answer to Exercise 5.19(b) is far from obvious. To help you out, consider constructing a table as in Figure 4.1 on page 80.

5.4 Base-rate neglect

One source of imperfect reasoning about probabilities is confusion between conditional probabilities $\Pr(A\,|\,B)$ and $\Pr(B\,|\,A)$. It might seem obvious that these two are distinct. As we know from Section 4.4, the probability that a randomly selected human being is a smoker is obviously different from the probability that a randomly selected smoker is a human being. However, there are contexts in which it is easy to mix these two up. In this section, we will consider some of these contexts.

Example 5.20 Mammograms Doctors often encourage women over a certain age to participate in routine mammogram screening for breast cancer. Suppose that from past statistics about some population, the following is known. At any one time, 1 percent of women have breast cancer. The test administered is correct in 90 percent of the cases. That is, if the woman does have cancer, there is a 90 percent probability that the test will be positive and a 10 percent probability that it will be negative. If the woman does not have cancer, there is a 10 percent probability that the test will be positive and a 90 percent probability that it will be negative. Suppose a woman has a positive test during a routine mammogram screening. Without knowing any other symptoms, what is the probability that she has breast cancer?

When confronted with this question, most people will answer close to 90 percent. After all, that is the accuracy of the test. Luckily, we do not need to rely on vague intuitions; we can compute the exact probability. In order to see how, consider Figure 5.2, in which C denotes the patient having cancer and P denotes the patient testing positive. Plugging the numbers into Bayes's rule (Proposition 4.42), we get:

$$\Pr(C|P) = \frac{\Pr(P|C) * \Pr(C)}{\Pr(P|C) * \Pr(C) + \Pr(P|\neg C) * \Pr(\neg C)}$$
$$= \frac{0.9 * 0.01}{0.9 * 0.01 + 0.1 * 0.99} \approx 0.08$$

Notice that the probability that somebody who has been identified as having cancer in fact has cancer is not equal to – in fact, not even remotely similar to – the accuracy of the test (which in this case is 90 percent). The probability that the woman has cancer is only about 8 percent – much lower than people think. This is paradoxical because we know that the test is reasonably good: it gives the correct outcome in 90 percent of cases. What we are forgetting is that relatively few people actually have cancer. Out of 1000 people, only about ten can be expected to have cancer. Of those, nine will test positive. Of the 990 women who do not have cancer, only 10 percent

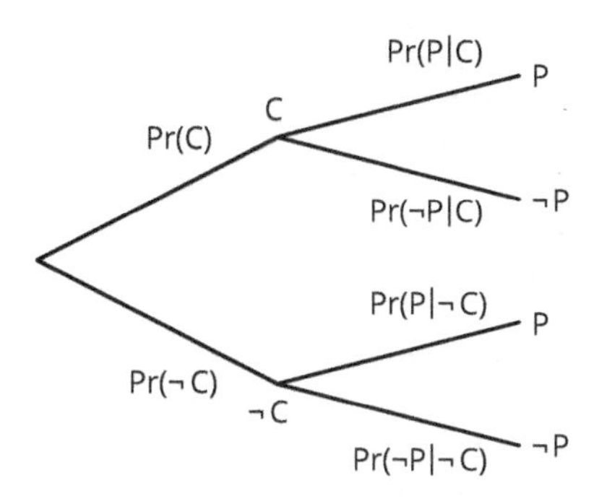

Figure 5.2 Breast cancer screening

will test positive, but that is still 99 people. So only nine of the 108 people who test positive actually have cancer, and that is about 8 percent. Notice that this case is similar to the frisbee case: although the new machine has a lower failure rate than the old one, the average frisbee produced in the factory is more likely than not to come from the new machine, simply because it produces so many more frisbees.

The fraction of all the individuals in the population who have cancer (or some other characteristic of interest) is called the **base rate**. In the cancer case, the base rate is only one percent. One way to diagnose the mistake that people make is to say that they fail to take the base rate properly into account. Thus, the mistake is sometimes referred to as **base-rate neglect** or the **base-rate fallacy**. The judgment that we make in these situations should reflect three different factors: first, the base rate; second, the evidence; third, the conditional probabilities that we would see the evidence when the hypothesis is true and when it is false. We commit the base-rate fallacy when we fail to take the first of these three factors properly into account.

Incidentally, this example makes it clear why younger women are not routinely tested for breast cancer. In younger women, the base rate would be even lower; so the ratio of true positives to all positives would be even lower. Notice that in the previous example, when a woman from the relevant population gets a positive result the probability that she has cancer only increases from one percent to about eight percent, which is not a very large increase. If the base rate were even lower, the increase would be even smaller, and the conditional probability $Pr(C \mid P)$ would not be very different from the base rate $Pr(C)$. When this is the case, the test does not give the doctor any additional information that is relevant when producing a diagnosis, and so the test is said to be **non-diagnostic**. If the base rate were very high, the test would still not be diagnostic. In order for a test to be diagnostic, it helps if the base rate is somewhere in the middle.

Exercise 5.21 Mammograms, cont. Men can get breast cancer too, although this is very unusual. Using the language of "base rates" and "diagnosticity," explain why men are not routinely tested for breast cancer.

Testimony can be non-diagnostic, as the following classic example illustrates.

Exercise 5.22 Testimony A cab company was involved in a hit-and-run accident at night. Two cab companies, the Green and the Blue, operate in the city. You are given the following data: 85 percent of the cabs in the city are Green, 15 percent are Blue. A witness identified the cab involved in the accident as Blue. The court tested the reliability of the witness under the same circumstances that existed on the night of the accident and concluded that the witness correctly identified each one of the two colors 80 percent of the time and failed 20 percent of the time. What is the probability that the cab involved in the accident was Blue rather than Green?

Exercise 5.23 Iron Bowl At an Auburn–Alabama game, 80 percent of attendees wore Alabama gear and 20 percent wore Auburn gear. During the game, one of the attendees apparently robbed a beer stand outside the stadium. A witness (who was neither an Alabama nor an Auburn fan) later told police that the robber wore Auburn gear. The witness, however, was the beer stand's best customer, and it was estimated that he would only be able to identify the correct gear about 75 percent of the time. What is the probability that the robber wore Auburn gear, given that the witness said that he did?

Here is a slightly different kind of problem.

Exercise 5.24 Down syndrome The probability of having a baby with Down syndrome increases with the age of the mother. Suppose that the following is true. For women 34 and younger, about one baby in 1000 is affected by Down syndrome. For women 35 and older, about one baby in 100 is affected. Women 34 years and younger have about 90 percent of all babies. What is the probability that a baby with Down syndrome has a mother who is 34 years or younger?

Base-rate neglect helps explain the planning fallacy: the fact that plans and predictions are often unreasonably similar to best-case scenarios (see Section 5.3). Notice that people who fall prey to the planning fallacy are convinced that their own project will be finished on time, even when they know that the vast majority of similar projects have run late.

From the point of view of the theory we explored in the previous chapter, the planning fallacy is surprising. If people updated their beliefs in Bayesian fashion, they would take previous overruns into account and gradually come up with a better estimate of future projects. We can, however, understand the optimistic estimates as a result of base-rate neglect. It is possible to think of the fraction of past projects that were associated with overruns as the base rate, and assume that people tend to ignore the base rate in their assessments.

The last problems in this section all relate to the war on terror.

Exercise 5.25 Jean Charles de Menezes In the aftermath of the July 21, 2005, terrorist attacks in London, British police received the authority to shoot terrorism suspects on sight. On July 22, plainclothes police officers shot and killed a terrorism suspect in the London Underground. Use Bayes's rule to compute the probability that a randomly selected Londoner, identified by the police as a terrorist, in fact is a terrorist. Assume that London is a city of 10 million people, and that ten of them (at any given time) are terrorists. Assume also that police officers are extraordinarily competent, so that their assessments about whether a given person is a terrorist or not are correct 99.9 percent of the time.

The suspect, Jean Charles de Menezes, 27, was shot seven times in the head and once in the shoulder. He was later determined to be innocent. Notice, again, that the probability that somebody who has been identified as a terrorist is in fact a terrorist is not equal to – in fact, not even remotely similar to – the accuracy of the police officers' assessments. Notice, also, the time line: it is as though the police went out of their way to prove as soon as possible that they cannot be entrusted with the authority to execute people on sight.

Exercise 5.26 Behavior detection The following passage is from *USA Today:*

> Doug Kinsey stands near the security line at Dulles International Airport, watching the passing crowd in silence. Suddenly, his eyes lock on a passenger in jeans and a baseball cap.
>
> The man in his 20s looks around the terminal as though he's searching for something. He chews his fingernails and holds his boarding pass against his mouth, seemingly worried.
>
> Kinsey, a Transportation Security Administration [TSA] screener, huddles with his supervisor, Waverly Cousins, and the two agree: The man could be a problem. Kinsey moves in to talk to him.
>
> The episode this month is one of dozens of encounters airline passengers are having each day - often unwittingly - with a fast-growing but controversial security technique called behavior detection. The practice, pioneered by Israeli airport security, involves picking apparently suspicious people out of crowds and asking them questions about travel plans or work. All the while, their faces, body language and speech are being studied.
>
> The TSA has trained nearly 2,000 employees to use the tactic, which is raising alarms among civil libertarians and minorities who fear illegal arrests and ethnic profiling. It's also worrying researchers, including some in the Homeland Security Department, who say it's unproven and potentially ineffectual.

The government did not publish data on the efficacy of this program, but we can make some reasonable assumptions. Every month, roughly 60 million people fly on US carriers. Let us imagine that 6 of them are terrorists. Let us also imagine that the TSA personnel are highly competent and will correctly identify a person as a terrorist or non-terrorist in 98/100 of cases. Questions:

(**a**) What is the probability that a passenger selected at random is a terrorist and is correctly identified as such by TSA personnel?

(**b**) What is the probability that a passenger selected at random is *not* a terrorist but is nevertheless (incorrectly) identified as a terrorist by TSA personnel?

(**c**) What is the probability that a passenger in fact is a terrorist conditional on having been identified as such by TSA personnel?

(**d**) Is this test diagnostic?

Notice that in the story above, the man was apparently a false positive, meaning that the story inadvertently ended up illustrating the lack of diagnosticity of the test.

Exercise 5.27 Diagnosticity Let us take it for granted that the behavior-detection test (from Exercise 5.26) is not diagnostic. The test may still be diagnostic in another setting, say, at a checkpoint at the US embassy in Kabul, Afghanistan. Explain how this is possible.

Recent evidence suggests that behavior-detection agents are not in fact very good at reading body language, in which case the program would not work in Kabul either. Critics call it "security theater," although this is insulting to real theater, which has artistic value.

On a related note: since 2004, the US Department of Homeland Security's US-VISIT Program, now called the Office of Biometric Identity Management, collects digital fingerprints and photographs from international travelers at US visa-issuing posts and ports of entry. The database now contains hundreds of millions of fingerprints. If a terrorist's fingerprint – recovered from a crime scene, perhaps – is found to match one in the database, what do you think the probability is that the match is the actual terrorist? If you find yourself caught up in this kind of dragnet, the only thing standing between you and the electric chair might be a jury's understanding of Bayes's rule. Good luck explaining it to them.

5.5 Confirmation bias

One striking feature of Bayesian updating in Section 4.6 is that John and Wes come to agree on the nature of the coin so quickly. As you will recall, after only about 15 flips of the coin, both assigned a probability of almost 100 percent to the possibility that the coin had two heads. People sometimes refer to this phenomenon as **washing out of the priors**. That is, after so many flips, John and Wes will assign roughly the same probability to the hypothesis, independently of what their priors used to be. This represents a hopeful picture of human nature: when rational people are exposed to the same evidence, over time they come to agree regardless of their starting point. (As is often the case, things get tricky when probabilities are zero; I continue to ignore such complications.)

In real life, unfortunately, people do not in general come to agree over time. Sometimes that is because they are exposed to very different evidence: conservatives tend to read conservative newspapers and blogs that present information selected because it supports conservative viewpoints; liberals tend to read liberal newspapers and blogs that present information selected because it supports liberal viewpoints. Yet sometimes people have access to the very same evidence presented in the very same way (as Wes and John do) but nevertheless fail to agree over time. Why is this?

Part of the story is a phenomenon that psychologists call **confirmation bias**: a tendency to interpret evidence as supporting prior beliefs to a greater extent than warranted. In one classic study, participants who favored or opposed the death penalty read an article containing ambiguous information about the advantages and disadvantages of the death penalty. Rather than coming to agree as a result of being exposed to the same information, both groups of people interpreted the information as supporting their beliefs. That is, after reading the article, those who were previously opposed to the death penalty were even more strongly opposed and those who favored it were even more in favor. In the presence of confirmation bias, then, the picture of how people's beliefs change as they are exposed to the evidence may look less like Figure 4.5 on page 91 and more like Figure 5.3.

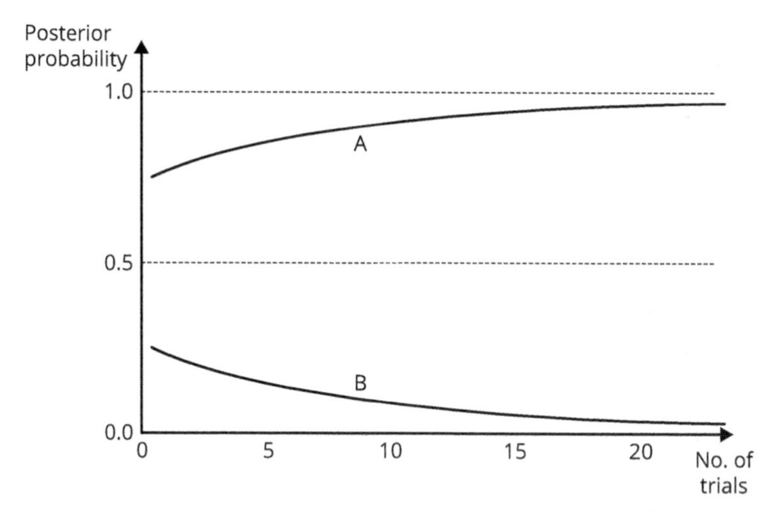

Figure 5.3 Confirmation bias

Exercise 5.28 Confirmation bias Imagine that John is suffering from confirmation bias. Which of the curves labeled A, B, and C in Figure 5.4 best represents the manner in which his probabilities change over time as the evidence comes in?

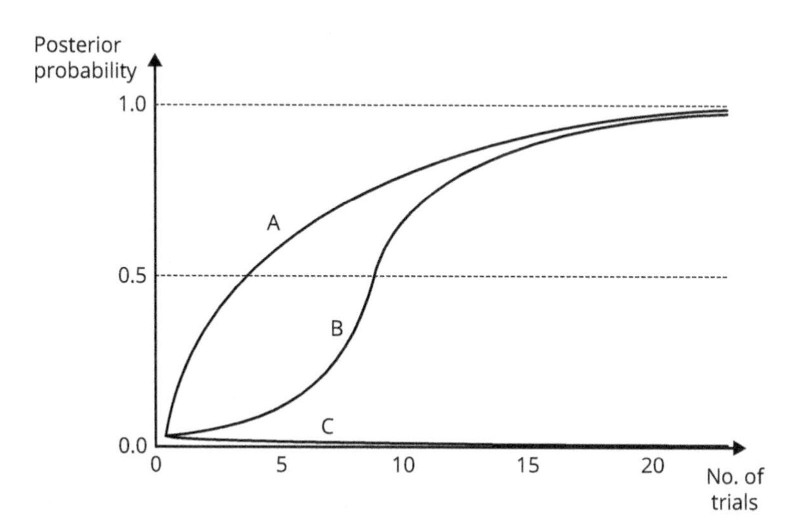

Figure 5.4 John's confirmation bias

Example 5.29 The jealous lover From literature or life, you may be familiar with the character of the jealous lover, who refuses to accept any evidence that his or her affections are reciprocated and who everywhere finds evidence fueling suspicions. As Marcel Proust, author of *In Search of Lost Time*, wrote: "It is astonishing how jealousy, which spends its time inventing so many petty but false suppositions, lacks imagination when it comes to discovering the truth." In more prosaic terms, the jealous lover exhibits confirmation bias.

Confirmation bias can explain a whole range of phenomena. It can explain why racist and sexist stereotypes persist over time. A sexist may dismiss or downplay evidence suggesting that girls are good at math and men are able to care for children but be very quick to pick up on any evidence that they are not. A racist may not notice all the people of other races who work hard, feed their families, pay their taxes, and do good deeds, but pay a lot of attention to those who do not. Confirmation bias can also explain why people gamble. Many gamblers believe that they can predict the outcome of the next game, in spite of overwhelming evidence that they cannot (they may, for example, have lost plenty of money in the past by mispredicting the outcomes). This could happen if the gambler notices all the cases when he did predict the outcome (and if the outcome is truly random, there will be such cases by chance alone) and fails to notice all the cases when he did not. The same line of thinking can explain why so many people think that they can beat the stock market, in spite of evidence that (in the absence of inside information) you might as well pick stocks randomly. Finally, confirmation bias can explain how certain conspiracy theories survive in spite of overwhelming contradictory evidence. The conspiracy theorist puts a lot of weight on morsels of evidence supporting the theory, and dismisses all evidence undermining it.

Exercise 5.30 Reputation The fact that people exhibit confirmation bias makes it very important to manage your reputation – whether you are a student, professor, doctor, lawyer, or brand. Why?

Scientists, by the way, are not immune from confirmation bias. Philosopher of science Karl Popper noted how some scientists find data supporting their theories everywhere. He describes his encounter with Alfred Adler, the pioneering psychoanalyst. In Popper's words:

> Once, in 1919, I reported to him a case which to me did not seem particularly Adlerian, but which he found no difficulty in analysing in terms of his theory of inferiority feelings, although he had not even seen the child. Slightly shocked, I asked him how he could be so sure. "Because of my thousandfold experience," he replied; whereupon I could not help saying: "And with this new case, I suppose, your experience has become thousand-and-one-fold."

Popper's description makes it sound as though Adler is suffering confirmation bias in a big way. Sadly, it is easy to find similar examples in economics. Because of how easy it is to "confirm" just about any theory, Popper ended up arguing that the hallmark of a scientific theory was not the fact that it could be confirmed, but rather that it could at least in principle be **falsified** – shown to be false by empirical observation. A good question to ask yourself and others is: "What sort of evidence would make you change your mind?" If you cannot think of anything short of divine intervention, you are almost surely suffering confirmation bias.

Problem 5.31 Confirmation bias among economists *(**a**) Name two famous economists who in your view are suffering confirmation bias. (**b**) Reflect upon the people you just named: if both are economists with whom you disagree, your response to (a) might itself be an expression of confirmation bias.*

Psychological research suggests that confirmation bias is due to a number of different factors. First, people sometimes fail to notice evidence that goes against their beliefs, whereas they quickly pick up on evidence that supports them. Second, when the evidence is vague or ambiguous, and therefore admits of multiple interpretations, people tend to interpret it in such a way that it supports their beliefs. Third, people tend to apply a much higher standard of proof to evidence contradicting their beliefs than to evidence supporting them.

Exercise 5.32 Destroying America Explain how book titles such as *Demonic: How the Liberal Mob Is Endangering America* and *American Fascists: The Christian Right and the War on America* contribute to political polarization.

Problem 5.33 Preventing confirmation bias *In matters of politics, philosophy, religion, and so on, do you expose yourself to the ideas of people "on the other side" as you do to the ideas of people "on your side"? Are you paying as much attention to what they say? Are you applying the same standards of evidence? If you can honestly answer yes to all these questions, you belong to a small but admirable fraction of the population. If not, you might give it a try; it is an interesting exercise.*

5.6 Availability

When physicians examine children with headaches, one of the things they look for is brain tumors. The base rate is very low. Children with brain tumors are virtually certain to have headaches, but there are of course many other causes of headaches. As it happens, a simple examination successfully identifies the children with tumors in very many of the cases. That is, of all the children who have been properly examined and judged not to have a tumor, very few actually do.

Exercise 5.34 CT scans In some populations, brain tumors in children are rare: the base rate is only about 1/10,000. A child with a tumor is very likely to have occasional headaches: 99 out of 100 do. But there are many other reasons a child can have a headache: of those who do not have a tumor, 1 in 10 have occasional headaches.

(**a**) Given that a child has occasional headaches (H), what is the probability that he or she has a brain tumor (T)?

(**b**) Let us say that among children with headaches, 999/1000 will ultimately be fine (F). Suppose that a physician using a simple test can correctly determine whether the child will be fine or not in 95/100 of children with headaches. Given that the doctor after performing the test gives the patient a green light (G), what is the probability that the child really will be fine?

As these exercises indicate, it is in fact very unlikely that the patient has a brain tumor provided that he or she has been properly examined. CT scans can determine almost conclusively whether the patient has a tumor, but they are prohibitively expensive. Knowing that patients who have been examined by a physician are unlikely to have a tumor, it is widely agreed that CT scans under these conditions are unjustified. However, once a physician happens to have a patient who turns out to have a tumor in spite of the fact that the examination failed to find one, the physician's behavior often changes dramatically. From then on, she wants to order far more CT scans than she previously did. Let us assume that the physician's experience with the last child did not change her values. Assuming that a drastic change in behavior must be due to a change in values or a change in beliefs, it follows that her beliefs must have changed. On the basis of what we know, did the physician update her beliefs rationally?

The story is, of course, far more complicated than it appears here. It is worth noticing, though, that the actual figures are widely known among medical doctors. This knowledge reflects the accumulated experience of far more cases than any one physician will see during her career. As a result, it seems unlikely that rational updating on the basis of one single case should have such a radical impact on a physician's behavior. So what is going on? Behavioral economists explain this kind of behavior in terms of **availability**: the ease with which information can be brought to mind when making a judgment. When the physician faces her next patient, though she at some level of consciousness still knows the figures that suggest a CT scan is uncalled for, chances are that the last case (in which she failed to find the tumor) will come to mind. It is particularly salient, in part because it happened recently, but also because it is highly emotionally loaded.

The **availability heuristic** is another prominent heuristic from the heuristics-and-biases program. When we rely on this heuristic, we assess the probability of some event occurring by the ease with which the event comes to mind. That is, the availability heuristic says that we can treat X as more likely than Y if X comes to mind more easily than Y. As pointed out in Section 3.6, heuristics are often perfectly functional. Suppose, for instance, that you happen to come across an alligator while walking to work. The chances are that images of alligators attacking other animals (including humans) will come to mind more easily than images of alligators acting cute and cuddly. If so, the availability heuristic tells you to assume the alligator is likely to be dangerous, which is obviously a helpful assumption under the circumstances. However, the availability heuristic (being a simple rule of thumb) can sometimes lead you astray, as in the case of the children with headaches. Thus, like anchoring and adjustment, availability can lead to bias.

Exercise 5.35 Contacts Your optometrist tells you that your new contacts are so good that you can wear them night and day for up to 30 days straight. She warns you that there is some chance that you will develop serious problems, but says that she has seen many clients and that the probability is low. After a week, you develop problems and see an ophthalmologist. The ophthalmologist tells you that he is the doctor who removes people's eyes when they have been injured as a result of improper contact use. He tells you that the probability of developing serious problems is high.

Use the concept of availability bias to explain how the optometrist and the ophthalmologist can report such different views about the likelihood of developing serious problems as a result of wearing contacts.

The availability heuristic can explain a variety of phenomena, including why people think violent crime is more common than other kinds of crime. Violent crime comes to mind so easily in part because images of violence can be particularly vivid, but also because it is covered so extensively in the press: "Area Man Not Shot Today" would make for a terrible headline. Availability can also explain why fears of airplane crashes, nuclear meltdowns, and terrorist attacks tend to increase dramatically shortly after such events occur, for the obvious reason that they come to mind particularly easily then. These considerations can help us see why anti-vaccination sentiments are so strong in spite of overwhelming evidence that vaccines are safe and effective. If you have a child, or even if you just hear of a child, who developed symptoms of autism shortly after being vaccinated, the dramatic series of events is likely to be highly salient and therefore strike you as more likely than it really is. Availability can also explain a variety of marketing practices. Advertising campaigns for grooming products, cigarettes, alcohol, and all sorts of other products depict users of the product as being particularly attractive and popular. If that is what comes to mind when potential buyers reflect on the consequences of using the products, availability might make attractiveness and popularity seem particularly likely outcomes. Availability can also explain why people repeat dangerous behaviors. If they once, for whatever reason, do something dangerous or reckless – drive drunk, do hard drugs, have unprotected sex, or whatever – and nothing bad happens, the salience of that event means that they might come to think the dangers have been exaggerated, which will make them more likely to engage in it again. Self-reinforcing cycles of this kind, especially when they involve multiple individuals whose beliefs and behaviors reinforce each other's, are called **availability cascades**.

The availability heuristic sheds light on the power of storytelling. As every writer knows, stories are often far more compelling than scientific data. If you doubt that, just ask a wolf. Wolves pose a trivial danger to humans: the number of verifiable, fatal attacks by wolves on humans is exceedingly low. And yet, fear of wolves runs deep. Part of the explanation is certainly that there are so many stories about big, bad wolves e.g., eating little girls' grandmothers. As a result of all these stories, the idea of wolves attacking humans is highly salient, which means that people treat it as likely – even though the data establish it is not. Far more dangerous organisms, such as the *Salmonella* bacterium that kills some 400 people per year in the US alone, do not figure in the public imagination in the same way and consequently are not as feared as they probably should be. The power of storytelling can be harnessed to communicate risk information very effectively, but it can also do immense harm. A single story about an illegal immigrant committing a heinous crime can generate strong anti-immigration sentiments in spite of evidence of the enormously beneficial aggregate welfare effects of migration.

Problem 5.36 Causes of death *According to the World Health Organization, the leading causes of death in the world are ischemic heart disease, stroke, lower respiratory infections, and chronic obstructive lung disease. This makes the leading causes of death quite different from the leading sources of fear. The effect is that people spend relatively much time thinking about threats they can do little about (terrorist groups across the world, for example) and relatively little time thinking about things they can (such as cardiovascular health, which can be improved by exercise and diet). Availability can explain why people overestimate the danger posed to them by the former and underestimate that by the latter. But a no less interesting question is this: Can the power of availability and other heuristics be harnessed to encourage people to think more about things such as cardiovascular health?*

Availability bias also helps explain the base-rate fallacy. Consider the cancer case. Even if you know that there are false positives, so that a positive test does not necessarily mean that you have the disease, the chances are that cases of true positives (people who were correctly diagnosed with cancer) are more likely to come to mind than cases of false positives (people who were incorrectly diagnosed with cancer). Insofar as you follow the availability heuristic, you will think of true positives as more likely than false positives. Because the actual probability of having the disease given a positive test is the ratio of true positives to all positives, an overestimation of the probability of a true positive will lead to an inflated probability of having the disease.

5.7 Overconfidence

Bayesian updating, as you know, can tell you precisely what probability you should assign to an event given the available evidence. Such probabilities are obviously relevant to statements of **confidence**: statements concerning how certain you are that various things will happen. There are many ways to express your confidence in a belief. After making an assertion, you can add: "… and I am 90 percent certain of that." Or, you can say: "It's 50–50," meaning that you are no more confident that the one thing will happen than you are that the other thing will happen. If you are in the business of providing professional forecasts, you may be used to offering 95 percent confidence intervals, which are ranges within which you expect the outcome to fall with 95 percent certainty. Thus, financial analysts might predict that a certain stock will reach $150 next year, and add that they are 95 percent certain that the value will be between $125 and $175; labor economists might predict that the unemployment rate will hit 6 percent and add that they are 95 percent sure that the actual unemployment range will fall in the 4–8 range. The more confident the analysts are in their predictive abilities, the narrower the confidence intervals will be.

When assessing statements of confidence, behavioral economists talk about **calibration**. Formally speaking, you are perfectly calibrated if, over the long run, for all propositions assigned the same probability, the proportion of the propositions that are true equals the probability you assigned. If you are calibrated and judge that something is 90 percent certain, nine times out of ten you will be right about that thing. Notice that you can be calibrated even if most of your predictions are wrong: if you are calibrated and 33 percent confident in your predictions, you will still be wrong two times out of three. Notice also that it is possible to be calibrated even when outcomes cannot be precisely predicted, e.g., because they are random. You do not know if an unbiased coin will come up heads or tails when you flip it; but if you predict with 50 percent confidence that it will come up heads, you will be perfectly calibrated.

All things equal, calibration seems to be a good thing. We certainly do expect competent people to be calibrated: if a structural engineer tells you he or she is 100 percent certain that a certain kind of house is safe to live in, you will be disappointed in him or her when half the houses of that kind collapse.

Yet, one of the most persistent findings in the literature is that people tend to exhibit **overconfidence**. That is, like the structural engineer in the previous paragraph, the certainty that people (including experts) express in their judgments tends to exceed the frequency with which those judgments are correct. In early studies, researchers asked undergraduates questions of the type "Absinthe is (a) a liqueur or (b) a precious stone?" and invited them to judge how confident they were that their answer was right. Participants who indicated that they were 100 percent certain that their answers were right were on the average correct 70–80 percent of the time. To test whether increased motivation would decrease the degree of overconfidence, the researchers asked participants to express their confidence in terms of odds (see text box on page 81) and offered participants to play a gamble based on those odds. When people said the odds that they were right were 100:1, in order to be well calibrated they should have said 4:1; when they said the odds were 100,000:1, they should have said 9:1. Though overconfidence was first studied in the lab, over time manifestations of systematic overconfidence have been found also among physicists, doctors, psychologists, CIA analysts, and others making expert judgments. Thus, overconfidence appears outside the laboratory, when knowledgeable judges make assertions within their field of specialization, and when they are motivated to provide accurate assessments.

Exercise 5.37 Meteorology Evidence suggests meteorologists are well calibrated and therefore an exception to the rule. This will strike many people as literally unbelievable. What heuristic might cause them to underestimate meteorologists' ability to offer calibrated predictions?

Studies indicate that overconfidence increases with confidence, and therefore is most extreme when confidence is high. Overconfidence is usually eliminated when confidence ratings are low, and when very low, people may even be underconfident. Overconfidence also increases with the difficulty of the judgment task. The more challenging the task is, the more likely a judge is to be overconfident; when it comes to very easy judgments, people may even be underconfident. Interestingly, overconfidence does not in general seem to decrease when people become more knowledgeable. In one famous study, the researcher asked participants in his study questions about the behaviors, attitudes, and interests of a real patient. As participants received more and more information about the patient's life, they assigned more and more confidence to their answers. Yet, their accuracy barely increased at all. Incidentally, the clinical psychologists who participated in the study – a majority of whom had PhDs – were no more accurate and no less confident than psychology graduate students and advanced undergraduates. The more educated people were actually *more* overconfident than the less educated people. It may well be true, as Alexander Pope said, that "[a] little learning is a dangerous thing" – and a lot of learning too.

Example 5.38 Apollo 11 On the 35th anniversary of the moon landing, CNN asked the crew of Apollo 11 what their biggest concern was at the time. Astronaut Neil Armstrong answered: "I think we tried very hard not to be overconfident, because when you get overconfident, that's when something snaps up and bites you. We were ever alert for little difficulties that might crop up and be able to handle those."

Is there anything you can do to be less overconfident and more calibrated? Research suggests that informing people about the prevalence of overconfidence makes little difference to their calibration. However, what does seem to help is to make highly repetitive judgments and to receive regular, prompt, and unambiguous feedback. (This is why meteorologists came to be so well calibrated.) Moreover, overconfidence can be reduced by considering reasons that you might be wrong.

How is it possible for overconfidence to be so prevalent? For starters, many of our judgments are not repetitive, and we do not receive feedback that is regular, prompt, and unambiguous. Furthermore, even in the presence of outcome feedback, learning from experience is more difficult than one might think. Confirmation bias (Section 5.5) makes us overweight evidence that confirms our predictions and underweight evidence that disconfirms them; in this way, confirmation bias makes us blind to our failures. Availability bias (Section 5.6) does not help either. If the image of a situation where you were right when others were spectacularly wrong comes to mind easily and often, you might end up overestimating the probability that that sort of thing will happen again. And a phenomenon referred to as **hindsight bias** – that is, the tendency to exaggerate the probability that an event would occur by people who know that it in fact did occur – plays other tricks with our minds. Victims of the hindsight bias may never learn that past predictions were no good, because they misremember what they in fact predicted and see no need to be less confident in the future. Finally, people are very good at explaining away false predictions, e.g., by arguing that they were *almost* right, or that any failures were due to inherently unpredictable factors.

The heuristics-and-biases program

According to the heuristics-and-biases program, we form judgments by using **heuristics** – functional but imperfect rules of thumb or mental shortcuts that help us form opinions and make decisions quickly. Here are four prominent heuristics:

- The **anchoring-and-adjustment** heuristic instructs you to pick an initial estimate (anchor) and adjust the initial estimate up or down (as you see fit) in order to come up with a final answer. Insufficient adjustment will lead to answers that track the anchors – even when irrelevant or uninformative.
- The **representativeness** heuristic tells you to estimate the probability that some outcome was the result of a given process by reference to the degree to which the outcome is representative of that process: the more representative the outcome, the higher the probability.
- The **availability heuristic** makes you assess the probability that some event will occur based on the ease with which the event comes to mind: the easier it comes to mind, the higher the probability.
- The **affect heuristic** gets you to assign probabilities to consequences based on how you feel about the thing they would be consequences of: the better you feel about it, the higher the probability of good consequences and the lower the probability of bad.

The heuristics-and-biases program does not say that people are dumb: to the contrary, it says that following heuristics is a largely functional way to make speedy decisions when it counts. Because the heuristics are imperfect, however, they can lead to **bias**, that is, systematic and predictable error.

Kahneman has proposed that the operation of heuristics can be understood in terms of **substitution**. When faced with a question we are unable to address directly, we sometimes replace it by an easier question and answer that one instead. Rather than addressing the question "How likely is this airplane to crash?" we substitute the question "How easily can I imagine this airplane crashing?" Substitution allows us to come up with quick answers. But because the question we are actually answering is different from the original one, the answer too might be different: while airplane crashes are unlikely, mental representations of airplane crashes are easy to conjure. Thus, we come to exhibit availability bias and exaggerate the probability of a crash. (Similar stories can be told about the other heuristics.) The tricky thing is that we are often unaware of the substitution and therefore of the difference between the answer we sought and the one we produced.

The overconfidence phenomenon receives indirect support from research on **competence**. For example, many studies suggest that people overestimate their competence in various practical tasks. The vast majority of drivers – in some studies, more than 90 percent – say that they are better than the median driver, which is a statistical impossibility. In one fascinating study, undergraduates whose test scores in grammar and logic put them in the bottom 25 percent of a group of peers, on the mean estimated that they were well above average. Even more surprising, perhaps, when participants received more information about their relative performance in the tests (by being asked to grade those of other participants), the strongest students became more calibrated, while the weakest students, if anything, became *less* calibrated. The results suggest the least competent are at a double disadvantage, in that their incompetence causes both poor performance and an inability to recognize their performance as poor: the *cognitive* skill required to perform a difficult mental task may well be tightly tied up with the *metacognitive* skill to assess the quality of our performance. This **Dunning–Kruger effect** may or may have been what comedian Ricky Gervais was getting at when he said: "When you are dead, you do not know you are dead. It is only painful for others. The same applies when you are stupid."

Exercise 5.39 Inevitability People think many things are inevitable. If you search for the expression "it was inevitable that" on Google News, you may get tens of thousands of hits. Which bias is reflected in the use of that expression?

Exercise 5.40 Adam Smith, once more What sort of phenomenon might Adam Smith have had in mind when he talked about the "over-weening conceit which the greater part of men have of their own abilities"?

5.8 Discussion

This chapter has explored phenomena that appear inconsistent with the theory of probabilistic judgment that we learned about in Chapter 4. While the theory of probability was never designed to capture the precise cognitive processes people use when forming judgments, there appears to be a wide range of circumstances under which people's intuitive probability judgments differ substantially, systematically, and predictably from the demands of the theory. As the examples have shown, deviations can be costly. The phenomena are typically construed as undercutting the adequacy of probability theory as a descriptive theory. Yet some of these phenomena can also be invoked when challenging the correctness of probability theory as a normative standard. The fact that living up to the theory is so demanding – surely, part of the explanation for why people fail to live up to it in practice – is sometimes thought to undercut its normative correctness.

We have also discussed some of the theoretical tools used by behavioral economists to capture the manner in which people actually make judgments. Thus, we explored further aspects of the heuristics-and-biases program, which is one prominent effort to develop a descriptively adequate theory of probabilistic judgment. Because heuristics are largely functional, it would be a mistake to try to eliminate reliance on them altogether. But heuristics can lead us astray, and an awareness of the conditions under which this may happen might reduce the likelihood that they do. As the marketing applications make particularly clear, knowledge of behavioral economics in general, and heuristics and biases in particular, permits us to influence other people's judgment, for good or for evil. But knowledge of behavioral economics also permits us to anticipate and to resist other people's efforts to influence our judgment.

Exercise 5.41 Misguided criticism Some critics of the heuristics-and-biases program attack it for saying that human beings are irredeemably stupid. Thus, "the heuristics-and-biases view of human irrationality would lead us to believe that humans are hopelessly lost in the face of real-world complexity, given their supposed inability to reason according to the canon of classical rationality, even in simple laboratory experiments." Explain where these critics go wrong.

Needless to say, the chapter does not offer a complete list of phenomena that are inconsistent with standard theory or of the theoretical constructs designed to explain them. One important heuristic that has not come up yet is the **affect heuristic**. This heuristic tells you to assign probabilities to consequences based on how you feel about the thing they would be consequences of: the better you feel about it, the higher the probability you assign to good consequences and the lower the probability you assign to bad. As a result, those who love guns will think of scenarios in which guns are beneficial as relatively more likely, and scenarios in which guns are harmful as relatively less likely, than those who hate guns. Like other heuristics, the affect heuristic is often functional: when you see something you really do not like the look of in the back of your fridge, you will think the probability that you will get sick from eating it as relatively high – which is likely the proper response. But the affect heuristic can also lead you astray, by making beliefs about benefits and risks reflect feelings rather than the other way around.

The next part of the book will explain how the theory of probability can be incorporated into a theory of rational decision, and explore its strengths and weaknesses.

Additional exercises

Exercise 5.42 Probability matching Imagine that your friend Anne has a coin that has a 2/3 probability of coming up heads (H) and a 1/3 probability of coming up tails (T). She intends to flip it three times and give you a dollar for every time you correctly predicted how the coin would come up. Would you be more likely to win if you predicted that the coin will come up HHH or HHT?

If your prediction was HHT (or HTH or THH) you might have engaged in **probability matching**: choosing frequencies so as to match the probabilities of the relevant events. Probability matching might result from the use of the representativeness heuristic, since an outcome such as HHT (or HTH or THH) might seem so much more representative for the random process than HHH or TTT. As the exercise shows, probability matching is a suboptimal strategy leading to bias.

Exercise 5.43 Gender discrimination, cont. In Exercise 4.51 on page 93, we computed the probability that an editorial board of 20 members is all male by chance alone. If the answer strikes a person as low, what fallacy may he or she have committed?

Exercise 5.44 IVF In vitro fertilization (IVF) is a procedure by which egg cells are fertilized by sperm outside the womb. Let us assume that any time the procedure is performed the probability of success (meaning a live birth) is approximately 20 percent. Let us also assume, though this is unlikely to be true, that the probabilities of success at separate trials are independent. Imagine, first, that a woman has the procedure done twice.

(a) What is the probability that she will have *exactly* two live births?

(b) What is the probability that she will have *no* live births?

(c) What is the probability that she will have *at least* one live birth?

(d) Imagine, next, that another woman has the procedure done five times.

(e) What is the probability that she will have at least one live birth?

Exercise 5.45 Mandatory drug testing In July 2011, the State of Florida started testing all welfare recipients for the use of illegal drugs. Statistics suggest that some 8 percent of adult Floridians use illegal drugs; let us assume that this is true for welfare recipients as well. Imagine that the drug test is 90 percent accurate, meaning that it gives the correct response in nine cases out of ten.

(a) What is the probability that a randomly selected Floridian welfare recipient uses illegal drugs and has a positive test?

(b) What is the probability that a randomly selected Floridian welfare recipient does not use illegal drugs but nevertheless has a positive test?

(c) What is the probability that a randomly selected Floridian welfare recipient has a positive test?

(d) Given that a randomly selected Floridian welfare recipient has a positive test, what is the probability that he or she uses illegal drugs?

(e) If a Florida voter favors the law because he thinks the answer to (d) is in the neighborhood of 90 percent, what fallacy might he be committing?

Exercise 5.46 CIA Intelligence services are deeply interested in how people think, both when they think correctly and when they think incorrectly. The following exercise is borrowed from the book *Psychology of Intelligence Analysis*, published by the US Central Intelligence Agency (CIA).

During the Vietnam War, a fighter plane made a non-fatal strafing attack on a US aerial reconnaissance mission at twilight. Both Cambodian and Vietnamese jets operate in the area. You know the following facts: (a) Specific case information: The US pilot identified the fighter as Cambodian. The pilot's aircraft recognition capabilities were tested under appropriate visibility and flight conditions. When presented with a sample of fighters (half with Vietnamese markings and half with Cambodian) the pilot made correct identifications 80 percent of the time and erred 20 percent of the time. (b) Base rate data: 85 percent of the jet fighters in that area are Vietnamese; 15 percent are Cambodian.

Question: What is the probability that the fighter was Cambodian rather than Vietnamese?

Exercise 5.47 Juan Williams In October 2010, National Public Radio (NPR) fired commentator Juan Williams after he made the following remark on Fox News: "When I get on a plane … if I see people who are in Muslim garb and I think, you know, they're identifying themselves first and foremost as Muslims, I get worried. I get nervous." Here, I will not comment on the wisdom of firing somebody for expressing such a sentiment or of expressing it in the first place. But we can discuss the *rationality* of the sentiment.

(**a**) In the United States, there are roughly 300 million people, of whom about 2 million are Muslims. Let us assume that at any one time there are ten terrorists able and willing to strike an airliner, and (implausibly) that as many as nine out of ten terrorists are Muslims. Under these assumptions, what is the probability that a randomly selected Muslim is a terrorist able and willing to strike an airliner?

(**b**) Use the notion of availability bias to explain how Juan Williams might overestimate the probability that a randomly selected Muslim is a terrorist able and willing to strike an airliner.

Exercise 5.48 Theories, theories Complete this sentence: "If all your observations support your scientific theories or political views, you are (probably) suffering…"

Exercise 5.49 Matthew Which heuristic is embodied in this line from Matthew 7:17–18: "So every good tree bears good fruit, but the bad tree bears bad fruit. A good tree cannot produce bad fruit, nor can a bad tree produce good fruit."

Exercise 5.50 Genetically modified organisms (GMOs) A person opposed to GMOs reads a compelling text about the benefits of such organisms and comes to quite like the thought of them. When asked about the risks, he says he has changed his mind and decided that not only are the benefits great, but the risks are trivial too – although he has acquired no new information about the risks. What heuristic might have been in play here?

Exercise 5.51 Schumpeter The Austrian economist Joseph Schumpeter claimed that he had set himself three goals in life: to be the greatest economist in the world, the best horseman in all of Austria, and the greatest lover in all of Vienna. He conceded that he had only reached two of the three goals. Suppose that he was wrong and had, in fact, reached none of them. Use each of the following ideas to explain how he might be so wrong: (**a**) confirmation bias, (**b**) availability bias, (**c**) overconfidence, and (**d**) conjunction fallacy.

Exercise 5.52 Match each of the vignettes below with one of the following phenomena: *availability bias, base-rate neglect, confirmation bias, conjunction fallacy, disjunction fallacy, hindsight bias,* and *overconfidence*. If in doubt, pick the best fit.

(**a**) Al has always been convinced that people of Roma descent are prone to thievery. In fact, several of his co-workers have a Roma background. But he knows his co-workers are not thieves, and he does not think twice about it. One day during happy hour, however, a racist acquaintance shares a story about two people "who looked like those people" stealing goods from a grocery store. "I knew it!" Al says to himself.

(**b**) Beth's car is falling apart. Her friends, who know these things, tell her that the car has a 10 percent probability of breaking down every mile. Beth really wants to go see a friend who lives about ten miles away, though. She ponders the significance of driving a car that has a 10 percent probability of breaking down each mile, but figures that the probability of the car breaking down during the trip cannot be much higher than, say, 15 percent. She is shocked when her car breaks down halfway there.

(**c**) Cecile is so terrified of violent crime that she rarely leaves her house, even though she lives in a safe neighborhood. She is out of shape, suffers from hypertension, and would be much happier if she went for a walk every so often. However, as soon as she considers going for a walk, images of what might happen to innocent people quietly strolling down the sidewalk come to mind, and she is sure something horrible is going to happen to her too. As a result, she goes back to watching reruns of *Law and Order*.

(**d**) David, who has never left the country, somehow manages to get tested for malaria. The test comes back positive. David has never been so depressed. Convinced that he is mortally sick, he starts to draft his will.

(**e**) Because she has trouble getting up in the morning, Elizabeth often drives too fast on her way to school. After getting a speeding ticket on Monday last week, she religiously followed the law all week. Only this week is she starting to drive faster again.

(**f**) Fizzy does not think the US will want to start another front in the war on terror, so she believes that it is quite unlikely that the US will bomb Iranian nuclear facilities. She does, however, think that it is quite likely that the US will withdraw all troops from Afghanistan. When asked what she thinks about the possibility that the US will bomb Iranian nuclear facilities and withdraw all troops from Afghanistan, she thinks that it is more likely than the possibility that the US will bomb Iranian nuclear facilities.

(**g**) Georgina has trouble imagining an existence without iPhones and iPads. Thus, she thinks it was inevitable that an inventor such as Steve Jobs would appear and design such things.

(**h**) Harry lost all his luggage the last time he checked it. He is never going to check his luggage again, even if it means having unpleasant arguments with flight attendants.

Problem 5.53 *Drawing on your own experience, make up stories like those in Exercise 5.52 to illustrate the various ideas that you have read about in this chapter.*

Further reading

A comprehensive introduction to heuristics and biases in judgment is Hastie and Dawes (2010). The gambler's fallacy and related mistakes are discussed in Tversky and Kahneman (1971). The conjunction and disjunction fallacies are explored in Tversky and Kahneman (1983) and in Tversky and Shafir (1992). The planning fallacy and the Sydney Opera House are discussed in Buehler et al. (1994, p. 366); Kahneman (2011) reports the data about rail projects. Base-rate neglect is examined in Bar-Hillel (1980). The *USA Today* story is Frank (2007). An extensive review of confirmation bias is Nickerson (1998); the quotation from *In Search of Lost Time* is Proust (2002 [1925], p. 402), and the study of confirmation bias in the context of the death penalty is Lord et al. (1979). Popper (2002 [1963], p. 46) describes meeting Adler. Availability is discussed alongside anchoring and adjustment and representativeness in Tversky and Kahneman (1974). The section on overconfidence draws on Angner (2006), which argues that economists acting as experts in matters of public policy exhibit significant overconfidence even within their domain of expertise. Neil Armstrong is cited in O'Brien (2004). The study by Fischhoff and colleagues is Fischhoff et al. (1977), the study involving a real patient is Oskamp (1982), and the study on competence is Kruger and Dunning (1999). Kahneman (2011, Chapter 9) discusses substitution and Smith (1976 [1776], p. 120) our "over-weening conceit." The critics of the heuristics-and-biases program are Gigerenzer and Goldstein (1996); the affect heuristic appears in Finucane et al. (2000). The CIA's intelligence analysis example comes from Heuer (1999, pp. 157–8) and the Juan Williams affair is described in Farhi (2010).

PART 3

CHOICE UNDER RISK AND UNCERTAINTY

6 RATIONAL CHOICE UNDER RISK AND UNCERTAINTY

Learning objectives

After studying this chapter you will:

- Master multiple principles of rational choice under uncertainty, including the maximin principle
- Know the theory of expected utility
- Understand how rational decision-making under risk depends on the shape of the utility function

6.1 Introduction

In Part 2, we left the theory of decision aside for a moment in order to talk about judgment. Now it is time to return to questions of decision, and specifically, rational decision. In this chapter we explore the theory of rational choice under risk and uncertainty. According to the traditional perspective, you face a **choice under uncertainty** when the probabilities of the relevant outcomes are completely unknown or not even meaningful; you face a **choice under risk** when the probabilities of the relevant outcomes are both meaningful and known. At the end of the day, we want a theory that gives us principled answers to the question of what choice to make in any given decision problem. It will take a moment to develop this theory. We begin by discussing uncertainty and then proceed to expected value, before getting to expected utility. Ultimately, expected-utility theory combines the concept of utility from Chapter 2 with the concept of probability from Chapter 4 into an elegant and powerful theory of choice under risk.

6.2 Uncertainty

Imagine that you are about to leave your house and have to decide whether to take an umbrella or to leave it at home. You are concerned that it might rain. If you do not take the umbrella and it does not rain, you will spend the day dry and happy; if you do not take the umbrella and it does rain, however, you will be wet and miserable. If you take the umbrella, you will be dry no matter, but carrying the cumbersome umbrella will infringe on your happiness. Your decision problem can be represented as in Table 6.1(a).

In a table like this, the leftmost column represents your menu, that is, the individual acts available to you. Other than that, there is one column for each thing that can happen. These things are referred to as **states of the world** or simply **states**, listed

Table 6.1 Umbrella problem

	Rain	No rain
Take umbrella	Dry, not happy	Dry, not happy
Leave umbrella	Wet, miserable	Dry, happy

(a) Payoffs

	Rain	No rain
Take umbrella	3	3
Leave umbrella	0	5

(b) Utility payoffs

	Rain	No rain
Take umbrella	0	2
Leave umbrella	3	0

(c) Risk payoffs

in the top row. The resulting matrix defines the outcome space. (Throughout, I will assume that the state of the world is independent of the acts chosen by the agent.) In this case, obviously, there are only two states: either it rains, or it does not. Nothing prevents expressing your preferences over the four outcomes by using our old friend the utility function from Section 2.7. Utility payoffs can be represented as in Table 6.1(b). Under the circumstances, what is the rational thing to do? Let us assume that you treat this as a choice under uncertainty. There are a number of different criteria that could be applied.

According to the **maximin criterion**, you should choose the alternative that has the greatest minimum utility payoff. If you take the umbrella, the minimum payoff is three; if you leave the umbrella at home, the minimum payoff is zero. Consequently, maximin reasoning would favor taking the umbrella. According to the **maximax criterion**, you should choose the alternative that has the greatest maximum utility payoff. If you take the umbrella, the maximum payoff is three; if you leave the umbrella at home, the maximum payoff is five. Thus, maximax reasoning would favor leaving the umbrella at home. The maximin reasoner is as cautious as the maximax reasoner is reckless. The former looks at nothing but the *worst* possible outcome associated with each act, and the latter looks at nothing but the *best* possible outcome associated with each act. Some of my students started referring to the maximax criterion as the **YOLO criterion** – from "You Only Live Once," for the reader unfamiliar with millennial.

Exercise 6.1 The watch Having just bought a brand-new watch, you are asked if you also want the optional life-time warranty.

(**a**) Would a maximin reasoner purchase the warranty?
(**b**) What about a maximax reasoner?

There are other criteria as well. According to the **minimax-risk criterion**, you should choose the alternative that is associated with the lowest maximum **risk** or **regret**. If you take the umbrella and it rains, or if you leave the umbrella at home and it does not rain, you have zero regrets. If you take the umbrella and it does not rain, your regret equals the best payoff you could have had if you had acted differently (five) minus your actual payoff (three), that is, two. By the same token, if you leave the umbrella at home and it does rain, your regret equals three. These "risk payoffs" can be captured in table form, as shown in Table 6.1(c). Since bringing the umbrella is associated with the lowest maximum regret (two, as opposed to three), minimax-risk reasoning favors taking the umbrella. The term **regret aversion** is sometimes used when discussing people's tendency to behave in such a way as to minimize anticipated regret. Regret aversion may be driven by loss aversion (see Section 3.5), since regret is due to the loss of a payoff that could have resulted from the state that obtains, had the agent acted differently. (We return to the topic of regret in Section 7.4.)

Exercise 6.2 Rational choice under uncertainty This exercise refers to the utility matrix of Table 6.2. What course of action would be favored by (**a**) the maximin criterion, (**b**) the maximax criterion, and (**c**) the minimax-risk criterion? As part of your answer to (c), make sure to produce the risk-payoff matrix.

Table 6.2 Decision under uncertainty

	S_1	S_2
A	1	10
B	2	9
C	3	6

Quite a number of authors have offered advice about how to minimize regret. In *The Picture of Dorian Gray*, Oscar Wilde wrote: "Nowadays most people die of a sort of creeping common sense, and discover when it is too late that the only things one never regrets are one's mistakes." And books with titles such as *The Top Five Regrets of the Dying* aspire to tell us what we too might one day come to regret: working too much, trying to please others, not allowing ourselves to be happy, and so on. Then again, Søren Kierkegaard – sometimes called "the father of existentialism" – altogether despaired of avoiding regret. "My honest opinion and my friendly advice is this: Do it or do not do it – you will regret both," he wrote in a book titled *Either/Or*.

Problem 6.3 The dating game under uncertainty *Imagine that you are considering whether or not to ask somebody out on a date. (**a**) Given your utility function, what course of action would be favored by (i) the maximin criterion, (ii) the maximax criterion, and (iii) the minimax-risk criterion? (**b**) In the words of Alfred, Lord Tennyson, "'Tis better to have loved and lost / Than never to have loved at all." What decision criterion do these lines advocate?*

Of all criteria for choice under uncertainty, the maximin criterion is the most prominent. It is, among other things, an important part of the philosopher John Rawls's theory of justice. In Rawls's theory, the principles of justice are the terms of cooperation that rational people would agree to follow in their interactions with each other if they found themselves behind a "veil of ignorance," meaning that they were deprived of all morally relevant information about themselves, the society in which they live, and their place in that society. Suppose, for example, that you have to choose whether to live either in a society with masters and slaves or in a more egalitarian society, without knowing whether (in the former) you would be master or slave. According to Rawls, the rational procedure is to rank societies in accordance with the worst possible outcome (for you) in each society – that is, to apply the maximin criterion – and to choose the more egalitarian option. Rawls took this to constitute a reason to think that an egalitarian society is more just than a society of masters and slaves.

Critics have objected to the use of maximin reasoning in this and other scenarios. One objection is that maximin reasoning fails to consider relevant utility information, since for each act, it ignores all payoffs except the worst. Consider the two decision problems in Table 6.3. Maximin reasoning would favor A in either scenario. Yet it does not seem completely irrational to favor A over B and B* over A, since B* but not B upholds the prospect of ten billion utiles.

Table 6.3 More decisions under uncertainty

	S_1	S_2
A	1	1
B	0	10

(a)

	S_1	S_2
A	1	1
B*	0	10^{10}

(b)

Another objection is that maximin reasoning fails to take into account the chances that the various states of the world will obtain. In a famous critique of Rawls's argument, Nobel Prize-winning economist John C. Harsanyi offered the following example:

Example 6.4 Harsanyi's challenge Suppose you live in New York City and are offered two jobs at the same time. One is a tedious and badly paid job in New York City itself, while the other is a very interesting and well-paid job in Chicago. But the catch is that if you wanted the Chicago job, you would have to take a plane from New York to Chicago (for example, because this job would have to be taken up the very next day). Therefore, there would be a very small but positive probability that you might be killed in a plane accident.

Assuming that dying in a plane crash is worse than anything that could happen on the streets of New York, as Harsanyi points out, maximin reasoning would favor the tedious NYC job, *no matter* how much you prefer the Chicago job and *no matter* how unlikely you think a plane accident might be. This does not sound quite right.

Perhaps there are scenarios in which the probabilities of the relevant outcomes are completely unknown or not even meaningful, and perhaps in those scenarios maximin reasoning – or one of the other criteria discussed earlier in this section – is appropriate. Yet the upshot is that, whenever possible, it is perfectly reasonable to pay attention to all possible payoffs as well as to the probabilities that the various states might obtain. When facing the umbrella problem in Table 6.1, for example, it seems rational to take into account all four cells in the payoff matrix as well as the probability of rain.

6.3 Expected value

From now on I will assume that it is both meaningful and possible to assign probabilities to outcomes; that is, we will be leaving the realm of choice under uncertainty and entering the kingdom of choice under risk. In this section, we explore one particularly straightforward approach – expected value – that takes the entire payoff matrix as well as probabilities into account.

The **expected value** of a gamble is what you can expect to win *on the average, in the long run*, when you play the gamble. Suppose I make you the following offer: I will flip a fair coin, and I will give you $10 if the coin comes up heads (H), and nothing if the coin comes up tails (T). This is a reasonably good deal: with 50 percent probability you will become $10 richer. This gamble can easily be represented in tree and table form, as shown in Figure 6.1. It is clear that on the average, in the long run, you would get $5 when playing this gamble; in other words, the expected value of the gamble is $5. That is the same as the figure you get if you multiply the probability of winning (1/2) by the dollar amount you stand to win ($10).

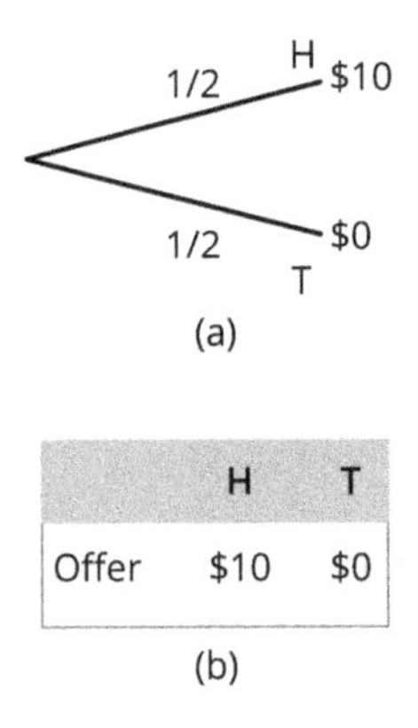

	H	T
Offer	$10	$0

Figure 6.1 Simple gamble

Exercise 6.5 Lotto 6/49 Represent the gamble accepted by someone who plays Lotto 6/49 (from Exercise 4.28 on page 82) as in Figure 6.1(a) and (b). Assume that the grand prize is a million dollars.

Example 6.6 Lotto 6/49, cont. What is the expected value of a Lotto 6/49 ticket, if the grand prize is a million dollars?

We know from Exercise 4.28 that the ticket is a winner one time out of 13,983,816. The means that the ticket holder will receive, on the average, in the long run, $1/13{,}983{,}816 * \$1{,}000{,}000$. You get the same answer if you multiply the probability of winning by the amount won: $0.000{,}000{,}07 * \$1{,}000{,}000 = \$0.07$. That is 7 cents.

Problem 6.7 *What would you pay to play this gamble? If you are willing to pay to play this game, what do you hope to achieve?*

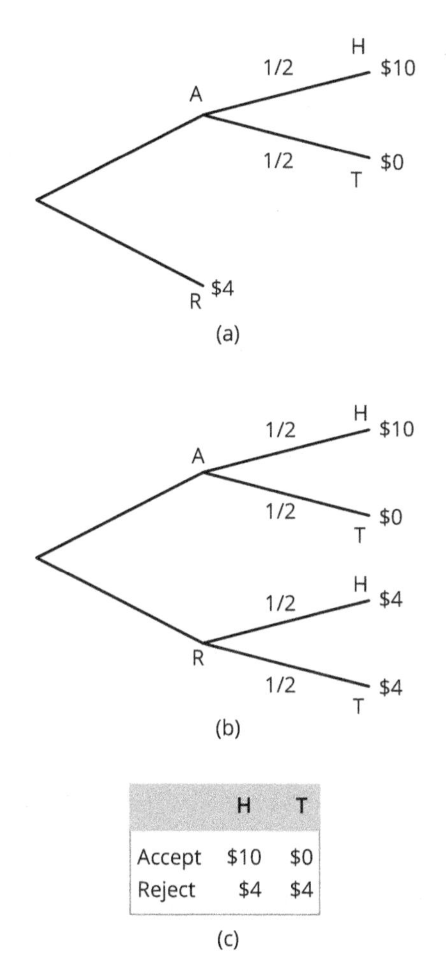

	H	T
Accept	$10	$0
Reject	$4	$4

Figure 6.2 Choice between gambles

Sometimes two or more acts are available to you, in which case you have a choice to make. Imagine, for instance, that you can choose between the gamble in Figure 6.1 and $4 for sure. If so, we can represent your decision problem in tree form as shown in Figure 6.2(a). We can also think of the outcome of rejecting the gamble as $4 no matter whether the coin comes up heads or tails. Thus, we can think of the gamble as identical to that in Figure 6.2(b). The latter decision tree makes it obvious how to represent this gamble in table form (see Figure 6.2(c)). The numbers in the row marked "Reject" represent the fact that if you reject the gamble, you keep the four dollars whether or not it turns out to be a winner.

Exercise 6.8 Expected value For the following questions, refer to Figure 6.2(c):
(**a**) What is the expected value of accepting this gamble?
(**b**) What is the expected value of rejecting it?

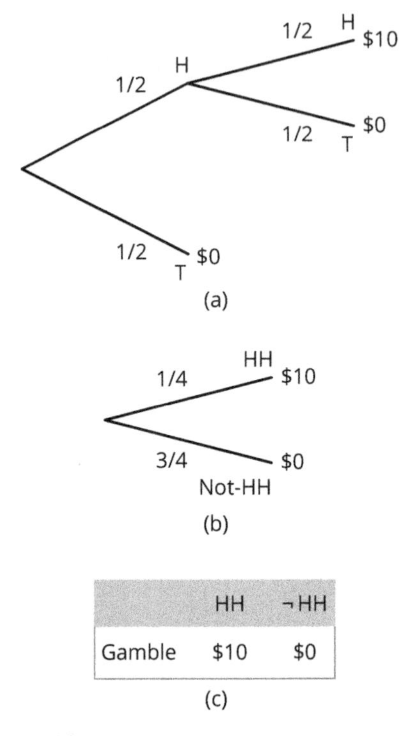

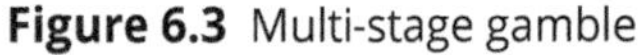

Figure 6.3 Multi-stage gamble

You may be wondering if all gambles can be represented in table form: they can. Consider, for instance, what would happen if you win the right to play the gamble from Figure 6.1 if you flip a coin and it comes up heads. If so, the complex gamble you are playing would look like Figure 6.3(a). The key to analyzing more complex, multi-stage gambles like this is to use one of the AND rules to construct a simpler one. In this case, the gamble gives you a 1/4 probability of winning $10 and a 3/4 probability of winning nothing. Hence, the gamble can also be represented as in Figure 6.3(b). This makes it obvious how to represent the complex gamble in table form, as in Figure 6.3(c). There can be more than two acts or more than two states. So, in general, we end up with a matrix like Table 6.4. By now, it is obvious how to define expected value.

Table 6.4 The general decision problem

	S_1	S_2	...	S_n
A_1	C_{11}	C_{12}	...	C_{1n}
⋮	⋮	⋮		⋮
A_m	C_{m1}	C_{m2}	...	C_{mn}

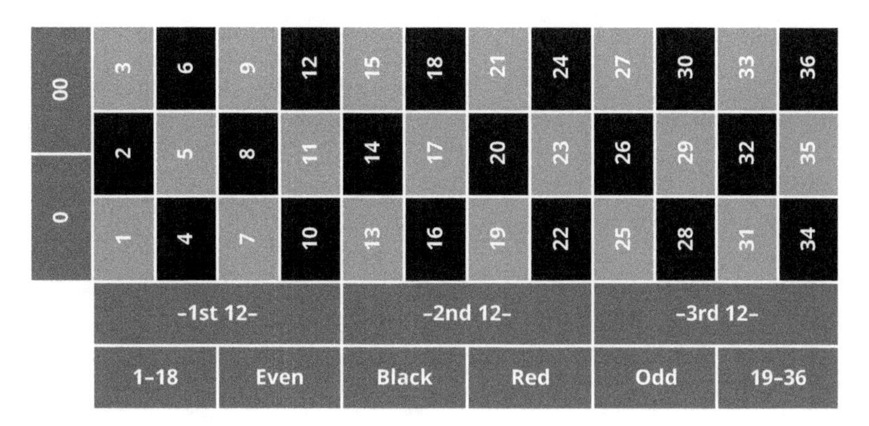

Figure 6.4 Roulette table

Definition 6.9 Expected value *Given a decision problem as in Table 6.4, the expected value $EV(A_i)$ of an act A_i is given by:*

$$EV(A_i) = \Pr(S_1) * C_{i1} + \Pr(S_2) * C_{i2} + \ldots + \Pr(S_n) * C_{in}$$
$$= \sum_{j=1}^{n} \Pr(S_j) C_{ij}$$

If this equation looks complicated, notice that actual computations are easy. For each state – that is, each column in the table – you multiply the probability of that state occurring with what you would get if it did; then you add all your numbers up. If you want to compare two or more acts, just complete the procedure for each act and compare your numbers. As you can tell, this formula gives some weight to each cell in the payoff matrix and to the probabilities that the various states of the world will obtain.

The table form is often convenient when computing expected values, since it makes it obvious how to apply the formula in Definition 6.9. The fact that more complex gambles can also be represented in table form means that the formula applies even in the case of more complex gambles. Hence, at least as long as outcomes can be described in terms of dollars, lives lost, or the like, the concept is well defined. To illustrate how useful this kind of knowledge can be, let us examine decision problems that you might encounter in casinos and other real-world environments.

Exercise 6.10 Roulette A roulette wheel has slots numbered 0, 00, 1, 2, 3, ..., 36 color-coded in red and black (see Figure 6.4). The players make their bets, the croupier spins the wheel, and depending on the outcome, payouts may or may not be made. Players can make a variety of bets. Table 6.5 lists the bets that can be made as well as the associated payoffs for a player who wins after placing a one-dollar bet. Fill in the table.

Exercise 6.11 Parking You are considering whether to park legally or illegally and decide to be rational about it. Use negative numbers to represent costs in your expected-value calculations.

Table 6.5 Roulette bets

Bet	Description	Payout	Pr(win)	Expected value
Straight Up	One number	$36		
Split	Two numbers	$18		
Street	Three numbers	$12		
Corner	Four numbers	$9		
First Five	0, 00, 1, 2, 3	$7		
Sixline	Six numbers	$6		
First 12	1–12	$3		
Second 12	13–24	$3		
Third 12	25–36	$3		
Red		$2		
Black		$2		
Even		$2		
Odd		$2		
Low	1–18	$2		
High	19–36	$2		

(**a**) Suppose that a parking ticket costs $30 and that the probability of getting a ticket if you park illegally is 1/5. What is the expected value of parking illegally?
(**b**) Assuming that you use expected-value calculations as a guide in life, would it be worth paying $5 in order to park legally?

It is perfectly possible to compute expected values when there is more than one state with a non-zero payoff.

Example 6.12 You are offered the following gamble: if a (fair) coin comes up heads, you receive $10; if the coin comes up tails, you pay $10. What is the expected value of this gamble?

The expected value of this gamble is $1/2 * 10 + 1/2 * (-10) = 0$.

Exercise 6.13 Suppose somebody intends to roll a fair die and pay you $1 if she rolls a one, $2 if she rolls a two, and so on. What is the expected value of this gamble?

Exercise 6.14 ***Deal or No Deal*** You are on the show *Deal or No Deal*, where you are facing so many boxes, each of which contains some (unknown) amount of money (see Figure 6.5). At this stage, you are facing three boxes. One of them contains

Figure 6.5 *Deal or No Deal*

\$900,000, one contains \$300,000, and one contains \$60, but you do not know which is which. Here are the rules: if you choose to open the boxes, you can open them in any order you like, but you can keep the amount contained in the *last* box only.

(**a**) What is the expected value of opening the three boxes?

(**b**) The host gives you the choice between a sure \$400,000 and the right to open the three boxes. Assuming you want to maximize expected value, which should you choose?

(**c**) You decline the \$400,000 and open a box. Unfortunately, it contains the \$900,000. What is the expected value of opening the remaining two boxes?

(**d**) The host gives you the choice between a sure \$155,000 and the right to open the remaining two boxes. Assuming you want to maximize expected value, which should you choose?

Expected-value calculations form the core of **cost–benefit analysis**, which is used to determine whether all sorts of projects are worth undertaking. Corporations engage in cost–benefit analysis to determine whether to invest in a new plant, start a new marketing campaign, and so on. Governments engage in cost–benefit analysis to determine whether to build bridges, railways, and airports; whether to incentivize foreign corporations to relocate there; whether to overhaul the tax system; and many other things. The basic idea is simply to compare expected benefits with expected costs: if the benefits exceed or equal the cost, the assumption is that it is worth proceeding, otherwise not.

So far, we have used our knowledge of probabilities to compute expected values. It is also possible to use Definition 6.9 to compute probabilities, provided that we know enough about the expected values. So, for example, we can ask the following question.

Example 6.15 Parking, cont. If a parking ticket costs \$30, and it costs \$5 to park legally, what does the probability of getting a ticket need to be for the expected value of parking legally to equal the expected value of parking illegally?

We solve this problem by setting up an equation. Assume, first of all, that the probability of getting a parking ticket when you park illegally is p. Assume, further, that the expected value of parking illegally equals the expected value of parking legally: $p * (-30) = -5$. Solving for p, we get that $p = -5/-30 = 1/6$. This means that if the probability of getting a ticket is 1/6, the expected values are identical. If p is greater than 1/6, the expected value of parking legally is greater than the expected value of parking illegally; if p is lower than 1/6, the expected value of parking legally is smaller than the expected value of parking illegally.

Exercise 6.16 Parking, cont. Assume that the cost of parking legally is still \$5.

(**a**) If the parking ticket costs \$100, what does the probability need to be for the expected value of parking legally to equal the expected value of parking illegally?

(**b**) What if the ticket costs \$10?

Problem 6.17 Parking, cont. *Given what you pay for parking and given what parking fines are in your area, what does the probability of getting a ticket need to be for the expected value of parking legally to equal the expected value of parking illegally?*

There is a whole field called **law and economics** that addresses questions such as this, exploring the conditions under which it makes sense for people to break the law, and how to design the law so as to generate the optimal level of crime.

Exercise 6.18 Lotto 6/49, cont. Suppose a Lotto 6/49 ticket costs $1 and that the winner will receive $1,000,000. What does the probability of winning need to be for this lottery to be **actuarially fair**, that is, for its price to equal its expected value?

Exercise 6.19 Warranties A tablet computer costs $325; the optional one-year warranty, which will replace the tablet computer at no cost if it breaks, costs $79. What does the probability p of the tablet computer breaking need to be for the expected value of purchasing the optional warranty to equal the expected value of not purchasing it?

As this exercise suggests, the price of warranties is often inflated relative to the probability that the product will break (for the average person, anyway).

Unfortunately, when used as a guide in life, expected-value calculations have drawbacks. Obviously, we can only compute expected values when consequences can be described in terms of dollars, lives lost, or similar. The definition of expected value makes no sense if the consequences C_{ij} are not expressed in numbers. Moreover, under many conditions expected-value considerations give apparently perverse advice, and therefore cannot serve as a general guide to decision-making in real life. Consider, for example, what to do if you have 30 minutes in a casino before the mafia comes after you to reclaim your debts. Assuming that you will be in deep trouble unless you come up with, say, $10,000 before they show up, gambling can be a very reasonable thing to do, even if the expected values are low. Consider, finally, the following famous example.

Example 6.20 St Petersburg paradox A gamble is resolved by tossing an unbiased coin as many times as necessary to obtain heads. If it takes only one toss, the payoff is $2; if it takes two tosses, it is $4; if it takes three, it is $8; and so forth (see Table 6.6). What is the expected value of the gamble?

Notice that the probability of getting heads on the first flip (H) is 1/2; the probability of getting tails on the first flip and heads on the second (TH) is 1/4; the probability of getting tails on the first two flips and heads on the third (TTH) is 1/8; and so on. Thus, the expected value of the gamble is:

$$1/2 * \$2 + 1/4 * \$4 + 1/8 * \$8 + \ldots =$$
$$\$1 + \$1 + \$1 + \ldots = \$\infty$$

Table 6.6 St Petersburg gamble

	H	TH	TTH	...
St Petersburg gamble	$2	$4	$8	...

In sum, the expected value of the gamble is *infinite*. This means that if you try to maximize expected value, you should be willing to pay any (finite) price for this gamble. That does not seem right, which is why the result is called the **St Petersburg paradox**. And it would not seem right even if you could trust that you would receive the promised payoffs no matter what happens.

6.4 Expected utility

Our calculations in Section 4.4 suggested that games like Lotto 6/49 are simply not worth playing (see Exercise 4.28 on page 82). But the story does not end there. For one thing, as our deliberations in this chapter have illustrated, the size of the prize matters (see, for instance, Example 6.6). Equally importantly, a dollar is not as valuable as every other dollar. You may care more about a dollar bill if it is the first in your pocket than if it is the tenth. Or, if the mafia is coming after you to settle a \$10,000 debt, the first 9999 dollar bills may be completely useless to you, since you will be dead either way, whereas the 10,000th can save your life.

To capture this kind of phenomenon, and to resolve the St Petersburg paradox, we simply reintroduce the concept of utility from Section 2.7. The utility of money is often represented in a graph, with money (or wealth, or income) on the x-axis and utility on the y-axis. In Figure 6.6, for example, the dashed line represents the expected value of money if $u(x) = x$; the solid line represents the utility of money if the mafia is coming after you; and the dotted line represents the case when a dollar becomes worth less and less as you get more of them, that is, when the marginal utility of money is diminishing. When the curve bends downwards when you move to the right, as the dotted line does, it is said to be **concave**.

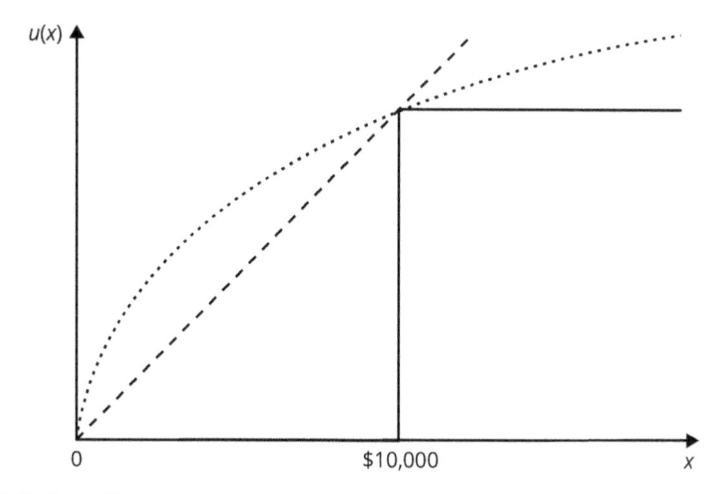

Figure 6.6 The utility of money

The marginal utility of most goods is probably diminishing. Back when your parents were young, it was common to buy newspapers out of boxes. You would just put a few coins in the slot, open the door, and grab a newspaper from the stack. Nothing prevented you from grabbing two or more copies, but most people did not. Why? Newspaper boxes work because the marginal utility of newspapers is sharply diminishing. While the first copy of the *Wall Street Journal* permits you to learn what is going on in the markets, subsequent copies are best used to make funny hats or to wrap fish. There are exceptions to the rule, however. Beer is not sold like newspapers, and for good reason: its marginal utility is not diminishing and may even be increasing. As you may have read in a book, after people have had a beer, a second beer frequently seems like an even better idea than the first one did, and so forth. This is why a beer – as in "Let's have a beer" – is a mythical animal not unlike unicorns.

Table 6.7 St Petersburg gamble with utilities

	H	TH	TTH	...
St Petersburg gamble	log(2)	log(4)	log(8)	...

How does this help? Consider the St Petersburg gamble from Section 6.3. Let us assume that the marginal utility for money is diminishing, which seems plausible. Mathematically, the utility of a given amount of money x might equal the logarithm of x, so that $u(x) = \log(x)$. If so, we can transform Table 6.6 into a table in which consequences are expressed in utilities instead of dollars (see Table 6.7). Now, we can compute the **expected utility** of the gamble. The expected utility of a gamble is the amount of utility you can expect to gain *on the average, in the long run*, when you play the gamble. In the case of the St Petersburg gamble, the expected utility is:

$$1/2 * \log(2) + 1/4 * \log(4) + 1/8 * \log(8) + \ldots = 2\log(2) \approx 0.602 < \infty$$

This way, the expected utility of the St Petersburg gamble is well defined and finite. (In Example 6.37 we will compute what the gamble is worth in dollars and cents.) Formally, this is how we define expected utility:

Definition 6.21 Expected utility *Given a decision problem like Table 6.4, the expected utility $EU(A_i)$ of an act A_i is given by*

$$\begin{aligned} EU(A_i) &= \Pr(S_1) * u(C_{i1}) + \Pr(S_2) * u(C_{i2}) + \ldots + \mathrm{r}(S_n) * u(C_{in}) \\ &= \sum_{j=1}^{n} \Pr(S_j) u(C_{ij}) \end{aligned}$$

Somebody who chooses that option with the greatest expected utility is said to engage in **expected-utility maximization**. Examples like the St Petersburg paradox suggest that expected-utility maximization is both a better guide to behavior, and a better description of actual behavior, than expected-value maximization. That is, the theory of expected utility is a better normative theory, and a better descriptive theory, than the theory of expected value. Computing expected utilities is not much harder than computing expected values, except that you need to multiply each probability with the utility of each outcome.

Example 6.22 Expected utility Consider, again, the gamble from Figure 6.2(c). Suppose that your utility function is $u(x) = \sqrt{x}$. Should you accept or reject the gamble?

The utility of rejecting the gamble is $EU(\text{R}) = u(4) = \sqrt{4} = 2$. The utility of accepting the gamble is $EU(\text{A}) = 1/2 * u(10) + 1/2 * u(0) = 1/2 * \sqrt{10} \approx 1.58$. The rational thing to do is to reject the gamble.

Exercise 6.23 Expected utility, cont. Suppose instead that your utility function is $u(x) = x^2$.

(a) What is the expected utility of rejecting the gamble?
(b) What is the expected utility of accepting the gamble?
(c) What should you do?

When the curve bends upwards as you move from left to right, like the utility function $u(x) = x^2$ does, the curve is said to be **convex.**

Exercise 6.24 Lotto 6/49, cont. Assume still that your utility function is $u(x) = \sqrt{x}$, that the probability of winning at Lotto 6/49 is one in 13,983,816, and that the prize is a million dollars.

(a) What is the expected utility of holding a Lotto ticket?
(b) What is the expected utility of the dollar you would have to give up in order to receive the Lotto ticket?
(c) Which would you prefer?

Exercise 6.25 Expected utility, again Suppose that you are facing three gambles. A gives you a 1/3 probability of winning \$9. B gives you a 1/4 probability of winning \$16. C gives you a 1/5 probability of winning \$25.

(a) What is the expected utility of each of these gambles if your utility function is $u(x) = \sqrt{x}$, and which one should you choose?
(b) What is the expected utility of each of these gambles if your utility function is $u(x) = x^2$, and which one should you choose?

Exercise 6.26 Expected value and expected utility Assume again that your utility function is $u(x) = \sqrt{x}$. Compute (i) the expected value and (ii) the expected utility of the following gambles:

(a) G: You have a 1/4 chance of winning \$25 and a 3/4 chance of winning \$1.
(b) G*: You have a 2/3 chance of winning \$7 and a 1/3 chance of winning \$4.

Another major advantage of the expected-utility framework is that it can be applied to decisions that do not involve consequences expressed in terms of dollars, lives lost, or the like. The expected-utility formula can be used quite generally, as long as it is possible to assign utilities to all outcomes. That is to say that expected utilities can be calculated whenever you have preferences over outcomes – which you do, if you are rational. Hence, expected-utility theory applies, at least potentially, to all decisions. The following exercises illustrate how expected-utility reasoning applies even when consequences are not obviously quantifiable.

Exercise 6.27 Hearing loss A patient with hearing loss is considering whether to have surgery. If she does not have the surgery, her hearing will get no better and no worse. If she does have the surgery, there is an 85 percent chance that her hearing will improve, and a five percent chance that it will deteriorate. If she does not have the surgery, her utility will be zero. If she does have the surgery and her hearing improves, her utility will be ten. If she does have the surgery but her hearing is no better and no worse, her utility will be minus two. If she does have the surgery and her hearing deteriorates, her utility will be minus ten.
(**a**) Draw a tree representing the decision problem.
(**b**) Draw a table representing the problem.
(**c**) What is the expected utility of not having the operation?
(**d**) What is the expected utility of having the operation?
(**e**) What should the patient do?

Exercise 6.28 Thanksgiving indecision Suppose you are contemplating whether to go home for Thanksgiving. You would like to see your family, but you are worried that your aunt may be there, and you genuinely hate your aunt. If you stay in town you are hoping to stay with your roommate, but then again, there is some chance that she will leave town. The probability that your aunt shows up is 1/4, and the probability that your roommate leaves town is 1/3. The utility of celebrating Thanksgiving with your family without the aunt is 12 and with the aunt is minus two. The utility of staying in your dorm without your roommate is three and with the roommate is nine.
(**a**) Draw a decision tree.
(**b**) Calculate the expected utility of going home and of staying in town.
(**c**) What should you do?

Exercise 6.29 Pascal's wager The seventeenth-century French mathematician and philosopher Blaise Pascal suggested the following argument for a belief in God. The argument is frequently referred to as **Pascal's wager**. Either God exists (G), or He does not ($\neg$G). We have the choice between believing (B) or not believing ($\neg$B). If God does not exist, it does not matter if we believe or not: the utility would be the same. If God does exist, however, it matters a great deal: if we do believe, we will enjoy eternal bliss; if we do not believe, we will burn in hell.
(**a**) Represent the decision problem in table form, making up suitable utilities as you go along.
(**b**) Let p denote the probability that G obtains. What is the expected utility of B and $\neg$B?
(**c**) What should you do?

Notice that it does not matter what p is. B **dominates** $\neg$B in the sense that B is associated with a higher expected utility no matter what.

Of course, Definition 6.21 can also be used to compute probabilities provided that we know enough about the expected utilities. So, for example, we can ask the following kinds of question.

Example 6.30 Umbrella problem, cont. This question refers to Table 6.1(b), that is, the umbrella problem from Section 6.2. If the probability of rain is p, what does p need to be for the expected utility of taking the umbrella to equal the expected utility of leaving it at home?

To answer this problem, set up the following equation: *EU* (Take umbrella) = *EU* (Leave umbrella). Given the utilities in Table 6.1(b), this implies that $3 = p * 0 + (1 - p) * 5$, which implies that $p = 2/5$.

Exercise 6.31 Indifference This question refers to Table 6.2. Let p denote the probability that S_1 obtains.

(**a**) If an expected-utility maximizer is indifferent between A and B, what is his p?

(**b**) If another expected-utility maximizer is indifferent between B and C, what is her p?

(**c**) If a third expected-utility maximizer is indifferent between A and C, what is their p?

Axiomatic expected-utility theory

Like the theory of rational choice under certainty, the theory of expected utility is axiomatic. It requires two further axioms, in addition to the ones introduced in Chapter 2.

Continuity The first additional axiom is similar to the continuity axiom of consumer-choice theory (cf. box on page 28), in that it says that people have similar preferences for similar lotteries. More specifically, the continuity axiom of expected-utility theory says that preferences are insensitive to tiny changes in the probabilities involved. If you prefer dating Sam to dating Robin, continuity guarantees that you will continue to do so even when there is some sufficiently small (but positive) probability that Sam is nuts. Thus, the continuity axiom of expected-utility theory excludes "jumps," where people have radically different preferences over lotteries with very similar probabilities.

Independence Take three lotteries L, L', and L'', and suppose the person weakly prefers lottery L to lottery L'. The independence axiom says that the preference will be preserved even if each of the lotteries L and L' are "mixed" with the third lottery L''. To say that two lotteries are mixed means that you get the one lottery with some probability α and the other lottery with probability $(1 - \alpha)$. Formally, then, the independence axiom says that:

$$L \succcurlyeq L' \Leftrightarrow \alpha L + (1-\alpha) L'' \succcurlyeq \alpha L' + (1-\alpha) L''$$

If you prefer dating Sam to dating Robin, independence guarantees that you will continue to do so even when there is some chance that you will end up dating nobody. That is, you will prefer dating Sam if a coin comes up heads and dating nobody if it comes up tails, to dating Robin if the coin comes up heads and nobody if it comes up tails. The independence axiom therefore excludes situations where the preference depends on the particular third lottery involved, in this case, the possibility of dating nobody.

Given the other axioms, the independence axiom is equivalent to the sure-thing principle, which will appear in Section 7.4.

6.5 Attitudes toward risk

As you may have noticed already, expected-utility theory has implications for attitudes toward risk. Whether you reject a gamble (as in Example 6.22) or accept it (as in Exercise 6.23) depends, at least to some extent, on the shape of your utility function. This means that we can explain people's attitudes toward risk in terms of the character of their utility function.

The theory of expected utility can explain why people often reject a gamble in favor of a sure dollar amount equal to its expected value. We simply have to add to the theory of expected utility the auxiliary assumption of diminishing marginal utility of money.

Example 6.32 Risk aversion Suppose you own \$2 and are offered a gamble giving you a 50 percent chance of winning a dollar and a 50 percent chance of losing a dollar. This decision problem can be represented as in Figure 6.7. Your utility function is $u(x) = \sqrt{x}$, so that marginal utility is diminishing. Should you take the gamble?

The problem can be represented as in Table 6.8. Expected-utility calculations show that you should reject the gamble, since $EU(\text{Accept}) = 1/2 * \sqrt{3} + 1/2 * \sqrt{1} \approx 1.37$ and $EU(\text{Reject}) = \sqrt{2} \approx 1.41$.

Table 6.8 Another gambling problem

	Win (1/2)	Lose (1/2)
Accept (A)	$\sqrt{3}$	$\sqrt{1}$
Reject (R)	$\sqrt{2}$	$\sqrt{2}$

An expected-value maximizer would have been indifferent between accepting and rejecting this gamble, since both expected values are \$2. Trivially, then, if your utility function is $u(x) = x$, you will be indifferent between the two options. For comparison, consider the following problem.

Exercise 6.33 Risk proneness Consider, again, the gamble in Figure 6.7. Now suppose that your utility function is $u(x) = x^2$. Unlike the previous utility function, which gets flatter when amounts increase, this utility function gets steeper. Compute the expected utilities of accepting and rejecting the gamble. What should you do?

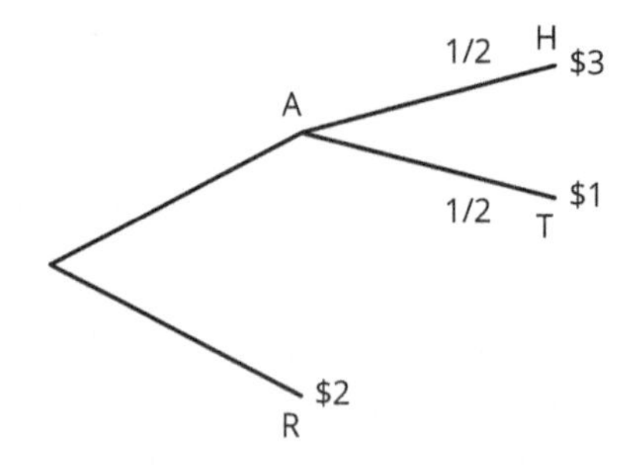

Figure 6.7 Risk aversion

As all these examples suggest, the shape of your utility function relates to your attitude toward risk – or your **risk preference** – in the following way. Whether you should reject or accept a gamble depends on whether your utility function gets flatter or steeper (bends downwards or upwards). In general, we say that you are **risk averse** if you would reject a gamble in favor of a sure dollar amount equal to its expected value, **risk prone** if you would accept, and **risk neutral** if you are indifferent. Thus, you are risk averse if your utility function bends downwards (as you move from left to right), risk prone if your utility function bends upwards, and risk neutral if your utility function is a straight line.

Notice that the theory itself does not specify what the shape of your utility function should be. Most of the time, economists will assume that utility for money is increasing, so that more money is better. But that is an auxiliary assumption, which is no part of the theory. The theory does not constrain your attitude toward risk; it does not even say that your attitude toward risk has to be the same when you get more (or less) money. For instance, you may have a utility function that looks like the solid line in Figure 6.8. Here, you are risk prone in the range below x^* and risk averse in the range above x^*. Or, you may have a utility function that looks like the dashed line. Here, you are risk averse in the range below x^* and risk prone in the range above x^*. The next exercise illustrates the manner in which attitudes toward risk are expressed in a variety of real-world behaviors.

Exercise 6.34 Attitudes to risk As far as you can tell, are the following people *risk prone, risk averse,* or *risk neutral*?

(**a**) People who invest in the stock market rather than in savings accounts.
(**b**) People who invest in bonds rather than in stocks.
(**c**) People who buy lottery tickets rather than holding on to the cash.
(**d**) People who buy home insurance.
(**e**) People who play roulette.
(**f**) People who consistently maximize expected value.
(**g**) People who have unsafe sex.

Sometimes it is useful to compute the **certainty equivalent** of a gamble. The certainty equivalent of a gamble is the amount of money such that you are indifferent between playing the gamble and receiving the amount for sure.

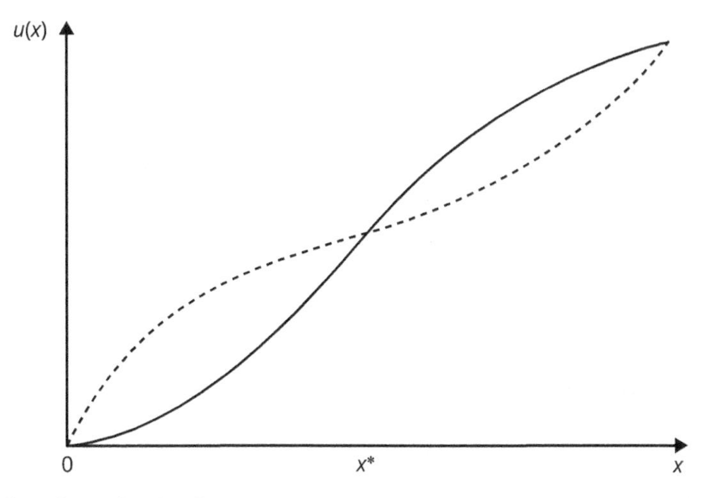

Figure 6.8 *S*-shaped utility functions

Definition 6.35 Certainty equivalent *The certainty equivalent of a gamble G is the number CE that satisfies this equation:* $u(CE) = EU(G)$.

The certainty equivalent represents *what the gamble is worth to you*. The certainty equivalent determines your willingness-to-pay (WTP) and your willingness-to-accept (WTA). In graphical terms, suppose that you have to find the certainty equivalent given a utility function like that in Figure 6.9. Suppose the gamble gives you a 50 percent chance of winning *A* and a 50 percent chance of winning *B*.

1. Put a dot on the utility curve right above *A*. This dot represents the utility of *A* (on the *y*-axis).
2. Put another dot on the utility curve right above *B*. This dot represents the utility of *B* (on the *y*-axis).
3. Draw a straight line between the two dots.
4. Put an X halfway down the straight line. The X represents the expected utility of the gamble (on the *y*-axis) and the expected value of the gamble (on the *x*-axis).
5. Move sideways from the X until you hit the utility curve.
6. Move straight down to the *x*-axis, and you have the certainty equivalent (on the *x*-axis).

The procedure is illustrated in Figure 6.9. The same procedure can also be used for gambles where the probabilities are not 50–50. The only thing that changes is the placement of the X on the straight line in Figure 6.9. If there is, say, a 3/7 probability of winning *A*, and a 4/7 probability of winning *B*, starting from the left, you put that X four-sevenths of the way up from *A* to *B*. As the probability of winning *B* increases, the X will move right toward *B* and the expected utility of the gamble will approach the utility of *B*; as the probability of winning *A* increases, the X will move left toward *A* and the expected utility of the gamble will approach the utility of *A*. This makes sense.

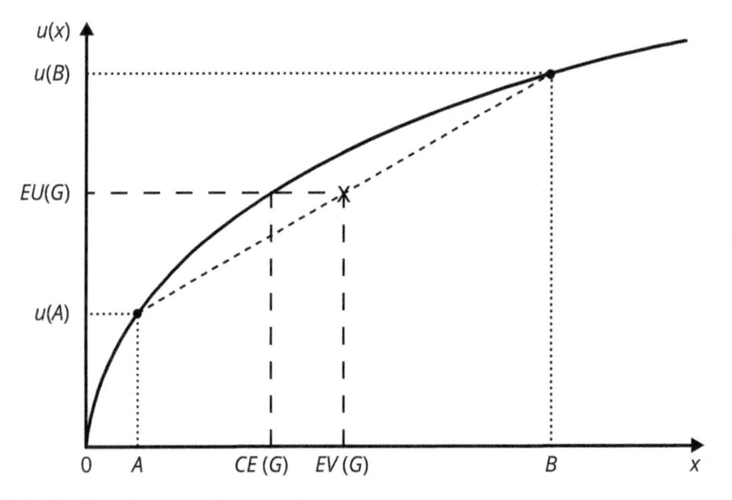

Figure 6.9 Finding the certainty equivalent

Exercise 6.36 Certainty equivalents Demonstrate how to find the certainty equivalent of the same gamble in the case when the utility function bends upwards. Confirm that the certainty equivalent is greater than the expected value.

Remember: when the utility function bends downwards, you are risk averse, the dashed line falls below the utility function, and the certainty equivalent is less than the expected value. When the utility function bends upwards, you are risk prone, the dashed line falls above the utility function, and the certainty equivalent is greater than the expected value. Read this paragraph one more time to make sure you understand what is going on.

In algebraic terms, you get the certainty equivalent of a gamble by computing the expected utility x of the gamble, and then solving for $u(CE) = x$. Thus, you get the answer by computing $CE = u^{-1}(x)$. As long as $u(\cdot)$ is strictly increasing in money, which it ordinarily will be, the inverse function is well defined. Consider the gamble in Table 6.8. We know that the expected utility of this gamble is approximately 1.37. You get the certainty equivalent by solving for $u(CE) = 1.37$. Given our utility function $u(x) = \sqrt{x}$, this implies that $CE = 1.37^2 \approx 1.88$. Because you are risk averse, the certainty equivalent of the gamble is lower than the expected value of 2.

Example 6.37 St Petersburg paradox, cont. In Section 6.4 we learned that for an agent with utility function $u(x) = \log(x)$, the expected utility of the St Petersburg gamble is approximately 0.602. What is the certainty equivalent of the gamble?

We compute the certainty equivalent by solving the following equation: $\log(CE) = 0.602$. Thus, the certainty equivalent $CE = 10^{0.602} \approx 4.00$. That is, the St Petersburg gamble is worth $4.

Exercise 6.38 Compute the certainty equivalent of the gamble in Figure 6.7, using the utility function $u(x) = x^2$.

We end this section with a series of exercises.

Exercise 6.39 Suppose that you are offered the choice between $4 and the following gamble, G: 1/4 probability of winning $9 and a 3/4 probability of winning $1.

(a) Suppose that your utility function is $u(x) = \sqrt{x}$. What is the utility of $4? What is the utility of G? What is the certainty equivalent? Which would you choose?

(b) Suppose instead that your utility function is $u(x) = x^2$. What is the utility of $4? What is the utility of G? What is the certainty equivalent? Which would you choose?

Exercise 6.40 Suppose that your utility function is $u(x) = \sqrt{x}$, and that you are offered a gamble which allows you to win $4 if you are lucky and $1 if you are not.

(a) Suppose that the probability of winning $4 is 1/4 and the probability of winning $1 is 3/4. What is the expected value of this gamble?

(b) Suppose that the probability of winning $4 is still 1/4 and the probability of winning $1 is 3/4. What is the expected utility of this gamble?

(c) Suppose that the probability of winning $4 is still 1/4 and the probability of winning $1 is 3/4. What is the certainty equivalent of the gamble; that is, what is the amount of money X such that you are indifferent between receiving $$X$ for sure and playing the gamble?

(d) Imagine now that the probability of winning $4 is p and the probability of winning $1 is $(1 - p)$. If the utility of the gamble equals 3/2, what is p?

Exercise 6.41 Suppose that your utility function is $u(x) = \sqrt{x}$, and that you are offered a gamble which allows you to win \$16 if you are lucky and \$4 if you are not.

(**a**) Suppose that the probability of winning \$16 is 1/4 and the probability of winning \$4 is 3/4. What is the expected utility of this gamble?

(**b**) Suppose that the probability of winning \$16 is still 1/4 and the probability of winning \$4 is 3/4. What is the certainty equivalent of the gamble?

(**c**) Imagine now that the probability of winning \$16 is p and the probability of winning \$4 is $(1 - p)$. If the expected utility of the gamble equals 9/4, what is p?

(**d**) Are you risk averse or risk prone, given the utility function above?

Exercise 6.42 Lotto 6/49, cont. Compute the certainty equivalent of the Lotto 6/49 ticket from Exercise 6.5 if $u(x) = \sqrt{x}$.

6.6 Discussion

In this chapter, we have explored principles of rational choice under risk and uncertainty. As pointed out in the introduction to this chapter, according to the traditional perspective, you face a choice under uncertainty when probabilities are unknown or not even meaningful. In Section 6.2, we explored several principles of rational choice that may apply under such conditions. When it is both meaningful and possible to assign probabilities to the relevant states of the world, it becomes possible to compute expectations, which permits you to apply expected-value and expected-utility theory instead.

The distinction between risk and uncertainty is far from sharp. In real life, it may not be obvious whether to treat a decision as the one or the other. Consider the regulation of new and unstudied chemical substances. Though there are necessarily little hard data on such substances, there is always some probability that they will turn out to be toxic. Some people argue that this means that policy-makers are facing a choice under uncertainty, that the maximin criterion applies, and that new chemicals should be banned or heavily regulated until their safety can be established. Others argue that we can and must assign probabilities to all outcomes, that the probability that new substances will turn out to be truly dangerous is low, and that expected-utility calculations will favor permitting their use (unless or until their toxicity has been established). Whether we treat a decision as a choice under uncertainty or under risk, therefore, can have real consequences. And it is not obvious how to settle such issues in a non-arbitrary way. (We will return to this topic under the heading of ambiguous probabilities in Section 7.5.)

One thing to note is that you cannot judge whether a decision was rational or not by examining the outcome alone. A rational decision, as you know, is a decision that maximizes expected utility given your beliefs and preferences at the time when you make the decision. Such a decision might lead to adverse outcomes. If something bad happens as a result of your decision, that does not mean you acted irrationally: you may just have been unlucky. Sometimes decision theorists use the term "right" to denote the decisions that lead to the best possible outcome. The fact that good decisions can have bad outcomes means that decisions can be rational but wrong. They can also be irrational but right, as when you do something completely reckless but see good results anyway; buying a lottery ticket as a means to get rich and winning truckloads of money might fall in this category. It goes without saying that we always want to make the right decision. The problem, of course, is that we do not know ahead of time which decision is the right one. That is why we aim for the *rational* decision – being the one with the greatest expectation of future utility.

Example 6.43 The rationality of having children It is sometimes argued that certain decisions cannot be made rationally. Philosopher L. A. Paul, for example, has argued that it is impossible to make a rational decision about having a child, because you cannot know ahead of time what it will be like, for you, to have a child. But such arguments may confuse the rational with the right. It is true that you cannot know what the right decision is: you may be very happy with a child, or you may be miserable. But you never know ahead of time what the right decision is.

Luckily, ignorance is no obstacle to making a rational decision, as "rationality" is understood here. Rationality does not require that you know what anything is like – only that you choose whatever option maximizes expected utility given your beliefs and preferences at the time you are making the decision.

We will return to the topic of the right and the rational in Section 7.7.

Our study of the theory of choice under risk sheds further light on the economic approach to behavior as understood by Gary Becker, and in particular on what he had in mind when talking about maximizing behavior (see Section 2.8). Recall (from Section 4.7) that the standard approach does not assume that people consciously or not perform calculations in their heads: all talk about maximization (or optimization) is shorthand for the satisfaction of preferences. Notice also that this approach does not assume that people are omniscient, in the sense that they know what state of the world will obtain. What it does assume is that people assign probabilities to states of the world, that these probabilities satisfy the axioms of the probability calculus, that people assign utilities to outcomes, and that they choose that alternative which has the greatest expected utility given the probabilities and utilities.

In the next chapter, we will consider some conditions under which these assumptions appear to fail.

Additional exercises

Exercise 6.44 The Precautionary Principle The Precautionary Principle enjoins us to avoid whatever course of action leads to the worst possible outcome. What principle of rational choice introduced in this chapter does this sound like?

Exercise 6.45 *Deal or No Deal*, cont. You are on *Deal or No Deal* again, and you are facing three boxes. One of the three contains \$1,000,000, one contains \$1000, and one contains \$10. Now the dealer offers you \$250,000 if you give up your right to open the boxes.

(**a**) Assuming that you use expected value as your guide in life, would you choose the sure amount or the right to open the boxes?

(**b**) Assuming that your utility function is $u(x) = \sqrt{x}$, and that you use expected utility as your guide in life, would you choose the sure amount or the right to open the boxes?
(**c**) Given the utility function, what is the lowest amount in exchange for which you would give up your right to open the boxes?

Exercise 6.46 The humiliation show You are on a game show where people embarrass themselves in the hope of winning a new car. You are given the choice between pressing a blue button and pressing a red button.
(**a**) If you press the blue button, any one of two things can happen: with a probability of 2/3, you win a toy frog (utility -1), and with a probability of 1/3 you win a bicycle (utility 11). Compute the expected utility of pressing the blue button.
(**b**) If you press the red button, any one of three things can happen: with a probability of 1/9 you win the car (utility 283), with a probability of 3/9 you win a decorative painting of a ballerina crying in the sunset (utility 1), and with a probability of 5/9 you end up covered in green slime (utility -50). Compute the expected utility of pressing the red button.
(**c**) What should you do?

Exercise 6.47 Misguided criticism Some critics attribute to neoclassical economists the view that human beings have the ability to compute solutions to every maximization problem, no matter how complicated, in their heads and on the fly. For example: "Traditional models of unbounded rationality and optimization in cognitive science, economics, and animal behavior have tended to view decision-makers as possessing supernatural powers of reason, limitless knowledge, and endless time." Explain why this criticism misses the mark.

See also Exercise 7.34 on page 173.

Further reading

The classic definition and discussion of choice under uncertainty is Luce and Raiffa (1957, Ch. 13); Rawls (1971) defends Rawls's theory of justice, and Harsanyi (1975) criticizes it. Helpful introductions to expected-utility theory include Allingham (2002) and Peterson (2009). The quip about not regretting one's mistakes appears in Wilde (1998 [1890], p. 34); *The Top Five Regrets...* is Ware (2012); and Kierkegaard (2000 [1843], p. 72) despaired of ever avoiding regret. The view that one cannot rationally decide whether to have children appears in Paul (2014). The critics are Todd and Gigerenzer (2000, p. 727).

7 DECISION-MAKING UNDER RISK AND UNCERTAINTY

Learning objectives

After studying this chapter you will:

- Be able to identify common choice patterns that violate expected-utility theory
- Know all the building blocks of prospect theory
- Apply the theory of expected utility in the real world – but also appreciate ways in which it may be inadequate

7.1 Introduction

The theory of expected utility combines the concept of utility from Chapter 2 with the concept of probability from Chapter 4 into an elegant and powerful theory of choice under risk. The resulting theory, which we explored in the previous chapter, is widely used. Yet there are situations in which people fail to conform to the predictions of the theory. In addition, there are situations in which it is seemingly rational to violate it. In this section we explore some such situations. We will also continue to explore what behavioral economists do in the face of systematic deviations from standard theory. To capture the manner in which people actually make decisions under risk, we will make more assumptions about the value function, which we first came across in Section 3.5, and introduce the probability-weighting function. Both these functions are essential parts of prospect theory, the most prominent behavioral theory of choice under risk.

7.2 Framing effects in decision-making under risk

For the next set of problems, suppose that you are a public health official.

Example 7.1 Pandemic problem 1 Imagine that the US is preparing for the outbreak of an unusual contagious disease, which is expected to kill 600 people. Two alternative programs to combat the disease have been proposed. Assume that the exact scientific estimate of the consequences of the programs are as follows: if Program A is adopted, 200 people will be saved; if Program B is adopted, there is a 1/3 probability that 600 people will be saved, and a 2/3 probability that no one will be saved. Which of the two programs would you favor?

When this problem was first presented to participants, 72 percent chose A and 28 percent chose B.

Example 7.2 Pandemic problem 2 Imagine that the US is preparing for the outbreak of an unusual contagious disease, which is expected to kill 600 people. Two alternative programs to combat the disease have been proposed. Assume that the exact scientific estimate of the consequences of the programs are as follows: if Program C is adopted 400 people will die; if Program D is adopted there is a 1/3 probability that nobody will die, and a 2/3 probability that 600 people will die. Which of the two programs would you favor?

When this problem was first presented to participants, only 22 percent chose C and 78 percent chose D.

The observed response pattern is puzzling. Superficial differences aside, option A is the same as option C, and option B is the same as option D.

Before discussing what may be going on, let us briefly explore why this response pattern is hard to reconcile with expected-utility theory. As we learned in the previous chapter, expected-utility theory does not by itself say whether you should choose the safe or the risky option: the theory does not specify what your risk preference should be. But the theory does say that your choice should reflect your utility function. What matters is whether the point marked X in Figure 7.1 falls above or below the utility function itself. If the X falls below the curve, you will choose the safe option. This will occur if your utility function is concave, like the curve marked 1, and you are risk averse. If the X falls above the curve, you will choose the gamble. This will occur if your utility function is convex, like the curve marked 2, and you are risk prone. If the X falls on the curve, you are indifferent between the two options. This will occur if your utility function is a straight line, like the dashed line

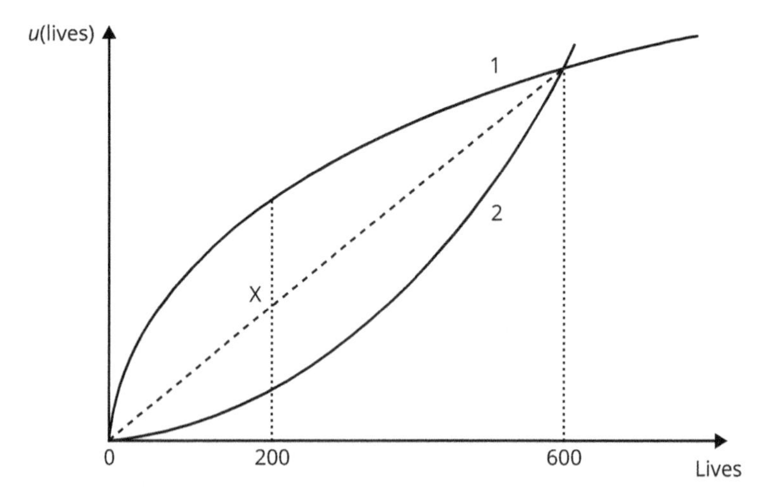

Figure 7.1 The utility of human lives

in the figure. The point is that as long as you act in accordance with expected-utility theory, you will prefer the safe option no matter how it is described, or you will prefer the risky option no matter how it is described, or you will be indifferent between the two. Your preference should definitely not depend on how the options are described.

So how do we account for the behavior exhibited in the study above? The key is to notice that the behavior can be interpreted in terms of framing. As you will recall from Section 3.5, framing effects occur when preferences and behavior are responsive to the manner in which the options are described, and in particular to whether the options are described in terms of gains or in terms of losses. Options A and B are both framed in a positive way, in terms of the lives that might be saved; that is, they are presented in a gain frame. Options C and D are both framed in a negative way, in terms of the lives that might be lost; that is, they are presented in a loss frame.

The talk about gains versus losses may remind you of our acquaintance the value function from prospect theory, which we used to model framing effects in Section 3.5. There, we learned that unlike the utility function, which ranges over *total* endowments, the value function ranges over *changes* relative to some reference point. We already know that many behavioral phenomena can be modeled by assuming a value function that is steeper for losses than for gains. Now we add the assumption that the value function has different curvatures for losses and for gains. In the realm of losses, we assume that the curve bends upwards (when moving from left to right), so that people are risk prone; in the realm of gains, we assume that the curve bends downwards, so that people are risk averse. In other words, the value function is convex in the realm of losses and concave in the realm of gains. This generates an *S*-shaped value function, as shown in Figure 7.2.

The curvature of the value function has interesting implications. For one thing, it entails that the absolute difference between $v(\pm 0)$ and $v(+10)$ is greater than the absolute difference between $v(+1000)$ and $v(+1010)$, and that this is true for losses as well as for gains. We can establish the result algebraically, as the next example shows.

Example 7.3 Curvatures An *S*-shaped value function $v(\cdot)$ can be defined by an expression that has two components: one corresponding to the realm of gains and one corresponding to the realm of losses. For example:

$$v(x) = \begin{cases} \sqrt{x/2} & \text{for gains } (x \geq 0) \\ -2\sqrt{|x|} & \text{for losses } (x < 0) \end{cases}$$

Using this equation, the value of ± 0 is $v(\pm 0) = 0$ while the value of $+10$ is $v(+10) = \sqrt{10/2} = \sqrt{5} \approx 2.24$. The difference is $v(+10) - v(\pm 0) \approx 2.24 - 0 = 2.24$. Meanwhile, the value of $+1000$ is $v(+1000) = \sqrt{+1000/2} \approx 22.36$ while the value of $+1010$ is $v(+1010) = \sqrt{+1010/2} \approx 22.47$. The difference is only $v(+1010) - v(1000) \approx 22.47 - 22.36 = 0.11$. The difference between $v(\pm 0)$ and $v(+10)$ is indeed greater than the difference between $v(+1000)$ and $v(+1010)$.

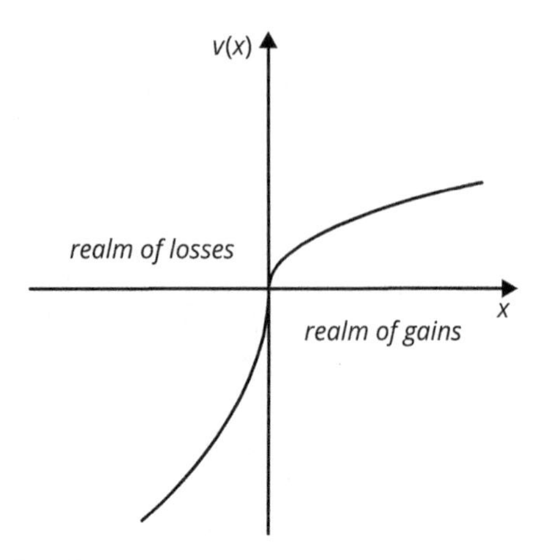

Figure 7.2 The value function

Because we are dealing with positive numbers throughout, we are in the realm of gains, which means that we are using the upper half of the equation. In the next exercise you will be using the lower half of the equation, since you will be dealing with negative numbers and the realm of losses. By the way, $|x|$ is the absolute value of x: that is, x with the minus sign removed (if there is one).

Exercise 7.4 Curvatures, cont. Given the same value function, which is greater: the absolute difference between $v(\pm 0)$ and $v(-10)$ or the absolute difference between $v(-1000)$ and $v(-1010)$?

Exercise 7.5 Jacket/calculator problem, again Consider again the classic jacket/calculator example from Section 3.2. Recall that many people were willing to make the drive when they could save \$5 on a \$15 calculator but not when they could save \$5 on a \$125 calculator. Using the same value function, show that the difference between $v(-10)$ and $v(-15)$ is much greater than the difference between $v(-120)$ and $v(-125)$.

The exercise shows that an *S*-shaped value function can in fact account for the observed behavior in the jacket/calculator case.

We are now in a position to return to the pandemic problem. The assumption that the value function is convex in the realm of losses and concave in the realm of gains helps account for the behavior of people facing this problem. The essential insight is that participants presented with the gain frame (as in Example 7.1) take their reference point to be the case in which no lives are saved (and 600 lost); participants presented with the loss frame (as in Example 7.2) take their reference point to be the case in which no lives are lost (and 600 saved). We can capture this in one graph by using two value functions to represent the fact that the two groups of participants use different outcomes as their reference point. In Figure 7.3, the concave value function on the top left belongs to people in the gain frame, while the convex value function on the bottom right belongs to people in the loss frame. As you can tell from the figure, people in the gain frame will prefer A to B, but people in the loss frame will prefer D to C.

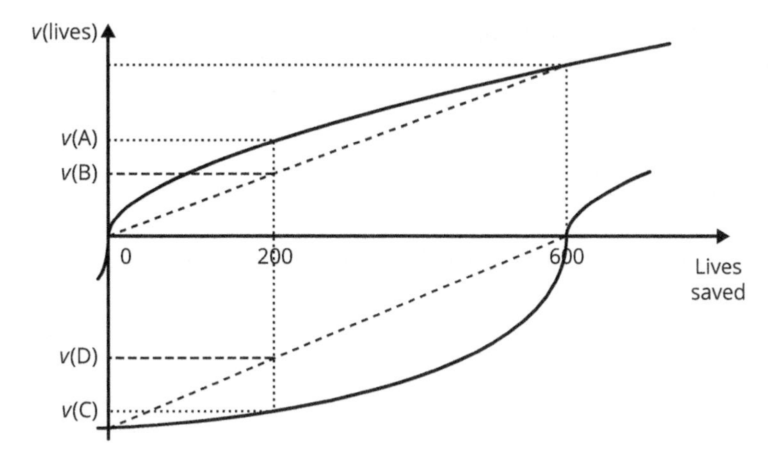

Figure 7.3 The value of human lives

Exercise 7.6 The ostrich farm Jen and Joe have an ostrich farm. They have just learned that the farm has been struck by an unusual virus. According to their vet, if they do nothing only 200 of the 600 animals will live. However, the vet offers an experimental drug. If this drug is used, the vet says there is a 2/3 chance that all the animals will die, but a 1/3 chance that all the animals will live. Jen says: "The drug isn't worth it. It's better to save 200 animals for sure than risk saving none." Joe says: "I think we should use the drug. Even if it's risky, that's the only way we have a chance of losing no animals at all. Taking the risk is better than losing 400 animals for sure." Draw a graph explaining how the two can come to such different realizations even though they have value functions with the same shape.

The following example is another nice illustration of the phenomenon.

Example 7.7 Prospect evaluation Consider the following two problems:

(**a**) In addition to whatever you own, you have been given $1000. You are now asked to choose between (A) a 50 percent chance of winning $1000 and (B) winning $500 for sure.

(**b**) In addition to whatever you own, you have been given $2000. You are now asked to choose between (C) a 50 percent chance of losing $1000 and (D) losing $500 for sure.

In terms of final outcomes, (A) is obviously equivalent to (C) and (B) to (D). Yet 84 percent of participants chose B in the first problem, and 69 percent chose C in the second.

The difference between Example 7.7(a) and (b) is that in the former, outcomes are described in a gain frame, whereas in the latter, they are described in a loss frame. Consequently, in (a) the gamble represents an opportunity to win the big prize, whereas in (b) the gamble represents an opportunity to prevent a loss. To show how the observed response pattern might emerge we can analyze the problem algebraically.

Exercise 7.8 Prospect evaluation, cont. This exercise refers to Example 7.7 above. Suppose that your value function $v(\cdot)$ is defined by $v(x) = \sqrt{x/2}$ for gains ($x \geq 0$) and $v(x) = -2\sqrt{|x|}$ for losses ($x < 0$).

(a) Draw the curve for values between 24 and 14. Confirm that it is concave in the domain of gains and convex in the domain of losses.

(b) Assuming that you have integrated the \$1000 into your endowment, what is the value of (A)?

(c) Assuming that you have integrated the \$1000 into your endowment, what is the value of (B)?

(d) Assuming that you have integrated the \$2000 into your endowment, what is the value of (C)?

(e) Assuming that you have integrated the \$2000 into your endowment, what is the value of (D)?

Notice how (B) turns out to be better than (A), but (C) better than (D).

The idea that people are risk averse in the domain of gains but risk prone in the domain of losses helps explain a range of phenomena. It can explain why some people are unable to stop gambling: once they find themselves in the red – which they will, soon enough, when playing games like roulette – they enter the domain of losses, where they are even more risk prone than they used to be. There is evidence that people betting on horses, etc., are more willing to bet on long shots at the end of the betting day. This phenomenon is often accounted for by saying that people who have already suffered losses are more prone to risk-seeking behavior. Analogously, the idea can explain why politicians continue to pursue failed projects and generals continue to fight losing wars: as initial efforts fail, the responsible parties enter the domain of losses, in which they are willing to bet on increasingly long shots and therefore take increasingly desperate measures. Somewhat paradoxically, then, this analysis suggests that people, countries, and corporations can be expected to be most aggressive when they are weakest – not when they are strongest, as one might think.

Here are some more exercises.

Exercise 7.9 A person's value function is $v(x) = \sqrt{x/2}$ for gains and $v(x) = -2\sqrt{|x|}$ for losses. The person is facing the choice between a sure \$2 and a 50–50 gamble that pays \$4 if she wins and \$0 if she loses.

(a) Show algebraically that this person is loss averse, in the sense that she suffers more when she loses \$4 than she benefits when she receives \$4.

(b) If she takes the worst possible outcome (\$0) as her reference point, what is the value of the sure amount and the gamble? Which would she prefer?

(c) If she takes the best possible outcome (\$4) as her reference point, what is the value of the sure amount and the gamble? Which would she prefer?

Exercise 7.10 Another person with the same value function is facing the choice between a sure \$2 and a 50–50 gamble that pays \$5 if he wins and \$1 if he loses.

(a) If he takes the worst possible outcome as his reference point, what is the value of the sure amount and the gamble? Which would he prefer?

(b) If he takes the best possible outcome as his reference point, what is the value of the sure amount and the gamble? Which would he prefer?

Exercise 7.11 Relative income It is well known that poor people, who can least afford to play the lottery, are most likely to do so. In a 2008 study, researchers wanted to know whether manipulating people's perceptions of their income can affect their demand for lottery tickets. Half of the participants were made to feel rich by answering a question about their yearly income on a scale from “<$10k,” “$10k–$20k,” and so on, to “>$60k.” The other half were made to feel poor by answering the same question on a scale from “<$100k,” “$100k–$200k,” and so on, to “>$1M.” At the conclusion of the study, participants who were made to feel relatively poor were more likely to choose lottery tickets than cash as a reward for their participation. Are these findings consistent with the analysis in this section or not?

Framing effects should not be confused with wealth effects, which occur when people's risk aversion changes when they go from being poor to being rich (or the other way around), and which can be represented using a single utility function. It would be normal, for instance, if your curve got flatter and flatter as your wealth increased, thereby making you less risk averse. Not all the data can be easily accommodated in this framework, however; much of it is better explained by a value function that is convex in the realm of losses and concave in the realm of gains. In the following section, we discuss other applications of these ideas.

7.3 Bundling and mental accounting

The fact that the value function is concave in the domain of gains and convex in the domain of losses has other interesting implications, one being that it matters how outcomes are **bundled** (see Section 3.5). Suppose that you buy two lottery tickets at a charity event, and that you win $25 on the first and $50 on the second. There are different ways to think of what happened at the event (see Figure 7.4). You can **integrate** the outcomes, and tell yourself that you just won $75, which in value terms would translate into $v(+75)$. Or you can **segregate** the outcomes and tell yourself that you first won $25 and then won $50, which in value terms would translate into $v(+25) + v(+50)$. Bundling can be seen as an instance of framing: at stake is whether you frame the change as one larger gain or as two smaller gains.

According to the standard view, bundling should not matter. The utility function ranges over total endowments, and no matter how you describe the various outcomes you end up with an additional $75 in your pocket. In utility terms, then, if you start off at $u(w)$, you will end up at $u(w + 25 + 50) = u(w + 75)$ either way.

According to prospect theory, however, bundling matters. Suppose that you start off with wealth w and that you take the status quo as your reference point. When the two gains are integrated, the value of winning $75 is represented in Figure 7.5(a). When the two gains are segregated, however, the situation looks different. This is so because you have the time to adjust your reference point before assessing the value of the second gain. When the two gains are segregated, the two separate gains are represented in Figure 7.5(b). It should be clear just from looking at these two figures that the value of a $25 gain plus the value of a $50 gain is greater than the value of a $75 gain; that is, $v(+25) + v(+50) > v(+75)$. This result follows from the value function

Figure 7.4 Integration vs. segregation of gains. Illustration by Cody Taylor

being concave in the domain of gains. The upshot is that people value two gains more when they are segregated than when they are integrated.

An analogy might help. If you are in a totally dark room and you turn on a light bulb, there is a huge difference: you can see, which is wonderful if you would rather not be in the dark. If you add a second light bulb of the same wattage, you will experience a small change in brightness, but the difference is not going to be that large. It is certainly not going to be as large as going from zero light bulbs to one: going from one to two makes a much smaller difference than going from zero to one. Something similar occurs with money. Winning a small amount is good. Winning ten times that amount is much better, obviously, but does not feel ten times as good. As a result, ten small gains are experienced as more impressive than the one gain that is ten times as large.

The fact that gains are valued more when segregated helps explain a variety of phenomena. For example, it explains why people do not put all their Christmas presents in one big box, even though that would save them time and money on wrapping: the practice of wrapping each present separately encourages the recipient to segregate the gains. The analysis also suggests that it is even better to give separate presents on separate nights, as on Hanukkah, rather than delivering all presents on Christmas Day, since this would do even more to encourage recipients to separate the gains. The analysis indicates that it is better to give people multiple small presents over the course of the year than to give them one big present once a year. While it is in good taste to give your spouse a present on your anniversary, you may wish to save some of the money and buy smaller presents during the rest of the year too. Similarly, segregation explains why fancy meals are served

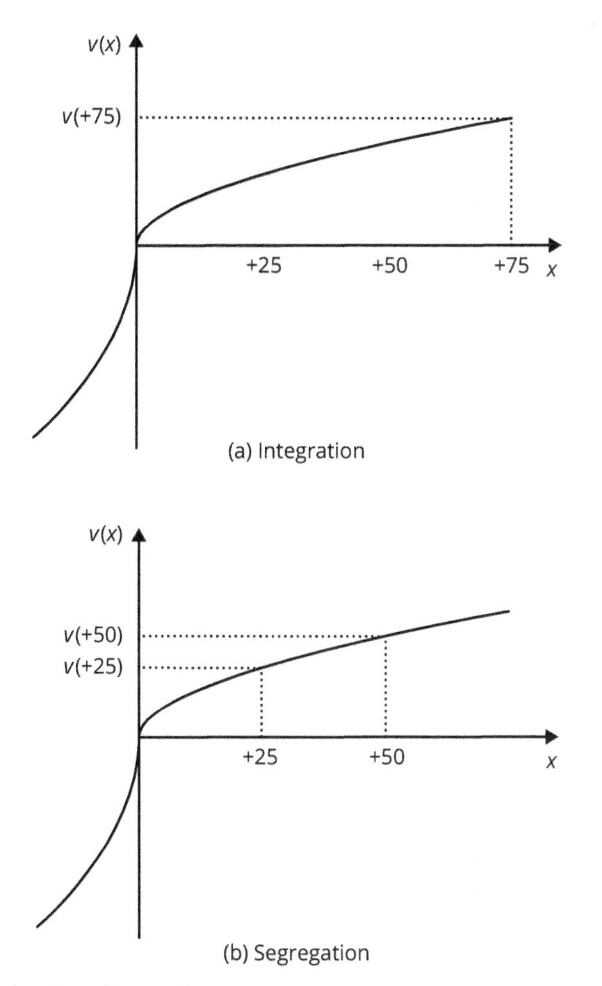

Figure 7.5 Evaluation of two gains

up dish-by-dish, rather than all at one time: chances are the eater will enjoy it more that way. In addition, the effect of segregating gains explains why workers receive end-of-year bonuses: receiving a $50k salary plus a $5k bonus encourages the segregation of gains in a manner that receiving a $55k salary does not. Finally, the value of segregated gains explains why people on daytime television try to sell pots and pans by offering to throw in lids, knives, cutting boards, and so on, rather than simply offering a basket consisting of all these goods. This practice too encourages the segregation of the gains.

Exercise 7.12 Evaluation of gains Yesterday, you had a decent day: you first received a $48 tax refund, and then an old friend repaid a $27 loan you had forgotten about. Suppose that your value function $v(\cdot)$ is defined by $v(x) = \sqrt{x/3}$ for gains ($x \geq 0$) and $v(x) = -3\sqrt{|x|}$ for losses ($x < 0$).

(**a**) If you integrate the two gains, what is the total value?

(**b**) If you segregate the two gains, what is the total value?

(**c**) From the point of view of value, is it better to integrate or to segregate?

Meanwhile, people are less dissatisfied when multiple losses are integrated than when they are segregated. By constructing graphs like those in Figure 7.5, you can confirm that, from the point of view of value, a $25 loss plus an additional $50 loss is worse than a $75 loss; that is, $v(-25) + v(-50) < v(-75)$. This result follows from the fact that the value function is convex in the domain of losses. Notice that if you do purchase the pots and pans from daytime television, and get all the other stuff in the bargain, your credit card is likely to be charged only once, meaning that costs are integrated. This is no coincidence: by encouraging customers to integrate the losses while segregating the gains, marketers take maximum advantage of these effects.

Example 7.13 Stalin Soviet dictator Joseph Stalin is alleged to have said: "The death of one man is a tragedy; the death of millions is a statistic." This line captures an important insight about integration: a million deaths is nowhere near as bad, from our subjective point of view, as a million times one death.

The fact that people experience less dissatisfaction when losses are integrated helps explain a variety of phenomena. It explains why sellers of cars, homes, and other pricey goods often try to sell expensive add-ons. Although you may never pay $1000 for a car radio, when it is bundled with a $25,999 car it may not sound like very much: for reasons explored in the previous section, a loss of $26,999 might not seem that much worse than a loss of $25,999. By contrast, since they are entirely different quantities, you might find it easy to segregate the car from the radio, which would further encourage you to accept the offer. Similarly, the wedding industry makes good use of integration by adding options that individually do not seem very expensive relative to what the hosts are already spending, but which jointly can cause major financial distress. "If you're spending $3000 on a dress, does it matter if you spend an additional $10 on each invitation?" The effects of integrating losses can also explain why so many people prefer to use credit cards rather than paying cash. When you use a credit card, though the monthly bill might be alarming, it only arrives once a month, thereby encouraging you to integrate the losses. By contrast, the satisfaction of receiving the things you buy is distributed throughout the month, thereby encouraging the segregation of gains.

Exercise 7.14 The opposite arrangement Suppose that the opposite were true: whenever you purchase something, you have to pay cash on the spot, but your purchases are not delivered until the end of the month in a giant box containing everything you bought in the last four weeks.
(**a**) Would you make more or fewer purchases this way?
(**b**) Use the language of integration and segregation to explain why.

Exercise 7.15 Air fares If you are old enough, you may remember the good old days when all sorts of conveniences were included in the price of an airline ticket. Under pressure to reduce the sticker price of their tickets, airlines have started charging less for the tickets themselves, but made it a habit to recover the losses by charging fees for everything from checked luggage to food and drink and early boarding. Perhaps they sell more tickets this way, but the effort is likely to sharply reduce customer satisfaction. Why?

This analysis might also explain why people hold on to cars even though taking cabs may be less expensive in the long run. The actual cost of owning a car for a month (including maintenance, car payments, insurance payments, gasoline, car washes, etc.) is so high that for many people it would make financial sense to sell the car and just hail cabs. The reason why people hesitate to do this (outside the major cities) may be related to the fact that car payments are made monthly or weekly, whereas taxi payments are made at the conclusion of every ride. Consequently, taxi companies could encourage more business by allowing people to run a tab and to pay it off once a month. The analysis can also explain why people prefer to pay a flat monthly fee for cell phone services, internet connections, and gym memberships: doing so permits them to integrate what would otherwise be many separate losses. The chances are you would not dream of joining a gym that charged you by the mile you ran on the treadmill. In part, this is for incentive-compatibility reasons: you do not want to join a gym that gives you a disincentive to exercise. But, in part, this may be because you wish to integrate the losses associated with your gym membership.

Exercise 7.16 Evaluation of losses Yesterday, you had a terrible day: you got a $144 speeding ticket on your way to the opera, and then had to pay $25 for a ticket you thought would be free. Suppose your value function remains that of Exercise 7.12.
(**a**) If you integrate the two losses, what is the total value?
(**b**) If you segregate the two losses, what is the total value?
(**c**) From the point of view of value, is it better to integrate or to segregate?

Example 7.17 Online booksellers If you look this book up at a large online retailer, you may see a little advertisement saying something like: "This book is frequently bought with Kahneman's *Thinking, Fast and Slow*. Buy both for twice the price!" Presumably the message contains no new information: you already knew that you could get more books for more money. How is this message supposed to work?

Related phenomena occur when experiencing a small loss in combination with a large gain or a small gain in combination with a large loss. People gain more value when they integrate a small loss with a large gain. That way, the loss is felt less intensely. This phenomenon is referred to as **cancellation**. Suppose that you win a million dollars, but have to pay a 10 percent tax on your winnings. The theory suggests that you will derive more value from telling yourself that you won $900,000 than by telling yourself that you won $1,000,000 and then lost $100,000.

Exercise 7.18 The pain of paying taxes The previous paragraph suggests that how you feel about paying your taxes will depend on whether you integrate that cost with the money you made or not.
(**a**) If you are a politician known for favoring high taxes, should you encourage voters to integrate or segregate? How should you shape your message?
(**b**) If you are a politician known for favoring low taxes, should you encourage voters to integrate or segregate? How should you shape your message?

Meanwhile, people gain more value when they segregate a large loss from a small gain. The small gain is often described as a **silver lining**. This analysis explains why some cars, apartments, and other big-ticket items sometimes come with cash-back offers. A customer may be more likely to buy a car with a \$27k price tag and a \$1k cash-back offer than to buy the very same car with a \$26k price tag; the \$1k gain, when segregated, helps offset the pain associated with the \$27k loss. The analysis also explains why credit-card companies frequently offer reward points for spending money on the card. Though the value of the reward points is small relative to monthly fees and charges, the company hopes that you will segregate and that the reward points therefore will serve to offset the larger loss. Finally, silver-lining phenomena may explain why rejection letters frequently include lines about how impressed the committee was with your submission or job application. The hope is that disingenuous flattery, when segregated, will to some extent offset the much greater perceived loss of the publication or job.

Exercise 7.19 Silver linings For this question, suppose your value function is $v(x) = \sqrt{x/2}$ for gains and $v(x) = -2\sqrt{|x|}$ for losses. Last night, you lost \$9 in a bet. There was a silver lining, though: on your way home, you found \$2 lying on the sidewalk.

(**a**) If you integrate the loss and the gain, what is the total value?
(**b**) If you segregate the loss and the gain, what is the total value?
(**c**) From the point of view of value, is it better to integrate or to segregate?

When do people integrate and when do they segregate? One possibility that might come to mind is that people bundle outcomes so as to maximize the amount of value that they experience. This is called the **hedonic-editing hypothesis**. According to this hypothesis, people will (1) segregate gains, (2) integrate losses, (3) cancel a small loss against a large gain, and (4) segregate a small gain from a large loss. Unfortunately, data suggest that the hypothesis is not in general true. In particular, it seems, people frequently fail to integrate subsequent losses. If you think about it, the failure of the hedonic-editing hypothesis is unsurprising in light of the fact that we need parents, therapists, boyfriends, and girlfriends to remind us of how to think about things in order not to be needlessly unhappy.

Bundling may be driven in part by **mental accounting**: people's tendency, in their minds, to divide money into separate categories. Mental accounting can be helpful in that it may prevent overspending. If the mental "clothing" account is seen as depleted, people may stop spending money on clothing even if there are funds left in some other mental account. But mental accounting can itself cause people to overconsume or underconsume particular kinds of goods. If the "entertainment account" is seen as having money left in it, but the "clothing account" is seen as overdrawn, people might spend more on entertainment even though they would maximize utility by buying clothes. This kind of behavior violates **fungibility**: the idea that money has no labels. Mental accounting might also affect the manner in which goods are bundled. For example, coding goods as belonging to the same category is likely to encourage integration, whereas coding goods as belonging to separate categories is likely to encourage segregation.

7.4 The Allais problem and the sure-thing principle

The following decision problem is called the **Allais problem.**

Example 7.20 Allais problem Suppose that you face the following options, and that you must choose first between (1a) and (1b), and second between (2a) and (2b). What would you choose?

(1a) $1 million for sure
(1b) An 89% chance of $1 million & a 10% chance of $5 million
(2a) An 11% chance of $1 million
(2b) A 10% chance of $5 million

A common response pattern here is (1a) and (2b). For the first pair, people may reason as follows: "Sure, $5 million is better than $1 million, but if I chose (1b) there would be some chance of winning nothing, and if that happened to me I would definitely regret not choosing the million dollars. So I'll go with (1a)." That is, a choice of (1a) over (1b) might be driven by regret aversion (see Section 6.2). For the second pair, people may reason in this way: "Certainly, an 11 percent chance of winning is better than a 10 percent chance of winning, but that difference is fairly small; meanwhile, 5 million dollars is a lot better than 1 million dollars, so I'll go with (2b)." For the second pair, the potential for regret is much less salient.

Unfortunately, this response pattern is inconsistent with expected-utility theory. To see this, consider what it means to prefer (1a) to (1b). It means that the expected utility of the former must be greater than the expected utility of the latter. Thus:

$$u(1M) > .89 * u(1M) + .10 * u(5M) \tag{7.1}$$

Preferring (2b) to (2a) means that the expected utility of the former must exceed the expected utility of the latter. Hence:

$$.10 * u(5M) > .11 * u(1M) \tag{7.2}$$

But because $.11 * u(1M) = (1 - .89) * u(1M) = u(1M) - .89 * u(1M)$, (7.2) is equivalent to

$$.10 * u(5M) > u(1M) - .89 * u(1M) \tag{7.3}$$

Rearranging the terms in (7.3), we get

$$.89 * u(1M) + .10 * u(5M) > u(1M) \tag{7.4}$$

But (7.4) contradicts (7.1). So the choice pattern that we are analyzing is in fact inconsistent with expected-utility theory.

There is another way of seeing why the choice pattern is inconsistent with expected-utility theory. Suppose that you spin a roulette wheel with 100 slots: 89 black, 10 red, and 1 white. This permits us to represent the four options in table form, as in Table 7.1.

Table 7.1 The Allais problem

	Black (89%)	Red (10%)	White (1%)
(1a)	$1M	$1M	$1M
(1b)	$1M	$5M	$0
(2a)	$0	$1M	$1M
(2b)	$0	$5M	$0

Let us begin by considering the first decision problem: that between (1a) and (1b). The table reveals that when black occurs, it does not matter what you choose; you will get a million dollars either way. In this sense, the million dollars if black occurs is a **sure thing**. The expression $.89 * u(1\text{M})$ appears in the calculation of the expected utility of (1a), of course, but because it also appears in the calculation of the expected utility of (1b), it should not affect the decision. Let us now consider the second decision problem. Again, the table reveals that when black occurs, you receive nothing no matter what you choose. So again, the $0 is a sure thing and should not affect the relative desirability of (2a) and (2b). Thus, what happens in the column marked "Black" should not affect your choices at all. Instead, your choices will be determined by what happens in the other two columns. But once you ignore the column marked "Black," (1a) is identical to (2a) and (1b) is identical to (2b): just compare the two shaded areas in Table 7.1. So if you strictly prefer (1a), you are rationally compelled to choose (2a); if you strictly prefer (1b), you are rationally compelled to choose (2b).

The **sure-thing principle** says that your decisions should not be influenced by sure things. This principle follows from expected utility theory (as we saw in the box on p. 143). The next exercise may help make the principle clearer.

Exercise 7.21 Sure-thing principle

(a) Suppose that you face the options in Table 7.2(a). Which state of the world does the sure-thing principle tell you to ignore?

(b) Suppose that you face the options in Table 7.2(b). What does the sure-thing principle tell you about this decision problem?

Table 7.2 Sure-thing principle

	A	B	C
(1)	2	1	4
(2)	3	1	3

(a)

	P	Q	R	S
(1)	3	2	4	1
(2)	3	1	4	2

(b)

	X	Y	Z
(1a)	80	100	40
(1b)	40	100	80
(2a)	40	0	80
(2b)	80	0	40

(c)

Exercise 7.22 Sure-thing principle, cont. Suppose that you face the options in Table 7.2(c) and that you must choose first between (1a) and (1b), and second between (2a) and (2b). What choice pattern is ruled out by the sure-thing principle?

As a normative principle, the sure-thing principle has its appeal, but it is not uncontroversial. Some people argue that violations of the sure-thing principle can be perfectly rational, and that there consequently is something wrong with expected-utility theory as a normative standard. Others insist that the sure-thing principle is a normatively correct principle. What is fairly clear, though, is that it is false as a description of actual behavior; people seem to violate it regularly and predictably. (We will return to this topic in the next section.)

One way to account for the Allais paradox is to say that people overweight outcomes that are certain, in the sense that they occur with a 100 percent probability. This tendency has been called the **certainty effect**. As suggested above, the certainty effect might result from regret aversion: whenever you forgo a certain option for a risky one, there is some chance that you will experience regret. Thus, a desire to minimize anticipated regret would lead to the rejection of the option that is not certain.

The certainty effect is apparent in slightly different kinds of context as well, as the following example shows.

Example 7.23 Certainty effect Which of the following options do you prefer: (A) a sure win of $30; (B) an 80 percent chance to win $45? Which of the following options do you prefer: (C) a 25 percent chance to win $30; (D) a 20 percent chance to win $45?

In this study, 78 percent of respondents favored A over B, yet 58 percent favored D over C.

The observed behavior pattern is an instance of the certainty effect, since a reduction from 100 percent to 25 percent makes a bigger difference to people than a reduction from 80 percent to 20 percent.

Exercise 7.24 Certainty effect, cont. Show that it is a violation of expected-utility theory to choose (A) over (B) and (D) over (C) in Example 7.23. Notice that (C) and (D) can be obtained from (A) and (B) by dividing the probabilities by four.

Does the certainty effect appear in the real world? It might. In a study of 72 physicians attending a meeting of the California Medical Association, physicians were asked which treatment they would favor for a patient with a tumor, given the choice between a radical treatment such as extensive surgery (options A and C), which involves a greater chance of imminent death, and a moderate treatment such as radiation (options B and D). They were presented with the following options:

A 80 percent chance of long life
(20 percent chance of imminent death)

B 100 percent probability of short life
(0 percent chance of imminent death)

C 20 percent chance of long life
(80 percent chance of imminent death)

D 25 percent chance of short life
(75 percent chance of imminent death).

The certainty effect was plainly visible: in violation of expected-utility theory, 65 percent favored B over A, yet 68 percent favored C over D. The fact that medical doctors exhibit the same behavior patterns as other people should not surprise us. It might be helpful to know this, whether or not you are a medical doctor.

7.5 The Ellsberg problem and ambiguity aversion

The following decision problem is referred to as the **Ellsberg problem**. It is due to Daniel Ellsberg, a US military analyst otherwise famous for releasing the so-called Pentagon Papers. Ellsberg was the subject of the 2009 documentary *The Most Dangerous Man in America*.

Example 7.25 Ellsberg problem Suppose that Dan shows you an urn with a total of 90 balls in it. There are three kinds of ball: red, black, and yellow. You know (from a trustworthy authority) that 30 are red, but you do not know how many of the remaining 60 are black and how many are yellow: there could be anywhere from 0 black and 60 yellow to 60 black and 0 yellow. The composition of the urn is illustrated by Table 7.3.

Table 7.3 Dan's urn

<table>
<tr><th></th><th>Red</th><th>Black</th><th>Yellow</th></tr>
<tr><td>Number of balls in urn</td><td>30</td><td colspan="2">60</td></tr>
</table>

Dan invites you to randomly draw a ball from the urn. He gives you the choice between two different gambles: (I) $100 if the ball is red, and (II) $100 if the ball is black. Which one would you choose? Next, Dan gives you a choice between the following two gambles: (III) $100 if the ball is red or yellow, and (IV) $100 if the ball is black or yellow. Which one would you choose?

When faced with the Ellsberg problem, many people will choose (I) rather than (II), apparently because they know that the chances of winning are 1/3; if they choose the other option the chances of winning could be anywhere from 0 to 2/3. Meanwhile, many people will choose (IV) rather than (III), apparently because they know the chances of winning are 2/3; if they choose the other option the chances of winning could be anywhere from 1/3 to 1.

However, and perhaps unfortunately, the choice of (I) from the first pair of options and (IV) from the second pair violates the sure-thing principle introduced in the previous section. The violation may be clearer if we represent the problem as in Table 7.4, which shows the payoffs for all four gambles and the three different outcomes.

As the table shows, what happens when a yellow ball is drawn does not depend on your choices. Whether you choose (I) or (II) from the first pair, when a yellow ball is drawn you will get nothing either way. The $0 when a yellow ball is drawn is a sure thing. Whether you choose (III) or (IV) from the second pair, when a yellow ball is drawn you will get \$100 either way. Again, the \$100 is a sure thing. Thus, the sure-thing principle says that your choices should not depend on what happens when you draw a yellow ball. That is, your choice should not depend on what is going on in the last column of Table 7.4. Your choice must reflect your evaluation of what is going on in the two columns to the left only. Ignoring the column marked "Yellow," however, you will see that (I) and (III) are identical, as are (II) and (IV): just compare the two shaded areas in the table. Hence, unless you are indifferent, you must either choose (I) and (III) or (II) and (IV).

Table 7.4 The Ellsberg problem

	Red (R)	Black (B)	Yellow (Y)
(I)	100	0	0
(II)	0	100	0
(III)	100	0	100
(IV)	0	100	100

There is another way of showing how the choice pattern (I) and (IV) is inconsistent with expected-utility theory. A strict preference for (I) over (II) entails that $EU(\text{I}) > EU(\text{II})$, which means that:

$$\Pr(R) * u(100) + \Pr(B) * u(0) + \Pr(Y) * u(0) >$$
$$\Pr(R) * u(0) + \Pr(B) * u(100) + \Pr(Y) * u(0)$$

Meanwhile, a strict preference for (IV) over (III) entails that $EU(\text{IV}) > EU(\text{III})$, which means that:

$$\Pr(R) * u(0) + \Pr(B) * u(100) + \Pr(Y) * u(100) >$$
$$\Pr(R) * u(100) + \Pr(B) * u(0) + \Pr(Y) * u(100)$$

Let us assume that $u(0) = 0$ and that $u(100) = 1$, which is only to say that you prefer \$100 over nothing. If so, these two expressions imply that the following two conditions must simultaneously be satisfied:

$$\Pr(R) > \Pr(B)$$
$$\Pr(B) > \Pr(R)$$

But that is obviously impossible. So again, the choice pattern we have been talking about is inconsistent with expected-utility theory.

How do we explain the fact that people exhibit this inconsistency? The two rejected options – (II) and (III) – have something in common, namely, that the exact probability of winning is unclear. We say that these probabilities are **ambiguous**. By contrast, the favored options – (I) and (IV) – are not associated with ambiguous probabilities. The observed choices seem to reflect an unwillingness to take on gambles with ambiguous probabilities. We refer to this phenomenon as **ambiguity aversion**. Some people have a greater tolerance for ambiguity than others, but any aversion to ambiguity is a violation of expected-utility theory. Insofar as people are in fact ambiguity averse, which they seem to be, expected-utility theory fails to capture their behavior. And insofar as it is rationally permissible to take ambiguity into account when making decisions, expected-utility theory does not capture the manner in which people should make decisions.

Exercise 7.26 The coins Suppose that you have the opportunity to bet on the outcome of a coin toss. If the coin comes up heads, you win; if it comes up tails, you lose. Suppose also that you are ambiguity averse. Would you rather bet on a fair coin (with equal probabilities of coming up heads and tails) or on a loaded coin with unknown, unequal probabilities of coming up heads and tails?

Exercise 7.27 Tennis You have been invited to bet on one of three tennis games. In game 1, two extraordinarily good tennis players are up against each other. In game 2, two extraordinarily poor tennis players are up against each other. In game 3, one very good and one very bad player are up against each other, but you do not know which is good and which is bad. As a result, as far as you are concerned, the probability that any given player will win is 50 percent. Suppose that you are ambiguity averse. Which of the three games would you be *least* likely to bet on? Why?

There is no principled reason why people cannot be **ambiguity prone** rather than ambiguity averse. In fact, evidence suggests that people's behavior in the face of ambiguous probabilities depends on the context. According to the **competence hypothesis**, for example, people are less averse to ambiguity in contexts where they consider themselves particularly knowledgeable. Thus, a football fan may be ambiguity averse in the Ellsberg case (where outcomes are completely random) but ambiguity prone when predicting the outcomes of football games (where he or she feels like an expert).

Exercise 7.28 Nevada's boom and bust Las Vegas entrepreneur Andrew Donner does not gamble at the casinos. Instead, he invests in real estate in the city's downtown. Interviewed on *Marketplace*, Donner said: "Well, you know casinos, you somewhat know the odds, and I think there's something beautiful about being somewhat ignorant of your odds out in the business marketplace. You keep working and hopefully you win more than you lose."
(**a**) Is Donner ambiguity averse or ambiguity prone?
(**b**) Are his attitudes consistent with the competence hypothesis or not?

Even so, the Ellsberg paradox and ambiguity aversion have potentially vast implications. What they suggest is that people do not in general assign probabilities satisfying the axioms of the probability calculus to events with ambiguous probabilities. And in the real world, ambiguous probabilities are common. The probability of bankruptcies, pandemics, and nuclear meltdowns can be estimated, but, outside games of chance, some ambiguity almost always remains. Thus, it is highly likely that people's choices *do* reflect the fact that people are ambiguity averse – or prone, as the case may be. And perhaps choices *should* reflect the ambiguity of the probabilities too.

Example 7.29 Known knowns In a 2002 press briefing, American Defense Secretary Donald H. Rumsfeld said: "[As] we know, there are known knowns; there are things we know we know. We also know there are known unknowns; that is to say we know there are some things we do not know. But there are also unknown unknowns – the ones we don't know we don't know." Rumsfeld was roundly satirized following the briefing, but the distinctions he was trying to draw may be very important.

7.6 Probability weighting

The idea that the value function is concave in the domain of gains and convex in the domain of losses helps us analyze a wide range of behaviors, as we saw in Sections 7.2 and 7.3. Yet there are widely observed behavior patterns that cannot be accommodated within this framework. Consider the fact that some people simultaneously gamble and purchase insurance. This is paradoxical from the point of view of expected-utility theory. If people are risk averse, they should buy insurance but not gamble; if they are risk prone, they should gamble but not buy insurance; and if they are risk neutral, they should do neither. It is theoretically possible that people have inverted-S-shaped utility functions, like the dashed line in Figure 6.8 on page 143, and that the inflection point (marked $x*$ in the figure) just happens to correspond to their present endowment. Yet it seems too much like a coincidence that this should be true for so many people.

Simultaneous gambling and insurance shopping are equally paradoxical from the point of view of the theory we have studied in this chapter so far. The fact that people are willing to accept a gamble in which they may win a large sum of money suggests that they are risk prone in the domain of gains, while the fact that they are willing to reject a gamble in which they may lose their house suggests that they are risk averse in the domain of losses. This would entail that their value function is convex in the domain of gains and concave in the domain of losses, which is the very opposite of what we have assumed to date. The only way to accommodate this behavior pattern within the framework above is to assume that people take the state when they win the grand prize as their reference point when gambling, and the state in which they lose things as their reference point when buying insurance. This seems artificial, however, in light of the other evidence that people otherwise frequently take their endowment as their reference point.

Another way to understand the observed behavior pattern is to think of those who gamble as well as those who buy insurance as prone to paying too much attention to unlikely events. The more weight you put on the probability of winning the lottery, the more likely you will be to gamble. And the more weight you put on the probability of losing your house, car, life, and limb, the more likely you will be to purchase insurance. This insight suggests a more systematic approach to explaining how people can simultaneously buy lottery tickets and insurance.

Prospect theory incorporates this kind of behavior by introducing the notion of **probability weighting**. We know from Definition 6.21 on page 138 that expected-utility theory says that the agent maximizes an expression of the following form:

$$EU(A_i) = \Pr(S_1) * u(C_{i1}) + \Pr(S_2) * u(C_{i2}) + \ldots + \Pr(S_n) * u(C_{in})$$

By contrast, prospect theory says that the agent maximizes an expression in which the value function $v(\cdot)$ is substituted for the utility function $u(\cdot)$, and in which probabilities are weighted by a **probability-weighting function** $\pi(\cdot)$.

Definition 7.30 Value *Given a decision problem as in Table 6.4 on page 132, the value (or weighted value) $V(A_i)$ of an act A_i is given by*

$$V(A_i) = \pi[\Pr(S_1)] * v(C_{i1}) + \pi[\Pr(S_2)] * v(C_{i2}) + \ldots + \pi[\Pr(S_n)] * v(C_{in})$$
$$= \sum_{j=1}^{n} \pi[\Pr(S_j)] v(C_{ij})$$

The probability-weighting function $\pi(\cdot)$ assigns weights, from zero to one inclusive, to probabilities. It is assumed that $\pi(0) = 0$ and that $\pi(1) = 1$. But, as shown in Figure 7.6, for values strictly between zero and one, the curve does not coincide with the 45-degree line. For low probabilities, it is assumed that $\pi(x) > x$, and for moderate and high probabilities, that $\pi(x) < x$.

The probability-weighting function can help resolve the paradox that some people simultaneously buy lottery tickets and insurance policies. A well-informed expected-utility maximizer weights the utility of winning the grand prize by the probability of winning it, which as we know from Section 4.4 is low indeed. Thus, expected-utility

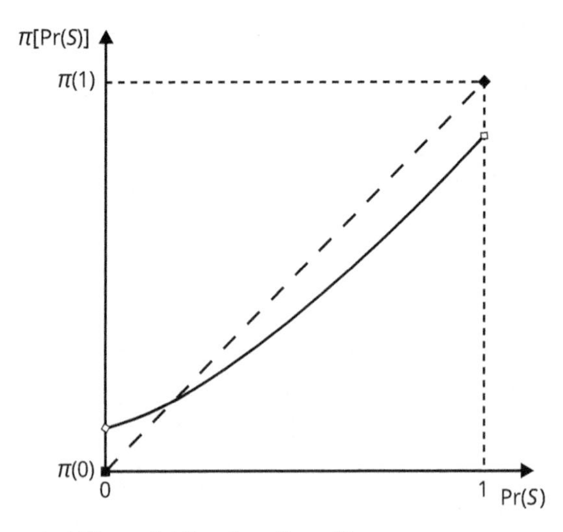

Figure 7.6 The probability-weighting function $\pi(\cdot)$

theory says that such outcomes should not loom very large in our decision-making. Prospect theory makes a different prediction. Winning the lottery and losing house, car, life, and limb are positive- but low-probability events, so the probability-weighting function implies that they would loom relatively large. And when such events loom large, people are willing to purchase lottery tickets and insurance policies. This helps explain why people fear airplane crashes, terrorist attacks, and many other unlikely things so much. Probability weighting also explains why people purchase extended warranties on equipment such as computers, in spite of the fact that simple expected-value calculations suggest that for most people extended warranties are not a very good deal (see Exercise 6.19 on page 136).

Prospect theory

Prospect theory describes the process of making decisions as having two phases. During the **editing phase** the decision maker edits the options to facilitate assessment. Editing may involve a number of different operations, including:

- **Coding:** Describing outcomes as gains or losses as compared with some reference point, which may be the current endowment, somebody else's current endowment, an expectation of a future endowment, or the like.
- **Combination:** Combining identical outcomes into a simpler one, so that a 25 percent chance of winning \$1 and another 25 percent chance of winning \$1 get represented as a 50 percent chance of winning \$1.
- **Simplification:** Simplifying the options, e.g., by rounding probabilities and outcomes. In particular, extremely small probabilities may be rounded down to zero and eliminated from consideration.

During the subsequent **evaluation phase** the decision maker assesses the edited options. The evaluation is based on two elements: the value function from Sections 3.5 and 7.2 and the probability-weighting function from Section 7.6. The value function $v(\cdot)$ is *S*-shaped, meaning convex in the realm of losses and concave in the realm of gains (see Figure 7.2 on page 152). The probability-weighting function $\pi(\cdot)$ normally satisfies the following conditions: $\pi(0) = 0$ and $\pi(1) = 1$; otherwise for low probabilities $\pi(x) > x$ and for moderate and high probabilities $\pi(x) < x$ (see Figure 7.6 on page 169). The value (or weighted value) $V(A_i)$ of an act A_i is evaluated in accordance with the formula

$$V(A_i) = \pi[\Pr(S_1)] * v(C_{i1}) + \pi[\Pr(S_2)] * v(C_{i2}) + \ldots + \pi[\Pr(S_n)] * v(C_{in})$$
$$= \sum_{j=1}^{n} \pi[\Pr(S_j)] v(C_{ij})$$

In the special case when $\pi(x) = x$ and $v(x) = u(x)$, the value of an option equals its expected utility (see Definition 6.21 on page 138).

Example 7.31 Freakonomics The book *Freakonomics* discusses the economics of crack-cocaine. Contrary to what many people think, the vast majority of crack dealers make little money – frequently less than the federally mandated minimum wage. They stay in the job, according to the *Freakonomics* authors, because of a small chance of joining the upper stratum of the organization, in which an exclusive "board of directors" makes decent money.

Even so, this does only seem to be part of the explanation. The directors do not make astronomical amounts of money, the probability of joining their ranks is small, and the probability of getting shot or landing in jail is high. We can augment the explanation by adding that aspiring directors might overweight the low probability of rising through the ranks and joining the board of directors. This is what prospect theory would predict.

As the authors suggest, the same analysis might apply to aspiring models, actors, concert pianists, and CEOs. The probability of succeeding in any one of these endeavors is low, yet people continue to bet that they will be the one who does. Their aspirations may, in part, be driven by the fact that they overweight the probability of success. (It can be argued that academia works the same way.)

The probability-weighting function can also account for the certainty effect: the tendency to overweight outcomes that are certain (see Section 7.4). As Figure 7.6 shows, there is a discontinuity at probabilities approaching one, so that as $x \to 1$, $\lim \pi(x) < \pi(1)$. Thus, events that are not certain (even when their probability is very high) will be underweighted relative to events that are certain.

The upshot is that people's behavior in the face of risk depends not just on whether outcomes are construed as gains or as losses relative to some reference point, but on whether the relevant probabilities are low or not. In the domain of losses, people tend to be risk prone, except for gambles involving a low-probability event of significant (negative) value, in which case they may be risk averse. In the domain of gains, people tend to be risk averse, except for gambles involving a low-probability event of significant (positive) value, in which case they may be risk prone. See Table 7.5 for a summary of these implications.

Table 7.5 Risk attitudes according to prospect theory

Probability	Domain	
	Losses	Gains
Low	risk averse	risk prone
Moderate	risk prone	risk averse
High	risk prone	risk averse

Example 7.32 Russian roulette Suppose that you are forced to play Russian roulette, but that you have the option to pay to remove one bullet from the loaded gun before pulling the trigger. Would you pay more to reduce the number of bullets in the cylinder from four to three or from one to zero? According to Kahneman and Tversky, if you are like most people, you would pay more to reduce the number from one to zero than from four to three. Why?

Reducing the number of bullets from four to three would reduce the probability of dying from 4/6 to 3/6. In this range, the probability-weighting function is relatively flat, meaning fairly unresponsive to changes in the underlying probability. Reducing the number of bullets from one to zero would reduce the probability of dying from 1/6 to 0. Here, there is a jump from $\pi(1/6) > 1/6$ to $\pi(0) = 0$. Thus, the value to you of reducing the number of bullets from one to zero exceeds that of reducing it from four to three.

Exercise 7.33 Lotteries as rewards Behavioral economists have found that using lotteries can be an effective way to incentivize behavioral change. Thus, a person may be more likely to fill in a survey or take a pill if offered a lottery ticket with a 1/1000 probability of winning $1000 than if offered a cash payment of $1. This might seem counterintuitive, given that people are often risk averse. Use the probability-weighting function to explain why lottery tickets can be so appealing.

7.7 Rabin's calibration theorem

The last challenge to expected-utility theory that we will explore in this chapter has a more mathematical flavor. According to the theory, as you know from Section 6.5, risk attitudes are determined by the shape of a utility function ranging over absolute levels of wealth (or whatever the agent cares about), not *changes* in wealth (or whatever). This idea, it turns out, has implications that are embarrassing for both the descriptive and normative interpretations of the theory. The implications are neatly brought out by a result referred to as **Rabin's calibration theorem**.

The theorem demonstrates that people who are moderately risk averse over small stakes must – as a matter of mathematical necessity – be absurdly risk averse over large stakes. For example, a person who rejects a 50–50 chance of winning $11 or losing $10 must reject a 50–50 chance of losing $100 and winning $1000, $1,000,000, *or any other amount of money*. In other words, there is no amount of money large enough such that this person will accept a 50 percent chance of losing $100 and a 50 percent chance of winning that amount. This is quite absurd. Most people fortunate enough to have a hundred bucks to invest would and should be quite pleased if offered a 50 percent chance of immediately making a bazillion dollars. Any real startup or asset you might otherwise invest in will have a much lower expected ROI (return on investment).

The theorem itself is complicated, but the implications easy to grasp. Table 7.6 shows some other implications of the theorem. For every line in this table, an expected-utility maximizer who turns down the bet on the left-hand side has no choice but to turn down the bet on the right-hand side. That is, if you turn down the bet on the left,

Table 7.6 Calibration theorem

If an expected-utility maximizer turns down this 50–50 bet...	... then she must also turn down this 50–50 bet
lose $10 / win $10.10	lose $1000 / win $∞
lose $10 / win $11	lose $100 / win $∞
lose $100 / win $101	lose $10,000 / win $∞
lose $100 / win $105	lose $2000 / win $∞
lose $100 / win $110	lose $1000 / win $∞

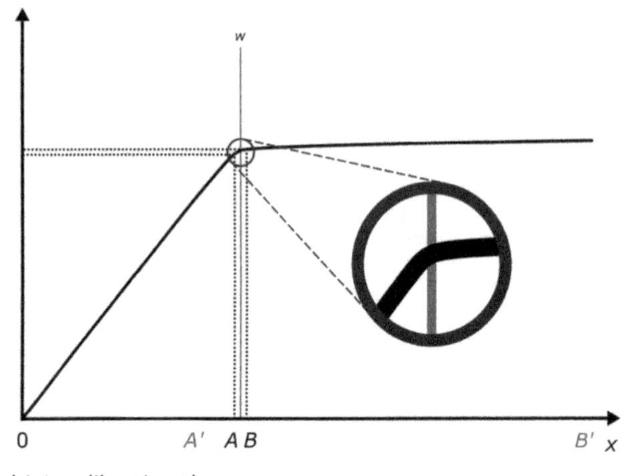

Figure 7.7 Rabin's calibration theorem

there is no amount of money, no matter how large, that will entice you to go for the bet on the right: a 50–50 chance of winning that amount or of losing the amount in the table.

One way to understand where these implications are coming from is to think of the problem in graphical terms. Consider again Figure 6.9 on p. 147. Here an expected-utility maximizer is facing a gamble that leaves her with A with some probability and with B otherwise. We know that her risk preference is entirely determined by the curvature of the utility function. And not only that: the only thing that matters is the curvature between A and B: the curvature to the left of A and the curvature to the right of B are irrelevant in the context. Now consider Figure 7.7, in which an agent with wealth w (represented by a grey line in the graph) is faced with a small-stakes lottery, say a 50–50 chance of winning \$11 or losing \$10. This means that the relevant outcomes are (A) a situation in which the agent ends up with $w - 10$ and (B) a situation in which the agent ends up with $w + 11$. For many people in developed nations w will be orders of magnitude greater than \$10–11, which means that the range from A to B will be a very narrow segment around w. If the person's wealth is \$10,000, for example, then the distance between A and B will be only $21/10{,}000 = 0.21$ percent of the distance between 0 and w. (If anything, Figure 7.7 exaggerates the distance between A and B.)

There is nothing in expected-utility theory that precludes having a curved utility function in the narrow range between A and B, as required to reject the small-stakes gamble. However, in order for the agent to be *sufficiently* risk averse to reject the small-stakes gamble, the function has to be *extremely* curved in this narrow range, as in the magnified segment of the curve. The extreme curvature within the A–B range has implications for the curvature outside of it. To the left of A, the curve has to be relatively straight and steep; to the right of B, the curve has to be relatively straight and flat. This, in turn, has implications for the levels of risk aversion permitted when it comes to large-stakes gambles, meaning gambles with outcomes outside of the range from A to B. Consider the same agent's attitude to this big-stakes gamble: a 50–50 chance of getting A' (just to the left of A) and B' (way to the right of B). Because the curve is relatively steep to the left of A, $u(A')$ will be considerably lower than $u(A)$, but because the curve is relatively flat to the right of B, $u(B')$ will only be marginally higher than $u(B)$. A 50–50 chance of winning A' or B' may well have a lower expected utility than the status quo $u(w)$.

Exercise 7.34 Show graphically that the expected value of the high-stakes gamble is lower than the utility of the status quo in Figure 7.7.

The theorem is quite general. For example, it does not assume any particular functional form.

The calibration theorem is a problem for expected-utility theory as a descriptive theory, because it is a fact that many people (maybe most?) exhibit moderate levels of risk aversion over all stakes, small and large. If you ask around, chances are many people will reject the small-stakes gamble but accept the large-stakes one, suggesting people you know are indeed moderately risk averse over the full range of gambles. But this apparently common pattern is inconsistent with the theory. People's risk attitudes, then, cannot be consistently captured by the curvature of a single utility function. To account for people's actual risk attitudes, we need to invoke other principles. Loss aversion, for example, could explain why people turn down small-stakes gambles involving a loss of some kind, even when the expected value is significantly positive.

The calibration theorem is a problem for expected-utility theory as a normative theory, because it clashes with a common (and deeply held) conviction that it is at least rationally permissible to be moderately risk averse for both small- and large-stakes gambles. If the conviction is correct, the theory cannot be normatively acceptable; if the theory is acceptable, the conviction has to be incorrect. If you want to avoid absurd levels of risk aversion when it comes to large stakes, while remaining true to expected-utility theory, you have to be effectively risk neutral when it comes to small stakes. It is important to note that the calibration theorem does not in and of itself *refute* expected-utility theory as a normative theory. But it does suggest, at a minimum, that we should be effectively risk neutral over small stakes.

7.8 Discussion

In this chapter we have discussed situations in which people appear to violate the standard implicit in the theory of expected utility outlined in Chapter 6. The problem is not that people fail to maximize some mathematical utility function in their heads. Rather, the problem is that people's observed choices diverge from the predictions of the theory. Though the divergences are not universal, they are substantial, systematic, and predictable, and they can have real, and sometimes adverse, effects on people's decision-making. Behavioral economists take the existence of divergences to undercut the descriptive adequacy of the theory of expected utility. Obviously, the chapter does not purport to offer a complete list of violations of expected-utility theory. We have also discussed situations where people's firmly held intuitions about the rational course of action differ from the recommendations of expected-utility theory, as in the presence of ambiguous probabilities. This raises deep issues about the nature of rationality.

In addition, we have explored more theoretical tools developed by behavioral economists to capture the manner in which people actually make decisions. Among other things, we studied other components of prospect theory, including the *S*-shaped value function and the probability-weighting function. This concludes our review of prospect theory (see text box on page 169). These tools can be used not just to explain and predict but also to influence other people's evaluations. It appears that, under certain conditions, a person's risk preferences can be reversed simply by changing the frame of the relevant options. This gives behavioral economists more levers to use, which is obviously important for therapists, marketers, public health officials, and others who hope to affect people's behavior. But again, knowledge of the tools also permits us to anticipate and prevent other people from employing them.

In Section 6.6, we discussed the distinction between the rational and the right, and the fact that you cannot judge the rationality of a decision by examining the outcome alone. It will come as no surprise to hear that people do, in fact, often judge the rationality of people's decisions by the outcome. This is called **outcome bias**, and the phenomenon appears pervasive. Sports commentary offers many examples: coaches' decisions tend to be praised just in case they have the intended result independently of whether their decisions were rational in the first place. (Of course, the fact that a person's decisions often turn out to be right is evidence in favor of their rationality.) One person who understood outcome bias was the early sixth-century Roman philosopher Boethius, who noted: "[The] world does not judge actions on their merit, but on their chance results, and they consider that only those things which are blessed with a happy outcome have been undertaken with sound advice." Boethius knew about judgment. Once a powerful public official, he wrote these lines on death row having been convicted of sorcery and other charges. He was bludgeoned to death shortly thereafter, but the work, *The Consolations of Philosophy*, became one of the most widely read texts of the Middle Ages.

In Part 4 we will add another layer of complexity to the analysis, by introducing the topic of time.

Additional exercises

Exercise 7.35 Savings decisions You are lucky enough to have a million dollars in the bank. You have decided that there are only three serious investment options: putting it in your mattress, investing in stocks, and investing in bonds. Your utility function over total wealth is $u(x) = \sqrt{x}$. There is no inflation.

(**a**) If you stick your money in the mattress (where we can assume that it will be perfectly safe), how much utility will you have at the end of the year?

(**b**) Bonds are in general very dependable, but markets have been jittery as of late. You estimate that there is a 90 percent chance that you will gain 4 percent, but there is a 10 percent chance that you will gain nothing (zero percent). What is the expected utility of investing the $1,000,000 in bonds?

(**c**) Stocks have recently been extremely volatile. If you invest in stocks, there is a 40 percent chance that you will gain 21 percent, a 40 percent chance that you will gain nothing (zero percent), and a 20 percent chance that you will lose ten percent. What is the expected utility of investing the $1,000,000 in stocks?

(**d**) Given that you are an expected-utility maximizer, what is the sensible thing to do with your money?
(**e**) If, instead of maximizing expected utility, you were exceedingly loss averse, what would you do?

Exercise 7.36 Zero expected value Prospect theory is consistent with the result that people frequently reject gambles with an expected value of zero. Suppose you are facing a gamble G with a 1/2 probability of winning \$10 and a 1/2 probability of losing \$10. According to prospect theory, would you prefer the gamble or the status quo? Assume that your value function is $v(x) = \sqrt{x/2}$ for gains and $v(x) = -2\sqrt{|x|}$ for losses.

Exercise 7.37 Life coaching Life coaches are people whose job it is to help you deal with challenges in your personal and professional life.
(**a**) If you spend too much money, life coaches will sometimes suggest that you cut up your credit cards and pay cash for everything. Use the language of integration and segregation to explain how this is supposed to help you rein in your spending.
(**b**) In order to help you spend money more wisely, life coaches sometimes suggest you allocate your money between a number of envelopes marked "food," "drink," "transportation," etc. What is the technical term for this kind of behavior?
(**c**) Though the envelope technique may have real benefits, it can also lead to problems. Identify some possible negative consequences of this technique.

Exercise 7.38 Match each of the vignettes below with one of the following phenomena: *ambiguity aversion, cancellation, certainty effect, competence hypothesis, silver lining,* and *mental accounting.* If in doubt, pick the best fit.
(**a**) Abraham is seriously depressed after his girlfriend of several years leaves him for his best friend. His therapist tells him that every time a door closes, another one opens. Abraham thinks about all the other potential girlfriends out there and finally feels a little better.
(**b**) Berit is trying to save money by limiting the amount of money she spends eating out. She tells herself she must never spend more than \$100 each week in restaurants. It is Sunday night, and Berit realizes that she has spent no more than \$60 eating out during the past week. She does not quite feel like going out, but tells herself that she must not miss the opportunity to have a \$40 meal.
(**c**) Charles is not much of a gambler and rarely accepts bets. The exception is politics. He considers himself a true policy wonk and is happy to accept bets when it comes to the outcome of local and national elections. His friends note that he still does not win the bets more than about half of the time.
(**d**) According to a common saying: "Better the devil you know than the devil you don't."
(**e**) Elissa very much wants to go to medical school, but cannot stand the thought of not knowing whether she will pass the rigorous curriculum. Instead, she decides to sign up for a less demanding physical therapy curriculum that she is confident that she can pass.

Problem 7.39 *Drawing on your own experience, make up stories like those in Exercise 7.38 to illustrate the various ideas that you have read about in this chapter.*

Further reading

Framing effects and probability weighting, which are part of prospect theory, are discussed in Kahneman and Tversky (1979); see also Tversky and Kahneman (1981), the source of the pandemic problem (p. 453). The example from the original prospect-theory paper appears in Kahneman and Tversky (1979, p. 273), and the two examples of the certainty effect in Tversky and Kahneman (1986, pp. S266–9); the study that manipulated people's perceptions of their income is Haisley et al. (2008). Bundling and mental accounting are explored in Thaler (1980, 1985), and hedonic editing in Thaler and Johnson (1990). The Allais problem is due to Allais (1953). The Ellsberg problem is due to Ellsberg (1961); the competence hypothesis is due to Heath and Tversky (1991). Donner is quoted in Gardner (2012), and Rumsfeld in US Department of Defense (2002). Probability-weighting is described in Kahneman and Tversky (1979); the roulette example (on p. 283) is attributed to Zeckhauser. *Freakonomics* is Levitt and Dubner (2005). Rabin's calibration theorem is discussed in Rabin and Thaler (2001), from which Table 7.6 is adapted. Outcome bias is the topic of Baron and Hershey (1988); the lines from *The Consolations of Philosophy* appear in Boethius (1999 [*c* 524], p. 14).

PART 4

INTERTEMPORAL CHOICE

8 THE DISCOUNTED UTILITY MODEL

Learning objectives

After studying this chapter you will:

- Recall how interest works
- Know the discounted utility model of intertemporal choice
- Understand a variety of perspectives on the rationality of time discounting

8.1 Introduction

So far, we have treated decision problems as though all possible consequences of actions were instantaneous. That is, we have assumed that all relevant consequences occur more or less immediately, or at least at one and the same time. There are cases when this is a perfectly reasonable assumption. If you are playing roulette once and have preferences over money, for example, whatever happens will happen more or less at the same time.

Very often, however, time is a factor. When you decide whether to purchase the one-year warranty for your new tablet computer (see Exercise 6.19 on page 136), you are not only choosing between a sure option (investing in the warranty) and a risky option (forgoing the warranty), but you are also choosing between a certain loss now (since the warranty has to be paid for now) and the possibility of suffering a loss later (the possibility of having to pay to replace a broken tablet some time down the road).

It may be that most decisions have consequences that occur at different points in time. Some decisions have immediate benefits and deferred costs: procrastination, for example, is a matter of favoring some immediate good (a dinner and movie with friends) over some later benefit (a clean house). Other decisions have immediate costs and deferred benefits: savings behavior, for example, is a matter of favoring some later benefit (a comfortable retirement) over some immediate good (a new video-game platform). In this chapter and the next, we will talk about how to model decisions when time is a factor.

8.2 Interest rates

Before we begin the theory, let us talk about **interest**. Much of this should be familiar, but knowing how to think about interest is so useful that it justifies a review. Evidence from studies such as that in Exercise 1.3(c) on page 7 indicates that many people have a limited grasp of how interest works.

Example 8.1 Interest Suppose you borrow $100 for a year at an annual interest rate of 9 percent. At the end of the process, how much interest will you owe the lender?

The answer is $\$100 * 0.09 = \$9$.

Slightly more formally, let r be the interest rate, P the **principal** (that is, the amount you borrow), and I the interest. Then:

$$I = Pr \tag{8.1}$$

This formula can be used to evaluate credit-card offers. Table 8.1 was adapted from a website offering credit cards to people with bad credit. Given this information, we can compute what it would cost to borrow a given amount by charging it to the card.

Table 8.1 Credit-card offers for customers with bad credit

Credit-card offer	APR	Fee
Silver Axxess Visa Card	19.92%	$48
Finance Gold MasterCard	13.75%	$250
Continental Platinum MasterCard	19.92%	$49
Gold Image Visa Card	17.75%	$36
Archer Gold American Express	19.75%	$99
Total Tribute American Express	18.25%	$150
Splendid Credit Eurocard	22.25%	$72

Example 8.2 Cost of credit Suppose you need to invest $1000 in a new car for one year. If you charge it to the Silver Axxess Visa Card, what is the total cost of the credit, taking into account the fact that you would be charged both interest and an annual fee?

Given that the annual percentage rate (APR) $r = 19.92\% = 0.1992$, you can compute $I = Pr = \$1000 * 0.1992 = \$199.20$. The annual fee is $48. The total cost would be the interest (I) plus the annual fee: $\$199.20 + \$48 = \$247.20$.

Expressed as a fraction of the principal, this is almost 25 percent. And that is not the worst offer, as the next exercise will make clear.

Exercise 8.3 Cost of credit, cont. What would it cost to borrow $1000 for one year using one of the other credit cards in Table 8.1? What if you need $100 or $10,000?

Fees and APRs fluctuate; never make decisions about credit cards without looking up the latest figures.

At the end of the year, the lender will want the principal back. In Example 8.1 above, the lender will expect the $100 principal as well as the $9 interest, for a total of $109. Let L be the **liability**, that is, the total amount you owe the lender at the end of the year. Then:

$$L = P + I \tag{8.2}$$

Substituting for I from (8.1) in (8.2), we get:

$$L = P + I = P + (Pr) = P(1 + r) \tag{8.3}$$

It is sometimes convenient to define R as one plus r:

$$R = 1 + r \tag{8.4}$$

Together, (8.3) and (8.4) imply that:

$$L = PR \tag{8.5}$$

Returning to Example 8.1, we can use this formula to compute the liability: $L = \$100 * (1 + 0.09) + \$109$, as expected. These formulas can also be used to compute interest rates given the liability and the principal.

Example 8.4 Implicit interest Suppose that somebody offers to lend you $105 on condition that you pay them back $115 one year later. What is the interest rate (r) implicit in this offer?

We know that $P = \$105$ and $L = \$115$. (8.5) implies that $R = L/P = \$115/\$105 = 1.095$. By (8.4), $r = R - 1 = 1.095 - 1 = 0.095$. Thus, the potential lender is implicitly offering you a loan at an annual interest rate of 9.5 percent.

Exercise 8.5 Payday loans Payday loan establishments offer short-term loans to be repaid on the borrower's next payday. Fees fluctuate, but such an establishment may offer you $400 on the 15th of the month, provided you repay $480 two weeks later. Over the course of the two weeks, what is the interest rate (r) implicit in this offer?

In some US states, the number of payday loan establishments exceeds the number of Starbucks and McDonald's locations combined. The answer to Exercise 8.5 suggests why. (See also Exercise 8.9.)

We can extend the analysis over longer periods of time. Broadly speaking, there are two kinds of scenario that merit our attention. Here is the first:

Example 8.6 Simple interest Imagine that you use a credit card to borrow $100, and that every month the credit-card company will charge you an interest rate of 18 percent of the principal. Every month, you pay only interest, but you pay it off in full. At the end of the year, you also repay the $100 principal. What is the total interest over the course of a year, expressed both in dollars and as a fraction of the principal?

Every month, by (8.1), you have to pay the credit-card company $I = Pr = \$100 * 0.18 = \$18$ in interest. Because you have to do this 12 times, the total interest will amount to $\$18 * 12 = \$216$. As a fraction of the $100 principal, this equals $\$216/\$100 = 2.16 = 216$ percent.

This is a case of **simple interest**. You can get the same figure by multiplying 18 percent by 12. Here is the other kind of case:

Example 8.7 Compound interest Imagine, again, that you use a credit card to borrow $100 and that the monthly interest rate is 18 percent. In contrast to the previous example, however, you do not make monthly interest payments. Instead, every month your interest is added to the principal. What is the total interest over the course of a year, expressed both in dollars and as a fraction of the principal?

At the conclusion of the first month, by (8.3), given that $r = 0.18$, you will owe $L = P(1 + r) = \$100 * 1.18 = \$118$. At the conclusion of the second month, your liability will be $L = \$118 * 1.18 = \$139.24$. Notice that you could have gotten the same answer by computing $L = \$100 * 1.18 * 1.18 = \$100 * 1.18^2$. At the conclusion of the third month, your liability will be $L = \$100 * 1.18 * 1.18 * 1.18 = \$100 * 1.18^3 \approx \$164.30$. At the conclusion of the 12th month, your liability will be $L = \$100 * 1.18^{12} \approx \$728.76$. The liability at the end of the year includes the \$100 principal, so your interest payments, by (8.2), only add up to $I = L - P \approx \$728.76 - \$100 = \$628.76$. As a fraction of the principal, this equals \$628.76/\$100 = 6.2876 = 628.76 percent.

The answer to Example 8.7 is so much higher than the answer to Example 8.6 because the former involves **compound interest**. Unlike the case of simple interest, here the interest accumulated during the first period is added to the principal, so that you will end up paying interest on the interest accumulated during previous periods. Notice that in cases of compound interest, you cannot simply multiply the interest accumulated during the first period by the number of periods, as we did in the case of simple interest. Instead, with compound interest your liability after t periods is computed as:

$$L = PR^t \tag{8.6}$$

This formula gives us the answer to Exercise 1.3(c), by the way: $\$200 * (1 + 0.10)^2 = \$242$.

Albert Einstein is sometimes quoted as having said that compound interest is one of the most powerful forces in the universe. This would have been a wonderful quotation, had he actually said it, which there is no evidence that he did. Even so, you can get the power of compounding to work in your favor by saving your money and allowing the interest to accumulate.

Exercise 8.8 Savings Suppose that you put \$100 into a savings account today and that your bank promises a 5 percent annual interest rate.
(**a**) What will your bank's liability be after 1 year?
(**b**) After 10 years?
(**c**) After 50 years?

Finally, let us return to the payday loan establishments.

Exercise 8.9 Payday loans, cont. Imagine that you borrow \$61 from a payday loan establishment. After one week, it wants the principal plus 10 percent interest back. But you will not have that kind of money; so, instead, you go to another establishment and borrow the money you owe the first one. You do this for one year. Interest rates do not change from one establishment to another, or from one week to another.
(**a**) How much money will you owe at the end of the year?
(**b**) What is the total amount of interest that you will owe at the end of the year, in dollar terms?
(**c**) What is the total amount of interest that you will owe at the end of the year, expressed as a fraction of the principal?
(**d**) What does this tell you about the wisdom of taking out payday loans?

Payday loan establishments have generated controversy, with some state legislatures capping the interest rates they may charge. Such controversy is not new. For much of the Christian tradition, it was considered a crime against nature to charge interest on loans, that is, to engage in **usury**. According to Dante's *Divina Commedia*, usurers are condemned to eternal suffering in the seventh circle of hell in the company of blasphemers and sodomites; you know where Dante would have expected to find payday loan officers. On the other hand, payday loan establishments provide a service that (rational and well-informed) people may have reason to demand. Parents may be willing to pay a premium to have money for Christmas presents at Christmas rather than in January, for example.

Either way, as these exercises show, knowing the basics about interest rates can be enormously useful. Now, let us return to the theory of decision.

8.3 Exponential discounting

Which is better, $100 today or $100 tomorrow? $1000 today or $1000 next year? The chances are you prefer your money today. There are exceptions – and we will discuss some of these in the next chapter – but typically people prefer money sooner rather than later. This is not to say that tomorrow you will enjoy a dollar any less than you would today. But it is to say that, *from the point of view of today*, the utility of a dollar today is greater than the utility of a dollar tomorrow.

There are many reasons to feel this way. The earlier you get your money, the more options will be available to you: some options may only be available for a limited period, and you can always save your money and go for a later option. In addition, the earlier you get your money, the longer you can save it, and the more interest you can earn. Whatever the reason, when things that happen in the future do not give you as much utility, from the point of view of today, as things that happen today, we say that you **discount the future**. The general term is **time discounting**. The extent to which you discount the future will be treated as a matter of personal preference. We will call it **time preference**.

There is a neat model that captures the basic idea that people prefer their money sooner rather than later: the model of **exponential discounting**. Suppose that $u > 0$ is the utility you derive from receiving a dollar today. From your current point of view, as we established, the utility of receiving a dollar tomorrow is less than u. We capture this by multiplying the utility of receiving a dollar now by some fraction. We will use the Greek letter delta (δ) to denote this fraction, which we call the **discount factor**. Thus, from your current point of view, a dollar tomorrow is worth $\delta * u = \delta u$. As long as $0 < \delta < 1$, as we generally assume, this means that $\delta u < u$. Hence, today you will prefer a dollar today to a dollar tomorrow, as expected. From the point of view of today, a dollar the day after tomorrow will be worth $\delta * \delta * u = \delta^2 u$. Because $\delta^2 u < \delta u$, today you prefer a dollar tomorrow to a dollar the day after tomorrow, again as expected.

In general, we want to be able to evaluate a whole sequence of utilities, that is, a **utility stream**. Letting t represent time, we will use $t = 0$ to represent today, $t = 1$ to represent tomorrow, and so on. Meanwhile, we let u_t denote the utility you receive at time t, so that u_0 represents the utility you receive today, u_1 represents the utility you receive tomorrow, and so on. We write $U^t(\mathbf{u})$ to denote the utility of stream $\mathbf{u}$ from time t. The number we seek is the utility $U^0(\mathbf{u})$ of the entire utility stream $\mathbf{u} = \langle u_0, u_1, u_2, \ldots \rangle$.

Definition 8.10 The delta function *According to the* ***delta function****, the utility $U^0(\mathbf{u})$ of utility stream $\mathbf{u} = \langle u_0, u_1, u_2, \ldots \rangle$ from the point of view of $t = 0$ is*

$$U^0(\mathbf{u}) = u_0 + \delta u_1 + \delta^2 u_2 + \delta^3 u_3 + \ldots$$
$$= u_0 + \sum_{i=1}^{\infty} \delta^i u_i$$

Thus, you evaluate different utility streams by adding the utility you would receive now, δ times the utility you would receive the next round, δ^2 times the utility you would receive in the round after that, and so on. The resulting model is called the **delta model.**

Utility streams can be represented in table form, as in Table 8.2. An empty cell means that the utility received at that time is zero. In order to compute the utility from the point of view of time zero, or any other point in time, you just need to know the discount factor δ. As soon as we are given the discount factor, we can use Definition 8.10 to determine which option you should choose.

Table 8.2 Simple time-discounting problem

	$t=0$	$t=1$	$t=2$
a	1		
b		3	
c			4
d	1	3	4

Example 8.11 Exponential discounting Suppose that $\delta = 0.9$, and that each utility stream is evaluated from $t = 0$. If so, $U^0(\mathbf{a}) = u_0 = 1$, $U^0(\mathbf{b}) = \delta u_1 = 0.9 * 3 = 2.7$, $U^0(\mathbf{c}) = \delta^2 u_2 = 0.9^2 * 4 = 3.24$, and $U^0(\mathbf{d}) = u_0 + \delta u_1 + \delta^2 u_2 = 1 + 2.7 + 3.24 = 6.94$. Hence, if given the choice between all four alternatives, you would choose **d.** If given the choice between **a**, **b**, and **c**, you would choose **c**.

Exercise 8.12 Exponential discounting, cont. Suppose instead that $\delta = 0.1$.
(**a**) Compute the utility of each of the four utility streams from the point of view of $t = 0$.
(**b**) What would you choose if given the choice between all four?
(**c**) What if you had to choose between **a**, **b**, and **c**?

As these calculations show, your discount factor can have a dramatic impact on your choices. If your discount factor is high (that is, close to one) what happens in future periods matters a great deal. That is to say that you exhibit **patience**: you do not discount your future very much. If your discount factor is low (that is, close to zero) what happens in the future matters little. That is to say that you exhibit **impatience**: you discount your future heavily. It should be clear how the value of δ captures your time preference.

Exercise 8.13 The ant and the grasshopper According to the fable, the grasshopper chirped and played all summer while the ant was collecting food. When winter came, the ant had plenty of food but the grasshopper died from hunger. What can you surmise about their deltas?

Economists believe discount factors can be used to explain a great deal of behavior (Figure 8.1). If your discount factor is low, you are more likely to spend money, procrastinate, do drugs, and have unsafe sex. If your discount factor is high, you are more likely to save money, plan for the future, say no to drugs, and use protection. Notice that this line of thought makes all these behaviors at least potentially rational. For somebody who discounts the future enough, there is nothing irrational about nurturing a crack-cocaine habit. In fact, Gary Becker (whom we came across in Section 2.8) is famous, among other things, for defending a theory of **rational addiction**.

Exercise 8.14 Discount factors For each of the following, identify whether the person's δ is likely to be high (as in close to one) or low (as in close to zero):
(**a**) A person who raids his trust fund to purchase a convertible.
(**b**) A person who enrolls in an MD/PhD program.
(**c**) A person who religiously applies sunscreen before leaving the house.
(**d**) A person who skips her morning classes to go rock climbing.
(**e**) The Iroquois Native American who requires that every deliberation must consider the impact on the seventh generation.

Figure 8.1 Time preference. Illustration by Cody Taylor

Exercise 8.15 The impartial spectator Adam Smith's *Theory of Moral Sentiments* made a big deal of the differences between an "impartial spectator" and our actual selves. An impartial spectator, Smith wrote, "does not feel the solicitations of our present appetites. To him the pleasure which we are to enjoy a week hence, or a year hence, is just as interesting as that which we are to enjoy this moment." When it comes to our actual selves, by contrast: "The pleasure which we are to enjoy ten years hence interests us so little in comparison with that which we may enjoy to-day." If we interpret the difference in terms of deltas, what would this entail when it comes to the deltas of (**a**) the impartial spectator, and (**b**) our actual selves?

Discounting can usefully be represented graphically. We put time on the x-axis and utility on the y-axis. A bar of height u at time t represents a reward worth u utiles to you when you get it at time t. A curve represents how much receiving the reward at t is worth to you from the point of view of times before t. As we know from Definition 8.10, that is δu at $t-1$, $\delta^2 u$ at $t-2$, and so on. As a result, we end up with a picture like Figure 8.2. As you move to the left from t, the reward becomes increasingly distant, and therefore becomes worth less and less to you in terms of utility.

We can use this graphical device to represent the difference between people with high and low δs. If δ is high, δu will not differ much from u, and the curve will be relatively flat: it will only approach the x-axis slowly as you move to the left in the diagram, like the dashed curve in Figure 8.2. If δ is low, δu will be much lower than u, and the curve will be relatively steep: it will approach the x-axis quickly as you move to the left in the diagram, like the dot-dashed curve in the figure. Again, if δ is high, the person does not discount the future very much, and the curve is flat; if d is low, the person does discount the future a great deal, and the curve is steep.

So far, we have used our knowledge of δ to determine a person's preferences over utility streams. Definition 8.10 also permits us to go the other way, as it were. Knowing the person's preferences over utility streams, we can determine the value of her discount factor.

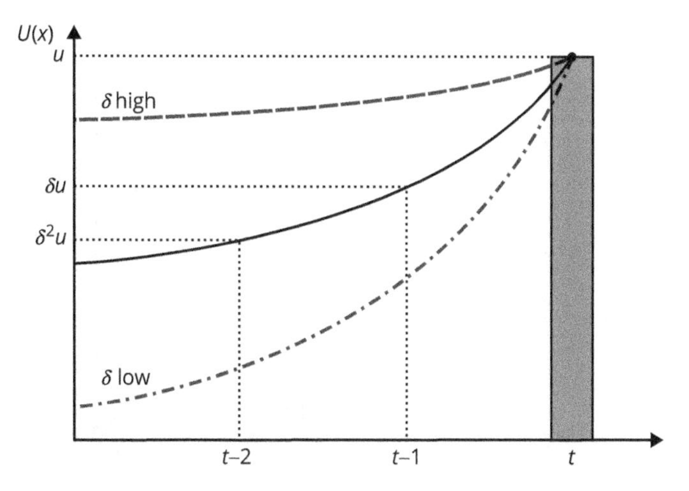

Figure 8.2 Exponential discounting

Example 8.16 Indifference Suppose that Alexandra, at time zero, is indifferent between utility streams **a** (2 utiles at $t = 0$) and **b** (6 utiles at $t = 1$). What is her discount factor δ?

Given that Alexandra is indifferent between **a** and **b** at time zero, we know that $U^0(\mathbf{a}) = U^0(\mathbf{b})$, which implies that $2 = 6\delta$ which is to say that $\delta = 2/6 = 1/3$.

When experimental economists study time discounting in the laboratory, they rely heavily on this kind of calculation. As soon as a laboratory subject is indifferent between an immediate and a delayed reward, his or her discount factor can easily be estimated.

Exercise 8.17 A stitch in time "A stitch in time saves nine," people say when they want you to do something now rather than later. But not everyone will be swayed by that sort of concern. Suppose that you can choose between one stitch at time zero and nine stitches at time one, and that each stitch gives you a utility of -1. What does it take for you to prefer the one stitch now?

Indifference can be represented graphically. Indifference between options **a** and **b** in Example 8.16 means that the picture would look like Figure 8.3. It can easily be ascertained that a strict preference for **a** over **b** would imply that $\delta < 1/3$ and that a strict preference for **b** over **a** would imply that $\delta > 1/3$.

Exercise 8.18 Use Figure 8.3 to answer the following questions:
(**a**) If $\delta < 1/3$, what would the curve look like?
(**b**) What if $\delta > 1/3$?

Exercise 8.19 This exercise refers to the utility streams in Table 8.2. For each of the following people, compute δ.
(**a**) At $t = 0$, Ahmed is indifferent between utility streams **a** and **b**.
(**b**) At $t = 0$, Bella is indifferent between utility streams **b** and **c**.
(**c**) At $t = 0$, Cathy is indifferent between utility streams **a** and **c**.
(**d**) At $t = 1$, Darrence is indifferent between utility streams **b** and **c**.

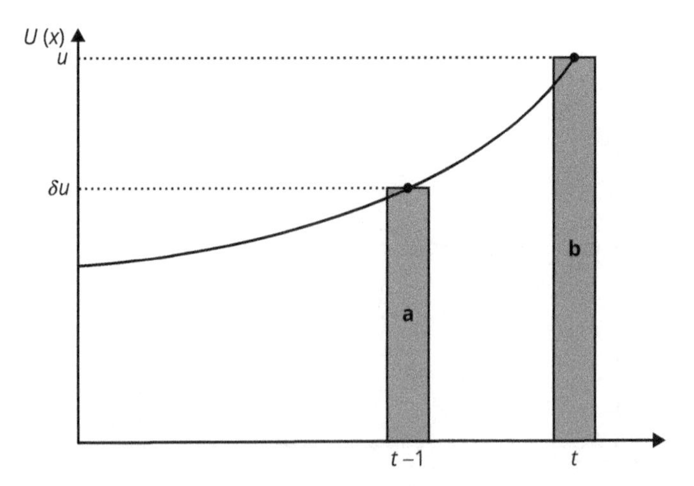

Figure 8.3 Exponential discounting with indifference

Exercise 8.20 For each of the three decision problems in Table 8.3, compute δ on the assumption that a person is indifferent between **a** and **b** at time zero.

Table 8.3 Time-discounting problems

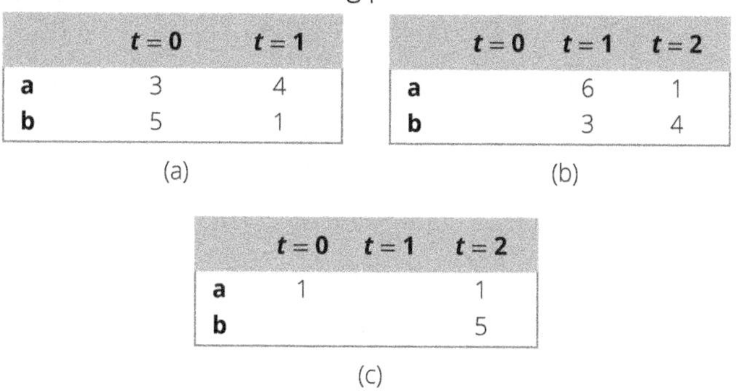

	$t = 0$	$t = 1$
a	3	4
b	5	1

(a)

	$t = 0$	$t = 1$	$t = 2$
a		6	1
b		3	4

(b)

	$t = 0$	$t = 1$	$t = 2$
a	1		1
b			5

(c)

Exercise 8.21 As a financial advisor, you offer your clients the possibility to invest in an asset that generates a utility stream of 1 utile this year ($t = 0$), 0 utiles next year ($t = 1$), and 1 utile the year after that ($t = 2$). For each of the following clients, determine their δ:

(**a**) Client P is indifferent between the investment and 2 utiles next year.
(**b**) Client Q is indifferent between the investment and 1 utile this year.
(**c**) Client R is indifferent between the investment and 1.25 utiles this year.

All the decision problems we have encountered so far in this chapter were defined by matrices of utilities. Very often, however, the relevant outcomes are given in dollars, lives saved, or the like. Given the appropriate utility function, however, we know how to deal with those problems too.

Example 8.22 Suppose you are indifferent between dollar streams **a** and **b** in Table 8.4(a). Your utility function is $u(x) = \sqrt{x}$. What is your δ?

Given the utility function, Table 8.4(a) can be converted into a matrix of utilities as in Table 8.4(b). We can compute δ by setting up the following equation: $3 + \delta 2 = 1 + \delta 5$. Or we can simply set up the following equation: $\sqrt{9} + \delta\sqrt{4} = \sqrt{1} + \delta\sqrt{25}$. Either way, the answer is $\delta = 2/3$.

Exercise 8.23 Suppose instead that the utility function is $u(x) = x^2$. What would Table 8.4(b) look like, and what would δ be?

Table 8.4 Time-discounting problem (in dollars and utiles)

	$t = 0$	$t = 1$
a	\$9	\$4
b	\$1	\$25

(a) In dollar terms

	$t = 0$	$t = 1$
a	3	2
b	1	5

(b) In utility terms

Discount rates

Sometimes discounting is expressed in terms of a **discount rate** r rather than a discount factor δ. The conversion is easy:

$$r = \frac{1-\delta}{\delta}$$

You can confirm that when $\delta = 1$, $r = 0$ and that as δ approaches zero, r increases. Knowing r, you can compute δ as follows:

$$\delta = \frac{1}{1+r}$$

In this text, I favor discount factors over discount rates. But it is useful to know what both are.

8.4 What is the rational delta?

In the above, I have followed what is common practice and treated the value of the discount factor δ (and discount rate r) as a mere preference. On this approach, a person's discount factor is just a personal matter, much like your preference for blueberry ice-cream over raspberry ice-cream, or your preference for ice-cream over LSD. In other words, rationality does not require you to have one delta rather than another. One implication of this analysis, as we saw in the previous section, is that destructive behaviors such as heavy drug use are perfectly consistent with rationality: if your discount factor is very low, aiming for instant gratification is perfectly rational in this analysis. Rationality does require you to consistently apply one discount factor, however. You cannot rationally plan for the future now and then act impulsively later; you cannot rationally act as though you have a high delta now and a low delta later. An inconsistent drug user is indeed irrational.

Historically, a number of thinkers have disagreed, and instead argued that time discounting (with $\delta < 1$ and $r > 0$) arises from some intellectual or moral deficiency. The economist A. C. Pigou, commonly considered the father of welfare economics, wrote that "this preference for present pleasures … implies only that our telescopic faculty is defective, and that we, therefore, see future pleasures, as it were, on a diminished scale." The polymath Frank P. Ramsey, who made path-breaking contributions to philosophy, statistics, and economics – his theory of saving is still taught in graduate-level macroeconomics courses – before dying at the age of 26, called time discounting "a practice which is ethically indefensible and arises merely from the weakness of the imagination." Some describe time discounting as a sort of crime committed by the current you against the future you – who is, they add, the person who should matter the most to you now. If these thinkers are right, repairing our intellectual and moral deficiencies would return us to a delta of one – and the behavior of the impartial spectator.

Others have advocated discounting the future. "Seize the day," or *carpe diem*, said the ancient poet Horace – and Robin Williams in *Dead Poets Society* – encouraging us to live in the present. Seneca, whom we came across in our discussion of the aspiration treadmill in Section 3.5, wrote:

> Can anything be more thoughtless than the judgement of those men who boast of their forethought? ... They organize their thoughts with the distant future in mind; but the greatest waste of life consists in postponement: that is what takes away each day as it comes, that is what snatches away the present while promising something to follow. The greatest obstacle to living is expectation, which depends on tomorrow and wastes today.

This passage appears in a treatise titled *On the Shortness of Life*, in which Seneca argues that life is long enough – the problem being that so much of it is squandered. In the same spirit it is not uncommon for people to volunteer the advice that you should live every moment as though it were your last. The late Apple founder Steve Jobs was one of them: "I have looked in the mirror every morning and asked myself: 'If today were the last day of my life, would I want to do what I am about to do today?'" Then again, Harvard psychologist Daniel Gilbert quips that such a comment "only goes to show that some people would spend their final [moments] giving other people dumb advice." (Some of these figures might have been talking about uncertainty too, since the future is not only removed in time, it is also as yet unrealized.)

Pioneering behavioral economist (and 1978 Nobel laureate) Herbert Simon offered another argument for discounting the future. In his view, the virtue of discounting is that it frees us from the burden of deliberating about remote events. When we discount the future, far-off events become immaterial to current decisions and can safely be ignored in our deliberations. And this is a good thing, according to Simon, in light of inevitable limitations of our cognitive powers. In his words: "If our decisions depended equally upon their remote and their proximate consequences, we could never act but would be forever lost in thought."

There is some empirical evidence that having a high discount factor is good for a person in the long term. The 2014 book *The Marshmallow Test* reviews a series of studies performed with children as young as four, who are given the choice between one marshmallow (or similar treat) now, or two such treats if they are able to wait for some 15 minutes. Since the first experiment in the series was performed in the 1960s, experimenters tracked the performance of the original participants in a variety of domains of life. Apparently, children who were able to defer gratification in preschool exhibited better concentration, intelligence, self-reliance, and confidence in adolescence and were better able to pursue and reach long-term goals, enjoyed higher educational attainment, and had a lower body-mass index in adulthood. (Replications have found smaller, but still significant, effects.) These findings may strike you as an argument for having a high delta. Then again, if your delta is low now, it is not as though you will care about adverse long-term consequences of time discounting – or anything else.

8.5 Discussion

In this chapter we have explored issues of interest – simple and compound – as well as exponential discounting. The model of exponential discounting has a remarkable feature in that it is the only model of discounting that entails time consistency. Consistency is an important topic in intertemporal decision-making, and we will return to it in Section 9.2. Though relatively simple, exponential discounting offers an extraordinarily powerful model. For this reason, it is critical to a variety of fields, including in cost–benefit analysis, to assess the costs and benefits of deferred outcomes, and in finance, to determine the present value of alternative investments. While the model might seem intuitively appealing and uncontroversial from both a descriptive and a normative perspective, it is anything but.

In Section 8.4, we saw that a person's discount factor delta is typically treated as a mere preference: as long as you are consistent in your attitudes to the future, rationality is consistent with any delta. In some contexts, however, we do not have the luxury of sidestepping the issue. When computing the costs and benefits of long-term investments, for example, some delta just has to be assumed and the specific value can matter hugely. Many investments (like railroads) have large upfront costs and benefits that extend into the indefinite future. If costs and benefits are computed with a low delta, the former will dominate and the project will be a no-go; if they are computed with a high delta, the latter will dominate, and the project will be a go. And there is simply no obvious, uncontroversial way to choose a number. Perhaps most prominently, the topic comes up in the discussion of climate change. On the assumption that taking action to prevent or mitigate the negative consequences of climate change would be costly in the short term but (at least potentially) beneficial in the long term, cost–benefit analysis with a high enough δ will favor taking action, but cost–benefit analysis with a low enough δ will favor not taking action. Thus, the rationality of taking action to prevent or mitigate the effects of climate change depends on the value of δ – and, again, there is no agreement whatsoever about the appropriate value of the parameter.

In the next chapter, we will discuss to what extent the exponential model of time discounting captures the manner in which people make decisions and whether it is appropriate as a normative standard.

Additional exercises

Exercise 8.24 Youth sports In a 2014 interview, basketball star Kobe Bryant discussed the importance of making sports fun for young people. "It's hard to tell a kid that you need to get out there and compete because it's going to decrease your chance of having diabetes 30–40 years from now," he said. In the language of time discounting, why are young people not motivated by the prospect of being healthy in 30–40 years?

Exercise 8.25 Credit scores A study in the journal *Psychological Science* by two economists from the Federal Reserve Bank in Boston found a connection between people's discount factor δ and their credit score. A credit score is a number representing a person's creditworthiness: the higher the score the better. Based on your understanding of discounting behavior:

(**a**) What is the connection? That is, do people with higher discount factors tend to have higher credit scores, or the other way around?

(**b**) Explain why that would be so.

Exercise 8.26 Heaven can wait Researchers have explored the possibility that a religious upbringing helps shape discount rates. A 2013 study starts off with the following stylized facts: (1) Calvinism discourages immediate consumption but encourages long-term accumulation, whereas Catholicism frowns equally upon both. (2) For Calvinists, even a single act of sin might signal that the sinner is not predestined to salvation, whereas the Catholic cycle of sin–confession–expiation underscores the possibility of forgiveness. The authors predicted that the discount rates of Dutch Calvinists, Italian Catholics, and atheists from both countries would differ significantly, and the prediction found support in the data. What did the researchers predict?

Exercise 8.27 The afterlife Some people believe in an afterlife where wrongs are righted and good behavior rewarded. What sort of time preference should we expect such believers to reveal in their behavior?

Exercise 8.28 Time discounting and interest rates Whether you should spend or save will depend not just on your time preference, but on the interest you can get when putting your money in a savings account. Suppose you have the option of spending $\$w$ now or saving it until next year. If you save it, the bank will pay you interest rate i.

(**a**) Suppose that your utility function is $u(x) = x$. What does your discount factor δ need to be for you to be indifferent between spending it now and saving it? What about your discount rate r (see the text box on page 188)?

(**b**) Suppose that your utility function is $u(x) = \sqrt{x}$. What does your discount factor δ need to be for you to be indifferent between spending it now and saving it? What about your discount rate r?

Further reading

Mas-Colell et al. (1995, Chapter 20) offer a more advanced discussion of intertemporal utility. The battle over payday loan establishments is covered by the *Wall Street Journal* (Anand, 2008). Smith (2002 [1759], pp. 221–2) compares our actual selves to the impartial spectator. Pigou (1952 [1920], p. 25) discusses our flawed telescopic faculty, and Ramsey (1928, p. 543) our weakness of imagination. Seneca (2007 [*c* 49], p. 148) and Jobs (2005) advocate living in the present, while Gilbert (2006, p. xiii) quips. *The Marshmallow Test* is Mischel (2014); the replication is due to Watts et al. (2018). Kobe Bryant is quoted in Shelburne (2014). The study from the Boston Fed is Meier and Sprenger (2012). The paper on religion and discounting is Paglieri et al. (2013).

9 INTERTEMPORAL CHOICE

Learning objectives

After studying this chapter you will:

- Be able to identify common behavior patterns that violate the discounted utility model
- Know multiple behavioral models of intertemporal choice, and when to apply them
- Apply the discounted utility model of intertemporal choice in the real world – but also understand its limitations as a model of rationality

9.1 Introduction

As the examples in the previous chapter suggest, the model of exponential discounting can be used to accommodate a variety of behavior patterns. For this reason, and because of its mathematical tractability, the model is heavily relied upon in a variety of disciplines. Yet it fails to capture some of the things people do. In this chapter we focus on a couple of phenomena that are not easily captured by the standard model. One is that people are time inconsistent; that is, their preferences appear to change for no reason other than the passing of time, as when a drug addict who woke up in the morning completely determined to clean up his act gives in and takes more drugs in the afternoon. Because people sometimes anticipate time-inconsistent behavior, they choose not to choose; that is, they take action intended to prevent themselves from taking action. Another phenomenon is the fact that people seem to have preferences over utility profiles; that is, they care about the shape of their utility stream, and not just about the (discounted) individual utilities. We will also study how behavioral economists go about capturing these phenomena. Time inconsistency will be captured by means of a model of hyperbolic discounting, a highly versatile model. We will see that the model of hyperbolic discounting does a good job of capturing time inconsistency, but that it is inadequate to account for preferences over profiles.

9.2 Hyperbolic discounting

As we saw in the previous chapter, the exponential discounting model can capture a great deal of behavior, including – perhaps surprisingly – forms of addiction. Yet the image of the rational addict does not sit well with observed behavior and first-person reports of many addicts. As the beat poet William S. Burroughs writes in the prologue to his autobiographical novel *Junky*:

> The question is frequently asked: Why does a man become a drug addict? The answer is that he usually does not intend to become an addict. You don't wake up one morning and decide to be a drug addict ... One morning you wake up sick and you're an addict.

Later in this chapter, we will discuss other behaviors that are hard to reconcile with the model of exponential discounting.

Exponential discounting implies the agent's behavior is **time consistent**, meaning that his or her preferences over two options do not change simply because time passes. If you are time consistent and feel (today) that **a** is better than **b**, then you felt the same way about **a** and **b** yesterday, and you will feel the same way about them tomorrow.

It is relatively easy to prove that anybody who discounts the future exponentially will be time consistent. But first we will need to review some notation. We continue to let $U^t(\mathbf{a})$ denote the utility from the point of view of time t of receiving some utility stream **a**. Let u_t refer to the utility received at time t. Then, $U^t(u_{t'})$ refers to the utility, from the point of view of t, of receiving $u_{t'}$ at time t'. As an example, if u_{tomorrow} refers to the utility you will receive tomorrow from eating ice-cream tomorrow, then $U^{\text{today}}(u_{\text{tomorrow}})$ is the utility, from the point of view of today, of eating ice-cream tomorrow. This number would normally be high, but not as high as $U^{\text{tomorrow}}(u_{\text{tomorrow}})$, which is the utility you will receive tomorrow when you eat ice-cream tomorrow.

Suppose you are facing two rewards **a** and **b**, as in Figure 8.3 on page 186. Let us say that **a** gives you u_t at time t, and that **b** gives you u_{t+1} at time $t+1$. Imagine that, from the point of view of time t, you strictly prefer **a** to **b**, that is, $U^t(\mathbf{a}) > U^t(\mathbf{b})$. If so, given that you are an exponential discounter, $U^t(\mathbf{a}) > U^t(\mathbf{b})$ implies that $u_t > \delta u_{t+1}$. Now let us look at what is going on before t, for example, at time $t-1$. Would it be possible to weakly prefer **b** to **a**? If you did, $U^{t-1}(\mathbf{b}) \geq U^{t-1}(\mathbf{a})$, and $\delta^2 u_{t+1} \geq \delta u_t$. Since $\delta > 0$, we can divide by δ on both sides, which gives us $\delta u_{t+1} \geq u_t$, which is a contradiction. So at $t-1$, you must strictly prefer **a** to **b**. What about $t-2$? The same conclusion obtains and for the same reason. What about $t-3$? We could go on.

In brief, if you discount the future exponentially, you must be time consistent. Graphically, what this means is that you either prefer **a** to **b** at all times (as in the dotted line in Figure 9.1), or you prefer **b** to **a** at all times (as in the dashed line in Figure 9.1), or you are indifferent between the two options (as in Figure 8.3 on page 186). At no point will the two lines cross. Your preference over **a** and **b** will never change simply because time passes.

The bad news, from the point of view of this model, is that people violate time consistency with alarming regularity. In the morning, we swear never to touch alcohol again; yet, by the time happy hour comes around, we gladly order another martini. On January 1, we promise our partners that we will stop smoking and start exercising; yet, when opportunity arises, we completely ignore our promises. Time inconsistency is nicely embodied in the figure of Ilya Ilyich Oblomov, who may be the most prominent procrastinator in all of literature. Here is how he appears on the first page of the eponymous book:

> [In] his dark-grey eyes there was an absence of any definite idea, and in his other features a total lack of concentration. Suddenly a thought would

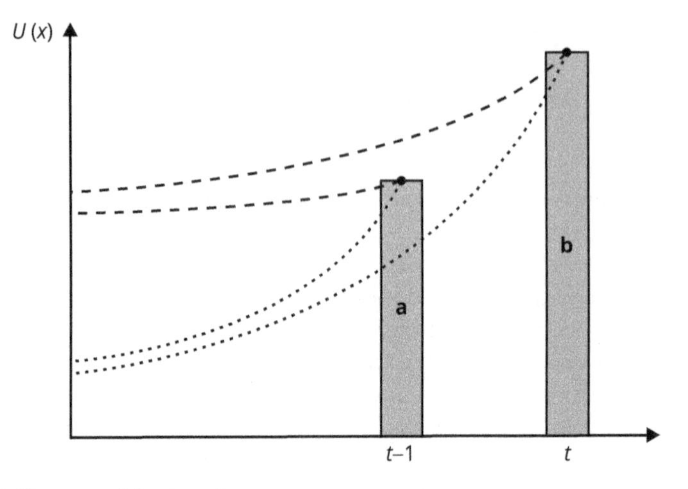

Figure 9.1 Time-consistent preferences

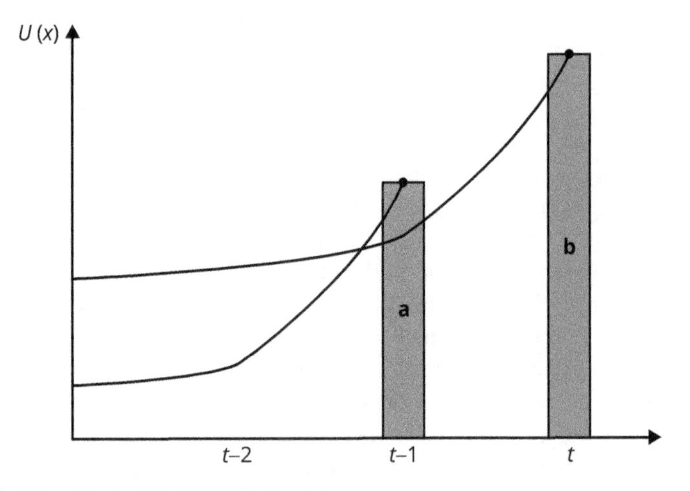

Figure 9.2 Time-inconsistent preferences

> wander across his face with the freedom of a bird, flutter for a moment in his eyes, settle on his half-opened lips, and remain momentarily lurking in the lines of his forehead. Then it would disappear, and once more his face would glow with a radiant insouciance which extended even to his attitude and the folds of his night-robe.

If this is unfamiliar, good for you! Otherwise, you will probably agree that time inconsistency is common. Graphically, we seem to discount the future in accordance with Figure 9.2. That is, at time $t-1$ (and possibly right before it), we want the smaller, more immediate reward; earlier than that, we want the larger, more distant reward.

It turns out that this kind of behavior can be usefully modeled with a slight variation of Definition 8.10 on page 183.

Definition 9.1 The beta–delta function *According to the* ***beta–delta function****, the utility* $U^0(\mathbf{u})$ *of utility stream* $\mathbf{u} = \langle u_0, u_1, u_2, \ldots \rangle$ *from the point of view of* $t = 0$ *is*

$$U^0(\mathbf{u}) = u_0 + \beta\delta u_1 + \beta\delta^2 u_2 + \beta\delta^3 u_3 + \ldots$$
$$= u_0 + \sum_{i=1}^{\infty} \beta\delta^i u_i$$

If you act in accordance with this formula, you evaluate utility streams by adding the utility you would receive now, $\beta\delta$ times the utility you would receive the next round, $\beta\delta^2$ times the utility you would receive in the round after that, and so on. The only difference relative to the exponential discounting function is that all utilities except u_0 are multiplied by an additional β, which is assumed to be a number such that $0 < \beta \leq 1$. Notice that while δ is raised to higher powers (δ, δ^2, δ^3, …) for later rewards, β is not. This form of discounting is called **quasi-hyperbolic discounting**. Here, I loosely refer to it as **hyperbolic discounting**. The resulting model is called the **beta–delta model**.

The introduction of the parameter β makes an interesting difference. When $\beta = 1$, an agent who discounts the future hyperbolically will behave exactly like an agent who discounts the future exponentially. For, if $\beta = 1$, the hyperbolic discounter will maximize the following expression:

$$U^0(\mathbf{u}) = u_0 + \beta\delta u_1 + \beta\delta^2 u_2 + \ldots = u_0 + \delta u_1 + \delta^2 u_2 + \ldots$$

which is identical to the delta function (Definition 8.10). When $\beta < 1$, however, things are different. In this case, all outcomes beyond the present time get discounted more than under exponential discounting, as shown in Figure 9.3. Compare this figure with Figure 8.2 on page 185. As you can tell, the hyperbolic curve is relatively steep between t and $t-1$, and relatively flat before $t-1$.

Exercise 9.2 The beta–delta function Suppose that you are facing a utility stream of 1 utile at $t = 0$, 3 utiles at $t = 1$, and 9 utiles at $t = 2$. For each of the following parameter values, apply the beta–delta function to determine the discounted utility of the stream.
(**a**) $\beta = 1/3$ and $\delta = 1$.
(**b**) $\beta = 1$ and $\delta = 2/3$.
(**c**) $\beta = 1/3$ and $\delta = 2/3$.

When an agent discounts the future hyperbolically, if given a choice between a smaller, earlier reward and a bigger, later reward, the picture may well end up looking like

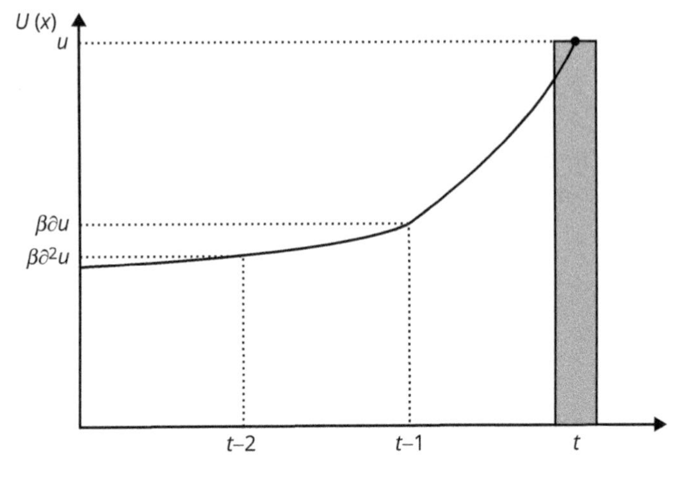

Figure 9.3 Hyperbolic discounting

Figure 9.2. The result is what Russians call "Oblomovitis" – in the novel, and sometimes in real life, a deadly condition. The fact that hyperbolic discounting may lead to time-inconsistent behavior can be shown algebraically, too.

Example 9.3 Dieting today and tomorrow Suppose that you are on a diet, but have to decide whether to have a slice of red-velvet cake at a party some random Saturday. Eating the cake would give you a utility of 4. If you have the cake, however, you will have to exercise for hours on Sunday, which would give you a utility of 0. The other option is to skip the cake, which would give you a utility of 1, and to spend Sunday relaxing in front of the television, for a utility of 6. Thus, you are facing the choice depicted in Table 9.1. You discount the future hyperbolically, with $\beta = 1/2$ and $\delta = 2/3$. Questions:

(**a**) From the point of view of Friday, what is the utility of eating the cake (**c**) and of skipping it (**d**)? Which would you prefer?

(**b**) From the point of view of Saturday, what is the utility of eating the cake and of skipping it? Which would you prefer?

Table 9.1 Red-velvet problem

	Saturday	Sunday
c	4	0
d	1	6

(**a**) From the point of view of Friday, Friday is $t = 0$, Saturday is $t = 1$, and Sunday is $t = 2$. From this point of view, eating the cake is associated with utility stream $\mathbf{c} = \langle 0, 4, 0 \rangle$ and not eating the cake is associated with utility stream $\mathbf{d} = \langle 0, 1, 6 \rangle$. Consequently, from the point of view of Friday, the utility of eating the cake is

$$U^0(\mathbf{c}) = 0 + 1/2 * 2/3 * 4 + 1/2 * (2/3)^2 * 0 = 4/3$$

Meanwhile, from the point of view of Friday, the utility of skipping the cake is

$$U^0(\mathbf{d}) = 0 + 1/2 * 2/3 * 1 + 1/2 * (2/3)^2 * 6 = 5/3$$

From the point of view of Friday, therefore, you will prefer to skip the cake and stick to your diet.

(**b**) From the point of view of Saturday, Saturday is $t = 0$ and Sunday is $t = 1$. From this point of view, eating the cake is associated with utility stream $\mathbf{c} = \langle 4, 0 \rangle$ and not eating the cake is associated with utility stream $\mathbf{d} = \langle 1, 6 \rangle$. Consequently, from the point of view of Saturday, the utility of eating the cake is

$$U^0(\mathbf{c}) = 4 + 1/2 * 2/3 * 0 = 4$$

Meanwhile, from the point of view of Saturday, the utility of skipping the cake is

$$U^0(\mathbf{d}) = 1 + 1/2 * 2/3 * 6 = 3$$

From the point of view of Saturday, therefore, you will prefer to eat the cake.

This example shows time inconsistency at work. Ahead of time, you prefer to stick to your diet and resolve to refrain from having the cake. And yet, when the opportunity arises, you prefer to ignore the diet and eat the cake. This means that you are exhibiting **impulsivity**. If impulsivity is not familiar to you, you belong to a small and lucky subsample of humanity. As the example shows, you can be impulsive and impatient at the same time. The next exercise illustrates the interaction between the two.

Exercise 9.4 Impulsivity and impatience Suppose you are offered the choice between option **a** (8 utiles on Thursday) and **b** (12 utiles on Friday).

(**a**) Assume that $\beta = 1$ and that $\delta = 5/6$. From the point of view of Thursday, which one would you choose? From the point of view of Wednesday, which one would you choose?

(**b**) Assume that $\beta = 1$ and that $\delta = 1/6$. From the point of view of Thursday, which one would you choose? From the point of view of Wednesday, which one would you choose?

(**c**) Assume that $\beta = 1/2$, and that $\delta = 1$. From the point of view of Thursday, which one would you choose? From the point of view of Wednesday, which one would you choose?

(**d**) Assume that $\beta = 1/2$, and that $\delta = 2/3$. From the point of view of Thursday, which one would you choose? From the point of view of Wednesday, which one would you choose?

Hyperbolic discounting can account not just for the fact that people emphasize their present over their future well-being, but also that they change their minds about how to balance the present versus the future. Thus, it can account for the fact that people fully intend to diet, stop smoking, do homework, and quit drugs, and then completely fail to do so. Here is another example.

Exercise 9.5 Cancer screening Most colon cancers develop from polyps. Because early screening can detect polyps before they become cancerous and colon cancer in its early stages, many doctors advise patients over a certain age to have a colonoscopy. Unfortunately, colonoscopies are experienced as embarrassing and painful. The typical person, when young, resolves to have a colonoscopy when older, but changes his or her mind as the procedure approaches. Assume that two patients, Abelita and Benny, have the choice between the following: (**a**) having a colonoscopy at time 1 (utility = 0) and being healthy at time 2 (utility = 18), and (**b**) avoiding the colonoscopy at time 1 (utility = 6) and be unhealthy at time 2 (utility = 0).

Abelita discounts the future exponentially. Her $\delta = 2/3$.

(**a**) At $t = 0$: What is her utility of **a**? What is her utility of **b**?
(**b**) At $t = 1$: What is her utility of **a**? What is her utility of **b**?

Benny discounts the future hyperbolically. His $\beta = 1/6$ and his $\delta = 1$.

(**c**) At $t = 0$: What is his utility of **a**? What is his utility of **b**?
(**d**) At $t = 1$: What is his utility of **a**? What is his utility of **b**?
(**e**) Who acts more like the typical patient?
(**f**) Who is more likely to end up with health issues?

The beta–delta function also permits you to go the other way. Knowing a person's preferences, the function permits you to compute their beta and/or delta. Consider the following exercise.

Exercise 9.6 Suppose that you discount the future hyperbolically, that is, in accordance with the beta–delta function, and that from the point of view of Thursday you are indifferent between options **a** (1 utile on Thursday) and **b** (3 utiles on Friday).

(**a**) If $\beta = 1/2$, what is δ?
(**b**) If $\delta = 4/9$, what is β?

Exercise 9.7 Suppose that you discount the future hyperbolically. Assume that both β and δ are strictly greater than zero but strictly smaller than one. At $t = 0$, you are given the choice between the following three options: **a** (1 utile at $t = 0$), **b** (2 utiles at $t = 1$), and **c** (3 utiles at $t = 2$). As a matter of fact, at $t = 0$ you are indifferent between **a** and **b** and between **b** and **c**.

(**a**) Compute β and δ.
(**b**) Suppose, in addition, that at $t = 0$ you are indifferent between **c** and **d** (x utiles at $t = 3$). What is x?

Exercise 9.8 Suppose that you discount the future hyperbolically. Assume that both β and δ are strictly greater than zero but strictly smaller than one. At $t = 0$, you are given the choice between the following three options: **a** (2 utiles at $t = 0$), **b** (5 utiles at $t = 1$), and **c** (10 utiles at $t = 2$). As a matter of fact, at $t = 0$ you are indifferent between **a** and **b** and between **b** and **c**. Compute β and δ.

Exercise 9.9 Wicksteed's blanket The theologian and economist Philip Wicksteed offers the following observation: "[We] lie awake (or what we call awake next morning) half the night consciously suffering from cold, when even without getting out of bed we could reach a blanket or a rug which would secure comfortable sleep for the rest of the night." Suppose that staying in a freezing bed gives Wicksteed 1 utile now

($t = 0$), 1 utile in the middle of the night ($t = 1$), and 1 utile in the early morning ($t = 2$). If he reached for the blanket, he would suffer 0 utiles now but enjoy 5 utiles in the middle of the night and in the early morning. A patient man, Wicksteed might have had a delta of one. What is his beta, given that he does *not* reach for the blanket?

Just one more thing, before we move on. Procrastination may not be all bad. As philosopher John Perry has pointed out, procrastinators put things off – but they are not idle. The problem is that they focus on marginally useful things like watering plants and sharpening pencils instead of whatever important task they are supposed to complete. Perry's "discovery," as he calls it, is that you can make this tendency work for you by structuring the tasks on your to-do list properly. The key to accomplishing task X is to always have another, more important one Y at the top of your list. Then you can use X as a means to avoid doing Y – thereby getting a lot done even while procrastinating. People often think they will get more done and miss fewer deadlines if they take on fewer tasks, but on Perry's analysis this gets things exactly backwards: the key to getting important stuff done is to take on even more, and more important tasks. For the discovery of **structured procrastination**, Perry was awarded the "Ig Nobel" Prize for Literature in 2011.

The next two sections contain more exercises on hyperbolic discounting.

9.3 Choosing not to choose

Another feature of human behavior is that we sometimes choose not to choose, in the sense that we take action to prevent ourselves from taking action. As recounted by Homer, Ulysses famously allowed himself to be tied to the mast of his ship so that he could listen to the sweet but seductive song of the sirens without risking doing anything stupid (see Figure 9.4). The rest of us are willing to pay a premium to buy snacks in small packages, soft drinks from overpriced vending machines, and beer in small quantities. Though we know that we could save money by buying in bulk, we fear that doing so would lead to overindulgence, leaving us overfed, drunk, and no better off financially. In a well-known study about procrastination and precommitment, executive-education students were allowed to set their own deadlines for three required papers. At the beginning of the term almost three-quarters, 73 percent, set deadlines that fell before the last week of class, even though they knew missed deadlines would lead to lower grades. Apparently the students were so afraid they would not get the work done without deadlines and external penalties that they were willing to take the risk.

This kind of behavior is theoretically puzzling. From the point of view of exponential discounting, choosing not to choose makes no sense. According to this model, if you do not want to indulge now, you will not want to indulge later either. But such behavior is not obviously entailed by the hyperbolic discounting model either. If you discount your future hyperbolically, you may very well plan not to overindulge but then overindulge anyway. Thus, neither one of the models that we have explored so far is appropriate for capturing common behaviors such as those above.

Behavioral economists approach the issue by drawing a distinction between **naive** and **sophisticated** hyperbolic discounters. When people are time inconsistent

Figure 9.4 Ulysses and the Sirens. Detail of mosaic from Dougga, Tunisia. Photograph by Dennis Jarvis. Used with permission

– meaning that they prefer x to y ahead of time, but y to x when the time arrives – they are said to have **self-control problems**. Naive time-inconsistent individuals – or *naifs*, for short – are unaware of their self-control problems. Naifs make their choices based on the inaccurate assumption that their future preferences will be identical to their current preferences. Sophisticated time-inconsistent individuals – or *sophisticates*, for short – are aware of their self-control problems. Sophisticates make their choices based on accurate predictions of their future behavior. An example might help.

It is October and you want to save for presents for the upcoming holiday season. You know that saving in October and November allows you to get some really nice presents for your loved ones in December. Saving and saving again is no fun, but your utility stream 3–3–27 ends on a high note when friends and family open their presents and love you again. You can also save in October, splurge in November, and still end up with decent presents in December, for a utility stream of 3–9–15. It goes without saying that you can just spend and spend, in which case your utility stream is 6–6–9. See Table 9.2(a) but ignore the last line for now.

Table 9.2 Layaway payoff matrices

	Oct	Nov	Dec
save-save	3	3	27
save-spend	3	9	15
spend-spend	6	6	9
layaway	2	3	27

(a) October

	Nov	Dec
save	3	27
spend	9	15

(b) November

Let us first assume that you are an exponential discounter with $\delta = 1$. In October, you will choose **save–save**:

$$U^{\text{Oct}}(\textbf{save–save}) = 3 + 3 + 27 = 33$$

$$U^{\text{Oct}}(\textbf{save–spend}) = 3 + 9 + 15 = 27$$

$$U^{\text{Oct}}(\textbf{spend–spend}) = 6 + 6 + 9 = 21$$

Let us now consider the possibility that you are a hyperbolic discounter with $\delta = 1$ and $\beta = 1/3$. If so, you will also choose **save–save**:

$$U^{\text{Oct}}(\textbf{save–save}) = 3 + 1/3(3 + 27) = 13$$

$$U^{\text{Oct}}(\textbf{save–spend}) = 3 + 1/3(9 + 15) = 11$$

$$U^{\text{Oct}}(\textbf{spend–spend}) = 6 + 1/3(6 + 9) = 11$$

So far so good. The problem is that having saved in October, you may be tempted to splurge in November; now you have savings, after all, that risk burning a hole in your pocket. You can continue to **save** in November and enjoy the expected 3–27 utility stream. Or you can **spend**, in which case you get 9–15. See Table 9.2(b) for the options available to you in November. The exponential discounter, being time consistent, would of course continue to **save**:

$$U^{\text{Nov}}(\textbf{save}) = 3 + 27 = 30$$

$$U^{\text{Nov}}(\textbf{spend}) = 9 + 15 = 24$$

A hyperbolic discounter, however, would **spend**:

$$U^{\text{Nov}}(\textbf{save}) = 3 + 1/3 * 27 = 12$$

$$U^{\text{Nov}}(\textbf{spend}) = 9 + 1/3 * 15 = 14$$

That is, come November, there is no way the hyperbolic discounter would stay on the **save–save** path charted in October. An unsophisticated, or naive, hyperbolic discounter is unable to anticipate his future behavior. Thus, he will get on the wagon in October, choose **save–save**, but fall off it and splurge in November; in spite of the very best intentions, therefore, he will end up with **save–spend**. A sophisticated hyperbolic discounter is able to anticipate her behavior and knows in October that **save–save** is not going to happen. Having eliminated **save–save** from her menu, the sophisticate might as well **spend–spend**, since she is indifferent between the remaining two options.

This is bad news not only for the hyperbolic discounter but also for purveyors of expensive goods; if customers are unable to save for the fancy stuff, nobody will be able to sell it. Department stores and other sellers therefore have every incentive to offer a **layaway** plan: for a small administrative fee of one, payable in October, stores will hold on to customers' savings in November, so as to make spending them then impossible. The fact that department stores offer a layaway plan does not eliminate any of the other options available in October; it merely adds one with a utility stream of 2–3–27, represented by the last line in Table 9.2(a).

No exponential discounter would choose **layaway**, since U^{Oct}(**layaway**) $= 2 + 3 + 27 = 32$, which is inferior to **save–save**. For our hyperbolic discounter, the discounted utility would be U^{Oct}(**layaway**) $= 2 + 1/3\ (3 + 27) = 12$. The unsophisticated hyperbolic discounter, who evaluates the options available in October without consideration of whether he can stick to the plan in November, will note that **save–save** looks better than **layaway** and choose **save–save** in October, fall off the wagon in November, and end up with **save–spend**. The sophisticated hyperbolic discounter, who eliminates **save–save** from her menu knowing that she will be unable to stick to it, finds that **layaway** beats both **save–spend** and **spend–spend**. She signs up for the **layaway** plan, is prevented from splurging in November, and gets the nice presents in December. Notice that the introduction of **layaway**, although inferior to **save–save** in terms of utilities, actually helps the sophisticated hyperbolic discounter do better than she would have done otherwise.

Here is another example of the behavior of an exponential discounter, a naive hyperbolic discounter, and a sophisticated hyperbolic discounter.

Example 9.10 Johnny Depp 1 Your local cinema theater offers a mediocre movie this week ($u_0 = 3$), a good movie next week ($u_1 = 5$), a great movie in two weeks ($u_2 = 8$), and a fantastic Johnny Depp movie in three weeks ($u_3 = 13$). Unfortunately, you must skip one of the four. For all questions below, suppose that $\delta = 1$ and $\beta = 1/2$. Will you skip **a** the mediocre, **b** the good, **c** the great, or **d** the fantastic movie?

If you are an exponential discounter, you will skip the worst movie. At $t = 0$, you know that U^0(**a**) $= 5 + 8 + 13 = 26$ is better than U^0(**b**) $= 3 + 8 + 13 = 24$, which is better than U^0(**c**), and so on. Because you are an exponential discounter, and therefore time consistent, you stick to your plan.

If you are a naive hyperbolic discounter, you will procrastinate until the very last moment and miss the fantastic movie. At $t = 0$, you choose between U^0(**a**) $= 1/2(5 + 8 + 13) = 13$, U^0(**b**) $= 3 + 1/2(8 + 13) = 13.5$, U^0(**c**) $= 3 + 1/2(5 + 13) = 12$, and U^0(**d**) $= 3 + 1/2\ 1/2(5 + 8) = 9.5$. You watch the mediocre movie, fully intending to skip the good one. But at $t = 1$, everything looks different. From there, you no longer have the option to skip the mediocre movie. You choose between U^1(**b**) $= 1/2(8 + 13) = 10.5$, U^1(**c**) $= 5 + 1/2 * 13 = 11.5$, and U^1(**d**) $= 5 + 1/2 * 8 = 9$. You watch the good movie, fully intending to skip the great movie. At $t = 2$, though, you choose between U^2(**c**) $= 1/2 * 13 = 6.5$ and U^2(**d**) $= 8$. Thus, you watch the great movie and at $t = 3$ have no choice but to skip the fantastic movie.

If you are a sophisticated hyperbolic discounter, by contrast, you will skip the good movie. Your sophistication allows you to predict that self-control problems at $t = 2$ would prevent you from watching the fantastic movie. Consequently, the choice at

$t = 1$ is between skipping the good movie for a utility of $U^1(\mathbf{b}) = 10.5$ or else end up with $U^1(\mathbf{d}) = 9$. So you know at $t = 0$ that at $t = 1$ you will choose **b**. At $t = 0$, therefore, the choice is between $U^0(\mathbf{a}) = 13$ and $U^0(\mathbf{b}) = 13.5$. Thus, you will watch the mediocre movie, skip the good one, and watch the great and fantastic ones.

This example shows how sophistication helps people anticipate the problems posed by time-inconsistent behavior. Behavior of this kind is probably common. At night, many people are determined to get up early the next morning, even though they anticipate that tomorrow morning they will want to sleep late. In order to prevent their morning self from sleeping late, therefore, they set alarms and put them on a window ledge across the room, behind a cactus. There are alarms that, when they go off, roll off the bedside table and under the bed or across the room, forcing your morning self to get up and chase them down, by which time (the idea is) you will be too awake to go back to bed. There are alarms that unless you get up and turn them off promptly will start shredding money. If any of these techniques sound familiar, or at least appealing, you may be a sophisticated hyperbolic discounter. Notice that the demand for schemes such as layaway plans and devices such as rolling alarm clocks shows not only that people exhibit time inconsistency, but also that they are fairly sophisticated in anticipating and subverting their own inconsistent behavior.

Bizarrely, however, sophistication can also exacerbate self-control problems. We have already come across one such case, namely, the sophisticated hyperbolic discounter who having eliminated **save–save** decides that she might as well **spend–spend**. She will therefore act even more myopically than the unsophisticated hyperbolic discounter, who attempts to **save–save** and at least ends up with **save–spend**. What follows is another, even more striking example of this phenomenon.

Exercise 9.11 Johnny Depp 2 This exercise refers to Example 9.10. Suppose instead that you can only watch one of the four movies. Will you watch **a** the mediocre, **b** the good, **c** the great, or **d** the fantastic movie?

(**a**) Show that an exponential discounter will watch **d** the fantastic movie.
(**b**) Show that a naive hyperbolic discounter will watch **c** the great movie.
(**c**) Show that a sophisticated hyperbolic discounter will watch **a** the mediocre movie.

The problem is that sophisticated hyperbolic discounters tend to **preproperate**, that is, doing something now when it would be better to wait. Preproperation in one sense is the very opposite of procrastination, which is a problem that naive hyperbolic discounters have. Paradoxically, then, there are situations in which naifs are better off than sophisticates – situations in which people do better when they do not attempt to anticipate their own behavior. Of course, people like Seneca and Simon, when encouraging you not to think so much about the future (see Section 8.4), already told you so.

9.4 Preferences over profiles

The model of hyperbolic discounting, especially when augmented with a story about naifs and sophisticates, can capture a number of phenomena that are simply inconsistent with the model of exponential discounting. Yet, there are many conditions under which both exponential and hyperbolic discounting fail to accurately capture people's actual behavior. The following exercise makes this clear.

Exercise 9.12 Cleaning the house It is Sunday morning ($t = 0$), and you are determined to accomplish two things today: cleaning the apartment and going to the movies. You can either clean during the morning (at $t = 0$) and go to the movies during the afternoon (at $t = 1$), or go to the movies during the morning (at $t = 0$) and clean in the afternoon (at $t = 1$). You hate cleaning: it only gives you a utility of 2. You love the movies: it gives you as much as 12.

For the first two questions, assume that you discount the future exponentially with $\delta = 1/2$. From the point of view of $t = 0$:

(**a**) What is the utility of cleaning first and going to the movies later?

(**b**) What is the utility of going to the movies first and cleaning later?

For the last two questions, assume that you discount the future hyperbolically with $\beta = 1/3$ and $\delta = 1/2$. From the point of view of $t = 0$:

(**c**) What is the utility of cleaning first and going to the movies later?

(**d**) What is the utility of going to the movies first and cleaning later?

What this exercise suggests is that whether you discount the future exponentially or hyperbolically, you will always schedule the pleasant experience first and the unpleasant one later.

This implication contrasts sharply with people's observed behavior. Personal experience suggests that, when choosing between sequences of events, people will make a point of scheduling the unpleasant experience first and the pleasant one later. "We must try to make the latter part of the journey better than the first, so long as we are en route," as the ancient Greek philosopher Epicurus said. In this case, personal experience and ancient wisdom are supported by evidence. In one study, the researchers presented people with verbal descriptions and graphical representations of increasing and decreasing salary profiles, and elicited preferences over the profiles. The authors conclude that all things equal, by and large, a large majority of workers prefer increasing wage profiles over flat or decreasing ones.

Such a **preference for increasing utility profiles** could in principle be captured by relaxing the assumption (which we made tentatively in Chapter 8) that δ is less than one. If δ exceeds one, a rational discounter will postpone pleasant events as much as possible. When $\delta > 1$, it follows that $r < 0$, which is why the resulting preference is called **negative time preference**. Yet, this solution is awkward, because the very same people who clean in the morning and go to the movies in the afternoon simultaneously discount the future with $\delta < 1$ and $r > 0$ (that is, exhibit **positive time preference**) in other contexts.

In addition, there is evidence that people also exhibit a **preference for spread**. That is, people sometimes like to distribute multiple desirable events over time. While some children eat all their Halloween candy in one sitting, others prefer to distribute the eating evenly over days or weeks. This kind of preference cannot be accounted for by either positive or negative time preference, at least on its own. Finally, people often exhibit a **preference for variation** over time, as they avoid choosing to consume the same good over and over again. "Variation in everything is sweet," the ancient Greek poet Euripides wrote. As a result, people diversify over time. (We will return to the theme of diversification in Section 9.6.)

All this suggests that people have **preferences over profiles**: they care about the *shape* of the utility stream as well as about (discounted) individual utilities. People often save the best for last: perhaps they want to end on a high note, hope to get the unpleasant experience over with, or rely on the prospect of a pleasant experience to motivate themselves to take care of the unpleasant one. People also wish to distribute pleasant and unpleasant events over time, and they value variety. Such preferences over profiles cannot be captured in the context of either one of the discounting models we have discussed so far. Yet preferences over profiles seem to be an important phenomenon.

Example 9.13 Economics professors Rumor has it that even economics professors frequently elect to receive their annual salary in twelve rather than nine installments, even though they would maximize their discounted utility by asking to be paid in nine. Obviously, they have the option of saving some of their money to smooth out consumption. A preference for a smooth income profile would explain this phenomenon.

The idea that the shape of an episode matters is familiar from literature and philosophy. American writer Kurt Vonnegut is most famous for the improbably funny novel *Slaughterhouse-Five*, about surviving the Allied fire bombings of Dresden as a prisoner of war. But he has written memorably about how good stories, and the stories people want to read, tend to have one of a small number of shapes. So for example:

> You will see this story over and over again. People love it and it's not copyrighted. The story is "Man in Hole," but the story needn't be about a man or a hole. It's: Somebody gets into trouble, gets out of it again ... It is not accidental that the [person's welfare] ends up higher than where it began. This is encouraging to readers.

The contemporary philosopher David Velleman has suggested that a person's well-being over some time period depends on the narrative or dramatic relations between the individual events that occur during that period – that is, on the manner in which the events hang together in a story. Velleman offers several "stories of a good life." One of them involves a character who suffers some misfortune, overcomes it, and learns important lessons in the process. You will notice immediately that this is Vonnegut's "Man in Hole." What makes this the story of a good life, according to Velleman, is the narrative or dramatic relations between the misfortune, what comes before it, and what comes after.

A synthesis of these ideas would suggest what we can call *narrativism*: the idea that people want to lead lives that are good stories, and that such lives are, other things equal, better than lives that are not. We want to live lives that include challenges and personal growth, obstacles and overcomings, sin and redemption, troubles and transformation, hardships and healing. And it is not just that such lives make for better stories, e.g., at funerals, although they do; it is also that they are stories of better lives for the people who live them. A desire for living a good story may explain why some people who seem to "have it all" recklessly throw it all away: the stable and high utility stream of the person who has it all just does not make for a very good story. The price of shaking things up can be high, however, costing people their property, marriages – sometimes even their lives. So it goes.

Example 9.14 Hunter S. Thompson The following passage, attributed to American writer Hunter S. Thompson, illustrates the power of stories:

> Life should not be a journey to the grave with the intention of arriving safely in a pretty and well-preserved body, but rather to skid in broadside in a cloud of smoke, thoroughly used up, totally worn out, and loudly proclaiming "Wow! What a ride!"

Economists have recently developed a whole new level of appreciation for the importance of narratives. The book *Narrative Economics*, by Nobel laureate Robert J. Shiller, argues that narratives are a principal driver of human behavior and economic events. Narratives, he writes, "are human constructs that are mixtures of fact, emotion, human interest, and other extraneous details that form an impression on the human mind." As such, narratives influence human behavior much more directly than, e.g., facts do. Moreover, narratives are contagious: they spread like viruses from one person to another, sometimes quickly and far. The fact that narratives can influence behavior in combination with the fact that they have a tendency to "go viral" means that they can have far-reaching economic consequences. Narrative economics, with its suggestion that economic behavior and events can be studied with the tools of epidemiology, opens up exciting new vistas for collaboration between economics, medicine, and the humanities.

The shape of utility profiles has received a lot of attention in the literature on the **peak–end rule**, which is used to assess the desirability of utility streams or "episodes." When people follow this rule, they consciously or unconsciously rank utility streams based on the average of the peak (that is, the maximum utility during the episode) and the end (that is, the utility near the end of the episode) and choose accordingly. Insofar as people act in accordance with the peak–end rule, the shape of the utility profile – and not just the (discounted) sum of utilities – will be critically important.

The peak–end rule has some interesting implications. Consider Figure 9.5. A person who applies the peak–end rule will assess episode (a) as superior to episode (b). If this is not immediately obvious, notice that the peak utility of the two episodes is identical and that episode (a) has a higher end utility than episode (b). If you apply the peak–end rule to make the choice between (a) and (b), therefore, you will choose (a). But there is something odd about this ranking: episode (b) has all the utility of episode (a) and then some. The peak–end rule entails **duration neglect**, meaning that the *length* of an episode will be relatively unimportant, contrary to exponential and hyperbolic discounting models.

Would anyone apply the peak–end rule? In a famous study of patients undergoing a colonoscopy – as in Exercise 9.5 – the researchers confirmed that retrospective

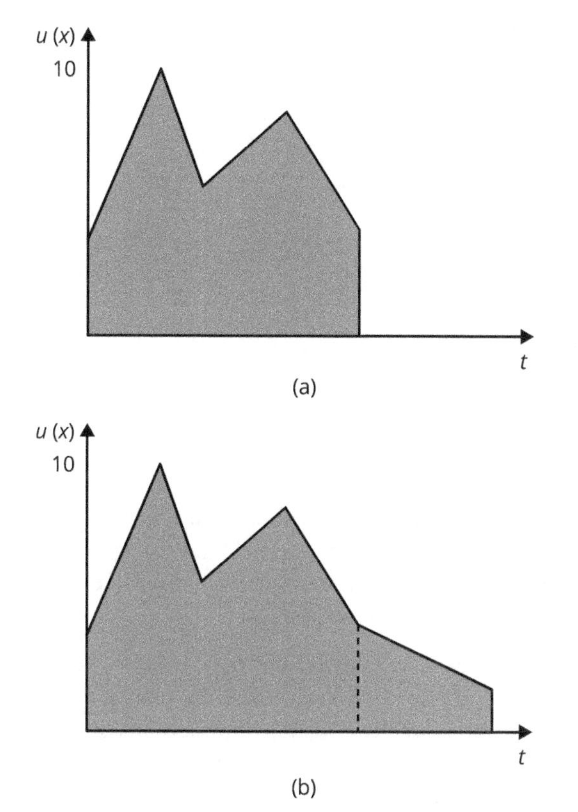

Figure 9.5 The peak-end rule

evaluations reflect the peak and end utility and that the length of the episode was relatively unimportant. Bizarrely, *adding* a painful tail to an already painful episode made people think of the episode as a whole as *less* painful and therefore *more* desirable.

Exercise 9.15 The peak–end rule Suppose you add a pleasant tail to an already pleasant episode. If people assess the episode as a whole in accordance with the peak–end rule, will this make people think of the episode as a whole as more or less pleasant?

Exercise 9.16 The peak–end rule, cont. This exercise refers to Figure 9.6. Would a person who follows the peak–end rule choose the episode represented by the solid line or the episode represented by the dashed line?

The peak–end rule can perhaps explain why people keep having children, even though systematic data suggest that parents are on the average less happy than non-parents, and that parents on the average are less happy when taking care of their children than when they are engaged in many other activities. As long as the most intense joy generated by children exceeds the most intense joy from other sources, and holding end experience constant, people will rank having children above other experiences. Duration neglect entails that long sleepless nights and the like will be relatively unimportant in the final analysis.

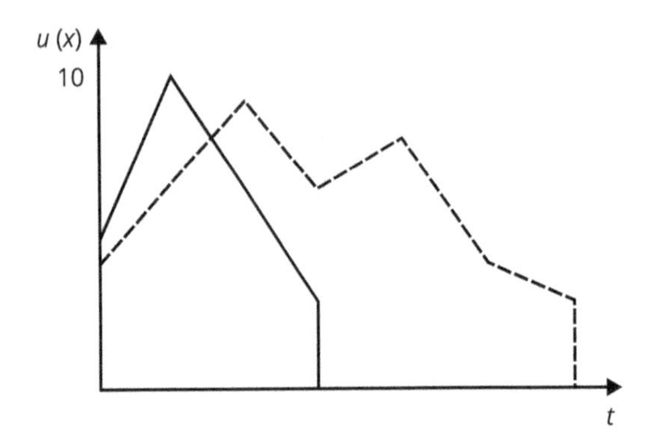

Figure 9.6 The peak–end rule, cont.

Exercise 9.17 College Many adults will tell you that their college years were the best years of their lives. This is puzzling since actual college students are not, on the average, fantastically happy. Use the concept of the peak–end rule to explain why people remember their college years so fondly.

There are choices that superficially look as if they must be the result of a preference over profiles but that really are not. For one thing, insofar as the anticipation of a pleasant event is itself pleasurable, you can rationally postpone the pleasant event. Suppose that it is Saturday, and that you are choosing whether to eat candy today or tomorrow. If you eat it today, you get 6 utiles today and 0 utiles tomorrow; if you postpone eating it until tomorrow, you get 2 utiles today (from the anticipation of the pleasant event) and 6 tomorrow (from the pleasant event itself). Given this sort of scenario, a rational discounter can postpone the eating of the candy until Sunday, even if he or she discounts the future somewhat. And this kind of narrative is consistent with standard theory.

9.5 Misprediction and miswanting

Implicit throughout this chapter and the last is the idea that many decisions depend on predictions of future preferences. When we go grocery shopping, we need to be mindful not only of the preferences we have when doing the shopping, but of the preferences that we will have when it is time to prepare and eat the food. When we choose a course of study, we have to consider the desires we will have while pursuing the degree, but also those we will have on the job for the rest of our careers. When we deliberate about whether to form a family and have children, we cannot ignore the preferences we will have 20, 40, or 60 years hence. And if we are time inconsistent and sophisticated about it, we have to anticipate how our preferences will change over time – or else we will not be able to choose not to choose as outlined in Section 9.3.

As it turns out, people are not very good at predicting their future preferences. This is to be expected. Many of the behavior patterns we have discussed in this book are surprising to social scientists and lay people alike. Consequently, there is little reason to think that a random decision-maker would be aware of them. Behavioral economists have identified a number of different ways in which people's predictions about their own preferences are systematically off target.

Underprediction of adaptation A number of studies suggest that people fail to appreciate the extent to which they will adapt to new conditions, such as a new endowment (see Section 3.5). Thus, people are unable to predict, ahead of time, just how attached they will be to an object after it has been incorporated in their endowment and loss aversion kicks in. In one study, participants who did not currently have a branded coffee cup reported that, if they had one, they would be willing to sell it for between $3 and $4; once they were awarded one, they said they would need between $4 and $6 to give it up.

Underprediction of adaptation helps explain marketing practices such as the no-questions-asked return policy from Section 3.5. If people underpredict the extent to which their preferences will change once some good has been incorporated into their endowment, they will underestimate the difficulties associated with returning the good once it has been brought home, and consequently help make the policy an effective tool for sellers.

Diversification bias Meanwhile, people often overestimate the degree to which their future selves will enjoy variety over time. In a famous study, undergraduate students were offered a choice of snacks to be delivered at the end of class on three subsequent class meetings over the course of as many weeks. When participants made separate choices on each of the three occasions, only 9 percent chose three different snacks. By contrast, when participants had to make all three choices during the first class meeting, 64 percent chose three different snacks. Such results suggest that people exaggerate the degree to which their future selves will favor variety and consequently diversify too much. Maybe variety in everything really is sweet – but not, apparently, as sweet as we think.

Projection bias You are probably familiar with projection: roughly, the unconscious process of attributing to others features of one's self. Projection bias is based on the idea that people project their current preferences onto their future selves. As a result, they will behave as though their future preferences are more like their current preferences than they actually are. The cliché according to which you should never go grocery shopping when extremely hungry or extremely full embodies the insight: when we are extremely hungry, we overestimate how much we will want to eat under more normal circumstances, and when we are extremely full, we underestimate it.

Another famous study asked office workers to choose between healthy and unhealthy snacks that would be delivered either at a time when they could expect to be hungry (late in the afternoon) or satiated (immediately after lunch). Some made the choice right after lunch; some in the late afternoon. Those who were scheduled to receive the snack at a time when they were likely to be hungry were more likely to opt for the unhealthy snack, presumably reflecting an increased taste for unhealthy snacks when hungry. However, those who were hungry when they made the decision were more likely to opt for unhealthy snacks than those who were satiated. Consistent with projection bias, people who were hungry when they made the decision acted as though they anticipated being more hungry when they actually received the snack a week later – like they would if they projected their current hunger onto their future self.

Example 9.18 Fasting Some religious traditions encourage fasting as a way to identify with less fortunate fellow human beings. The underlying insight is consistent with projection bias: it is only when we ourselves are hungry that we can begin to appreciate what it is like to actually starve.

Projection bias can explain underprediction of adaptation: we underestimate the extent to which our preferences will change after we incorporate something into our endowment and loss aversion kicks in.

Hot–cold empathy gaps The term "hot–cold empathy gap" refers to our inability when in a "hot" emotional state to empathize with people (ourselves or others) when in a "cold" state, and *vice versa*. The result is that when we are in a hot state – whether we are experiencing hunger, thirst, anger, embarrassment, or sexual arousal – we tend to underestimate how different our preferences are when we are in a cold state, and the other way around. Hot–cold empathy gaps occur both prospectively and retrospectively, and also across individuals. Empathy gaps are a direct result of projection bias.

One pioneering study found that sexual arousal dramatically increased college-aged males' reported willingness to engage in immoral and risky sexual behavior. Researchers compared students' answers to questions such as "Would you keep trying to have sex after your date says 'no'?" and "Would you always use a condom if you didn't know the sexual history of a new sexual partner?" when in a normal (presumably not aroused) and in a sexually aroused state (while masturbating). The results suggest that young men when in a cold state underestimate how willing they are, when in a hot state, to engage in immoral and risky sexual behavior. As the authors put it: "people seem to have only limited insight into the impact of sexual arousal on their own judgments and behavior" – which is important not the least from a prevention standpoint.

Miswanting Many of our preferences are based on affective forecasts: predictions of how we would feel under various circumstances. For example, it is not uncommon to form preferences over careers, spouses, cars, or whatever based on predictions of how happy we would be if we had this or did that. Unfortunately, there is sometimes a mismatch between wanting and liking – between what we want because we think that we will like it when we get it and what we in fact like when we get it. When this happens, we are victims of miswanting. A major source of miswanting is **impact bias**: the tendency to overestimate the enduring impact of future events on our emotional lives. While many of us imagine that winning a million dollars would make us lastingly happy and becoming disabled lastingly sad, a classic study argued the effects of lottery winnings and sudden disability are attenuated over time. Similarly, marriage, divorce, fame, and fortune have a much smaller effect on our emotional lives, including our happiness, than people tend to imagine.

Many subsequent studies have found that people adapt to illness and disability to a remarkable degree. The finding is often surprising to modern readers. It would not, however, have been surprising to Adam Smith, who wrote:

> By the constitution of human nature ... agony can never be permanent; and if [the disabled] survives the paroxysm, he soon comes, without any effort, to enjoy his ordinary tranquillity. A man with a wooden leg suffers, no doubt, and foresees that he must continue to suffer during the remainder of his life, a very

> considerable inconveniency. He soon comes to view it, however, exactly as every impartial spectator views it, as an inconveniency under which he can enjoy all the ordinary pleasures both of solitude and of society.

Impact bias may be driven in part by underprediction of adaptation: part of the story, certainly, is that lottery winners, the suddenly disabled, and everybody else adapt to changing conditions to a much greater degree, and sooner, than they anticipate. Another source of miswanting is the **focusing illusion**: the tendency for whatever you are attending to seem more important than it is. A person thinking about money, or a new car, or a bigger house, is likely to overestimate the emotional impact of getting more money, a new car, or a bigger house. But as Kahneman puts it: "Nothing in life is as important as you think it is when you are thinking about it."

Exercise 9.19 Sidgwick What kind of misprediction is described in the following passage, by the nineteenth-century moral philosopher Henry Sidgwick?

> In estimating for practical purposes the value of different pleasures open to us, we commonly trust most to our prospective imagination: we project ourselves into the future, and imagine what a particular pleasure will amount to under hypothetical circumstances. This imagination seems to be chiefly determined by our experience of past pleasures ... but partly also by the state of our mind or nerves at the time.

Exercise 9.20 Buffet lines Perhaps you too have the following experience when picking up dinner from a buffet, where you can serve yourself exactly what kind of food you want. At the end of the meal, people find there is food on their plate they have no interest in eating. This is puzzling: if people were able to correctly predict their preferences, all the food would get eaten. Use the concepts of (**a**) diversification bias, (**b**) projection bias, and (**c**) hot–cold empathy gaps to explain how a person can choose food he or she will later have no interest in eating.

Systematic misprediction and miswanting have many adverse consequences. The phenomena discussed in this section make us strangers to ourselves, in that they may leave us shocked and surprised at the bizarre tastes of the person we thought we knew the best – namely, ourselves. To the extent that the right decision depends on accurate predictions, moreover, misprediction and miswanting can lead to bad decisions – even by our own lights. Impact bias and the focusing illusion, for example, can make us overestimate the degree to which wealth will allow us to satisfy our preferences, as a result of which we risk working too much and spending too little time with family and friends.

Then again, Adam Smith believed that exactly this kind of mistake is a driver of economic progress. Smith describes a "poor man's son, whom heaven in its anger has visited with ambition," and who imagines that great wealth will lead to great happiness. Only in "the last dregs of life," Smith continues, does he learn that the pursuit is all in vain and wealth and greatness mere "trinkets of frivolous utility." But here is the kicker:

> And it is well that nature imposes upon us in this manner. It is this deception which rouses and keeps in continual motion the industry of mankind.

> It is this which first prompted them to cultivate the ground, to build houses, to found cities and commonwealths, and to invent and improve all the sciences and arts, which ennoble and embellish human life.

As if "by an invisible hand," Smith concludes, the poor man's pursuit for wealth and greatness, though misguided, ends up promoting the greater good.

If you want to avoid misprediction and miswanting, Sidgwick proposed keeping close tabs on your and others' emotional reactions. He advocated what he called the **empirical–reflective** method:

> [We] must substitute for the instinctive, implicit inference just described a more scientific process of reasoning: by deducing the probable degree of our future pleasure or pain under any circumstances from inductive generalizations based on a sufficient number of careful observations of our own and others' experience.

Sidgwick's empirical–reflective method has recently been endorsed by leading psychologists, although they do not call it that. Sonja Lyubomirsky, for example, tells us to "shelve" our gut reactions and instead systematically think things through based on our own and other people's experiences. Gilbert has even suggested that we refrain from trying to imagine what doing something will be like, for us, before we do it. He believes we can make far more accurate predictions by simply asking somebody else who is currently doing that thing to report what it feels like in the present. If you want to know what it is like to have children (cf. Exercise 6.43 on page 147), he believes you should not try to imagine what it would be like, for you, to have children; you should ask people who have children what it is like for them right now. You might, for example, just ask your parents, if they are still alive. That is probably the last piece of advice many people want to hear, but Gilbert swears by it.

Exercise 9.21 Happiest professions Table 9.3 shows the result of a survey about the happiness of people in various professions: the middle column lists the five happiest professions, and the right-hand column the five unhappiest. You might think that you would be happiest if you were a lawyer or a doctor or something. Although he (probably) does not know you, Gilbert thinks you are wrong. What profession does he think would make you the happiest? Explain why he believes that he knows more about how happy you would be in each of the professions than you do.

Table 9.3 The happiest and unhappiest jobs of 2015

	Happiest	Unhappiest
1	Principal	Security guard
2	Executive chef	Merchandiser
3	Loan officer	Salesperson
4	Automation engineer	Dispatcher
5	Research assistant	Clerk

9.6 Discussion

In this chapter, we have discussed the manner in which people make choices when time is a factor. We discovered several phenomena that are difficult to reconcile with the model of exponential discounting which we explored in Chapter 8. For many of these phenomena, the divergence appears significant and systematic, and therefore predictable. Again, knowledge of these phenomena permits us not only to explain and predict other people's behavior, but also to influence it – and to resist other people's efforts to influence ours.

There are many other phenomena that are at odds with the exponential discounting model we learned in Section 8.3. The **sign effect** says that gains are discounted at a higher rate than losses. The **magnitude effect** says that large outcomes are discounted at a lower rate than small ones. Consider also the likelihood that preferences change as a result of becoming sleepy, hungry, thirsty, excited, or old. Such changes are important not just because they mean that people's behavior can be expected to change over time in predictable ways, but also because people can be expected to respond to their own predictions about how their preferences will change. For example, many people know that they will buy too much food when shopping on an empty stomach and try to modulate their behavior accordingly. Economists have built models in which preference change is endogenous, but it may be more parsimonious to postulate the existence of a preference relation that evolves over time.

Behavioral economists take these phenomena to provide evidence that the exponential-discounting model and the assumption of stable preferences are descriptively inadequate. Some of these phenomena also cast doubt on the normative correctness of the model. It is true that some violations of exponential discounting – involving serious procrastination, extreme impulsivity, or similar – can harm the individual. While some degree of sophistication can help mitigate the effects of hyperbolic discounting, it can also hurt, as Exercise 9.11 showed. But then, there is real disagreement about the rationality of time discounting (see Section 8.4). The discussion about utility profiles in the previous section helps underscore these concerns. It can be argued that it is perfectly rational to desire a life of increasing utility, of ups-and-downs, or any other shape or story. If so, models of exponential and hyperbolic discounting both fail as a normative standard. And if exponential discounting is not the uniquely rational way to assess deferred outcomes, this would put in question its widespread use in disciplines such as cost–benefit analysis and finance.

In Part 5 we will consider strategic interaction, which will add yet another layer of complexity to the analysis.

Additional exercises

Exercise 9.22 Retirement savings When young, many people fully intend to save for retirement. However, when they start making money after college, they are often tempted to spend it immediately. Assume that Ximena and Yves have the choice between the following two options: (**a**) saving for retirement at time 1 ($u_1 = 0$) and retiring in style at time 2 ($u_2 = 12$), and (**b**) having more disposable income at time 1 ($u_1 = 6$) and retiring poor at time 2 ($u_2 = 0$).

Ximena is an exponential discounter. Her $\delta = 2/3$.
(a) At $t = 0$: What is her utility of **a**? What is her utility of **b**?
(b) At $t = 1$: What is her utility of **a**? What is her utility of **b**?
Yves is a naive hyperbolic discounter. His $\beta = 1/3$ and his $\delta = 1$.
(c) At $t = 0$: What is his utility of **a**? What is his utility of **b**?
(d) At $t = 1$: What is his utility of **a**? What is his utility of **b**?
(e) Who is more likely to experience regret?
(f) Who is more likely to retire in style?

Exercise 9.23 Addiction Suppose that life has three periods: youth, middle age, and old age. In every period you decide whether to do drugs ("hit") or not ("refrain"). The utility of hitting depends on whether you are addicted ("hooked") or not. If you are not hooked, the utility from hitting is 10 and the utility from refraining is 0. If you are hooked, the utility from hitting is −8 and the utility from refraining −25. In youth, you are not hooked; after that, you are hooked just in case you hit in the preceding period. Assume $\delta = 1$ and $\beta = 1/2$. What do you do if you are **(a)** time consistent, **(b)** time-inconsistent and naive, and **(c)** time-inconsistent and sophisticated?

Exercise 9.24 Orpheus and Eurydice Superstar singer–songwriter Orpheus messed up. Long story, but tl;dr he had to go get his main squeeze, Eurydice, back from the underworld. Things went well at first, but the king of the underworld, Hades, told Orpheus absolutely, positively not to look back, and what did he do? He looked back – and Eurydice disappeared. What kind of discounter do you think Orpheus was?

Exercise 9.25 From the underground The book *Notes from the Underground*, by Russian novelist Fyodor Dostoyevsky, is often described as the first existentialist novel. Here's a sample:

> Now I ask you: what can be expected of man since he is a being endowed with strange qualities? Shower upon him every earthly blessing, drown him in a sea of happiness, so that nothing but bubbles of bliss can be seen on the surface; give him economic prosperity, such that he should have nothing else to do but sleep, eat cakes and busy himself with the continuation of his species, and even then out of sheer ingratitude, sheer spite, man would play you some nasty trick. He would even risk his cakes and would deliberately desire the most fatal rubbish, the most uneconomical absurdity, simply to introduce into all this positive good sense his fatal fantastic element.

Can the strange behavior Dostoyevsky describes be explained in terms of **(a)** hyperbolic discounting, **(b)** the peak–end rule, and/or **(c)** narrativism?

Exercise 9.26 Weights and shackles Seneca offered the following advice:

> [Reflect] that prisoners at first find the weights and shackles on their legs hard to bear, but subsequently, once they have determined to endure them rather than chafe against them, necessity teaches them to bear them bravely, habit to bear them easily. In whatever life you choose you will find

> there are delights and relaxations and pleasures, if you are willing to regard your evils as light rather than to make them objects of hatred. In no respect has Nature done us a greater service, who, as she knew into what tribulations we were born, devised habit as a means of alleviating disasters, swiftly making us grow accustomed to the worst sufferings.

This passage suggests Seneca was well aware of at least one phenomenon discussed in Section 9.5. Which one?

Exercise 9.27 In much of the modern world, people are supposed to decide themselves whom to marry. The decision is often (expected to be) made when young, in love, and/or sexually aroused. (**a**) Give at least two reasons to think that people's happiness predictions under the circumstances might be off. (**b**) How could people improve the quality of their predictions?

Exercise 9.28 Match each of the vignettes below with one of the following phenomena: *hyperbolic discounting, preference over profiles, choosing not to choose,* and *misprediction / miswanting.* If in doubt, pick the best fit.

(**a**) Allie goes to bed at night fully intending to get up at 5 am and study hard before noon. When the alarm goes off, she smacks it hard and goes straight back to sleep.

(**b**) Bert wants to save more, but simply does not feel that he has enough money left at the end of the month. To encourage himself to save, he sets up an automatic transfer – on the first of each month – from his checking account to a savings account from which withdrawing money is a pain.

(**c**) Cherry can only afford a really nice meal at a restaurant once every semester. She makes sure to schedule it at the very end of the semester, so that she has something to look forward to when eating her ramen noodles.

(**d**) Darius knows that his wife would be so much happier, and his marriage so much healthier, if he spent more time cleaning the house. He keeps thinking that it would be great to clean the house. Yet, when it comes down to it, there is always something on TV that prevents him from doing his part.

(**e**) Epicurus, or one of his followers, wrote: "Life is ruined by procrastination, and every one of us dies deep in his affairs."

(**f**) Unlike some people she knows, Filippa will not finish a whole tub of ice-cream in one sitting. Rather, she will allow herself exactly one spoonful every day.

(**g**) A website helps you pursue your goals by requesting your credit-card number and the name of the non-profit organization you find most odious. If you fail to reach whatever goals you set for yourself, it will charge your card and send the money to the non-profit you despise.

(**h**) As Henry Sidgwick wrote: "In fact there is scarcely any point upon which moralizers have dwelt with more emphasis than this, that man's forecast of pleasure is continually erroneous."

Problem 9.29 *Drawing on your own experience, make up stories like those in Exercise 9.28 to illustrate the various ideas that you have read about in this chapter.*

Further reading

The quotation from *Junky* at the beginning of Section 9.2 is from Burroughs (1977 [1953], p. xv). A helpful review of violations of the exponential discounting model is Frederick et al. (2002), reprinted as Chapter 1 of Loewenstein et al. (2003). The hyperbolic discounting model is due to Ainslie (1975); *Oblomov* is Goncharov (1915 [1859], p. 7) and Wicksteed (2003 [1933], pp. 29–30) describes his blanket-related dilemma. Naive versus sophisticated hyperbolic discounting is discussed in O'Donoghue and Rabin (2000), from which the Johnny Depp-related example and exercise were adapted (pp. 237–8). The study of procrastination and precommitment is Ariely and Wertenbroch (2002), and the layaway example was inspired by Tabarrok (2013). The advice about making the latter part of the journey better than the first appears in Epicurus (2012 [*c* 300 BCE], p. 182), and the study of workers' preferences is Loewenstein and Sicherman (1991); the line from Euripides is cited in Aristotle (1999 [*c* 350 BCE], p. 119). Vonnegut (2006, p. 25) describes "Man in Hole," and Velleman (1991) discusses well-being over time. *Narrative Economics* is Shiller (2019); the quotation is from p. 65. The peak–end rule is discussed in Kahneman et al. (1997), and the colonoscopy study in Redelmeier and Kahneman (1996). Loewenstein and Angner (2003) examine preference change from a descriptive and normative perspective. The data in Section 9.5 come from Loewenstein and Adler (1995) on underprediction of adaptation, Simonson (1990) on diversification bias, Read and van Leeuwen (1998) on projection bias, and Ariely and Loewenstein (2006) on hot–cold empathy gaps (the conclusion cited appears on page 95). Gilbert et al. (2002) explain impact bias, Smith (2002 [1759], p. 172) the constitution of human nature, and Kahneman (2011, pp. 402–6) the focusing illusion. Sidgwick (2012 [1874]) appears to have discovered projection bias (p. 121), and he names and describes the empirical–reflective method (pp. 111, 122); Lyubomirsky (2013, p. 11) and Gilbert (2006, Ch. 11) endorse something similar. Smith (2002 [1759], pp. 214–5) says "it is well that nature imposes upon us in this manner." The data on the happiest and unhappiest professions appear in Adams (2015). The addiction example is due to O'Donoghue and Rabin (2000, pp. 240–1); Dostoyevsky (2009 [1864], p. 23) describes "man's strange qualities"; Seneca (2007 [*c* 49], pp. 127–8) offers advice; Epicurus (2012 [*c* 300 BCE], p. 181) claims life is ruined by procrastination; and Sidgwick (2012 [1874], p. 121) discusses "man's forecast of pleasure."

PART 5

STRATEGIC INTERACTION

10 ANALYTICAL GAME THEORY

Learning objectives

After studying this chapter you will:

- Know the difference between parametric and strategic decision-making
- Be able to find Nash equilibria (and some refinements) in a variety of games
- Understand how analytical game theory can be used for descriptive and normative purposes – and what its limitations are

10.1 Introduction

So far, we have assumed that the outcomes that you enjoy or suffer are determined jointly by your choices and the state of the world, which may be unknown to you at the time of the decision. Such a situation calls for **parametric** decision-making. For many real-world decisions, however, this is far from the whole story. Instead, many of the decision problems that you face in real life have an *interactive* or *strategic* nature. This means that whatever happens depends not just on what you do but also on what other people do. If you play chess, whether you win or lose is determined not just by your moves but also by your opponent's. If you invest in stock, whether or not you make money depends not only on your choice of stock but also on whether the stock goes up or down. And that is a function of supply and demand, which is a function of whether other people buy or sell the stock. Such situations call for **strategic** decision-making.

The presence of strategic interaction adds a whole new layer of complexity to the analysis. If you are a defense attorney, the outcome of your case depends not only on your actions but also on the actions of the prosecutor. Since you are well aware of this, you do your best to anticipate her decisions. Thus, your decisions will depend on what you think her decisions will be. Her decisions will reflect, in part, what she thinks you will do. So your decisions will depend on what you think she thinks you will do. But what she thinks you will do depends in part on what she thinks you think she will do … and so on. It is obvious that the correct analysis of strategic interaction is not obvious at all.

The analysis of strategic interaction is the province of **game theory**. In this chapter, I offer a brief overview of the standard theory, sometimes referred to as **analytical game theory**.

10.2 Nash equilibrium in pure strategies

The following story is a true internet legend.

Example 10.1 The makeup exam One year there were two students taking Chemistry. They both did so well on quizzes, midterms, and labs that they decided to leave town and go partying the weekend before the exam. They mightily enjoyed themselves. However, much like a scene in *The Hangover: Part IV*, they overslept and did not make it back to campus in time for the exam.

So they called their professor to say that they had got a flat tire on the way to the exam, did not have a spare, and had to wait for a long time. The professor thought about it for a moment, and then said that he was glad to give them a makeup exam the next day. The two friends studied all night.

At the assigned time, the professor placed them in separate rooms, handed them the exams, and asked them to begin. The two friends looked at the first problem, which was a simple one about molarity and solutions and was worth 5 points. "Easy!" they thought to themselves. Then, they turned the page and saw the second question: "(95 points) Which tire?"

This example illustrates the interactive or strategic nature of many decision problems. Here, the final grade of either friend will depend not just on his answer to the question, but on the other friend's answer too. The two will get A's whenever they give the same answer to the question and F's whenever they do not.

More formally speaking, you are playing a **game** whenever you face a decision problem in which the final outcome depends not just on your action, and on whatever state of the world obtains, but also on the actions of at least one other agent. According to this definition, the two friends are in fact playing a game against each other. And this is true whether or not they think of it as a game. Notice that you can play a game in this sense without competing *against* each other. Here, the name of the game is cooperation – and coordination. The agents involved in games are called **players**. A **strategy** is a complete plan of action that describes what a player will do under all possible circumstances. In the case of the makeup exam, each friend has four strategies to choose from: he can write "Front Left (FL)," "Front Right (FR)," "Rear Left (RL)," or "Rear Right (RR)."

Given a number of players, a set of strategies available to each player, and a set of payoffs (rewards or punishments) corresponding to each possible combination of strategies, a game can be represented using a **payoff matrix**. A payoff matrix is a table representing the payoffs of the players for each possible combination of strategies. The payoff matrix of the game played by the two friends can be represented as in Table 10.1. A **strategy profile** is a vector of strategies, one for each player. ⟨FL, RR⟩ is a strategy profile; so is ⟨RL, RL⟩. Thus, the payoff matrix shows the payoffs resulting from each strategy profile. Of course, the payoff matrix looks much like the tables representing non-strategic decision problems, except for the fact that each column represents a choice by the other player rather than a state of the world.

Analytical game theory is built around the concept of an equilibrium. The most prominent equilibrium concept is that of **Nash equilibrium**.

Table 10.1 The makeup exam

	FL	FR	RL	RR
FL	A	F	F	F
FR	F	A	F	F
RL	F	F	A	F
RR	F	F	F	A

Definition 10.2 Nash equilibrium *A Nash equilibrium is a strategy profile such that each strategy in the profile is a best response to the other strategies in the profile.*

To say that you are playing a "best response" means that you can do no better by switching to another course of action – given what the other players are up to. In other words, it means that there is no alternative strategy available to you that would give you a strictly higher payoff – again, given what the other people do. In a Nash equilibrium, therefore, everyone does as well as they can – given what the others are doing. This is different from saying that the outcome will be as good as it can get. As we will see, being in a Nash equilibrium does not mean that the outcome is particularly good for anyone.

In the makeup-exam game from Example 10.1, ⟨FL, FL⟩ is a Nash equilibrium: given that Player I plays FL, FL is a best response for Player II, and given that Player II plays FL, FL is a best response for Player I. In equilibrium, given the other players' strategies, no one player can improve his or her payoff by unilaterally changing to another strategy. By contrast, ⟨FL, RR⟩ is not a Nash equilibrium: given that Player I plays FL, Player II can do better than playing RR, and given that Player II plays RR, Player I can do better than playing FL. In this section, we will limit our analysis to Nash equilibria in pure strategies: Nash equilibria in which each player simply plays one of the individual strategies available to him or her (compare Section 10.3). In all, there are four Nash equilibria in pure strategies, one for each tire.

Example 10.3 Coffee shops You and your study partner are planning to meet at noon at one of two coffee shops, Lucy's Coffee and Crestwood Coffee. Unfortunately, you failed to specify which one, and you have no way of getting in touch with each other before noon. If you manage to meet, you get a utility of 1; otherwise, you get a utility of 0. Draw the payoff matrix and find the Nash equilibria in pure strategies.

The payoff matrix is Table 10.2. The convention is for the first number in each cell to represent the payoff of Player I, whose strategies are listed in the leftmost column; the second number in each cell represents the payoff of Player II, whose strategies are listed in the top row. The Nash equilibria in pure strategies are ⟨Lucy's, Lucy's⟩ and ⟨Crestwood, Crestwood⟩.

Table 10.2 A pure coordination game

	Lucy's	Crestwood
Lucy's	1,1	0,0
Crestwood	0,0	1,1

The coffee-shop game is an example of a **pure coordination game**: a game in which the players' interests are perfectly aligned. The makeup-exam game, obviously, is also a pure coordination game. In some coordination games, however, interests fail to align perfectly. The point is typically made by means of the politically incorrectly named **battle of the sexes**.

Example 10.4 Battle of the sexes A husband and wife must decide whether to have dinner at the steak house or at the crab shack. All things equal, both would rather dine together than alone, but the man (Player I) prefers the steak house and the woman (Player II) prefers the crab shack. The man gets 2 units of utility if both dine at the steak house, 1 if both dine at the crab shack, and 0 if they dine apart; the woman gets 2 units of utility if both dine at the crab shack, 1 if both dine at the steak house, and 0 if they dine apart. Draw the payoff matrix and find the Nash equilibria in pure strategies.

The payoff matrix is Table 10.3. There are two Nash equilibria in pure strategies. ⟨Steak House, Steak House⟩ is one. Because this is Player I's best outcome, he cannot improve his payoff by changing strategies. Although Player II would prefer it if *both* switched to Crab Shack, she cannot improve her payoff by *unilaterally* deviating: if she plays Crab Shack when Player I plays Steak House, she will end up with a payoff of 0 rather than 1. Of course ⟨Crab Shack, Crab Shack⟩ is the other Nash equilibrium in pure strategies.

Table 10.3 An impure coordination game

	Steak House	**Crab Shack**
Steak House	2,1	0,0
Crab Shack	0,0	1,2

Because Player I prefers the one equilibrium and Player II prefers the other, the battle of the sexes, sometimes euphemistically called "Bach or Stravinsky," is an example of an **impure coordination game**. Here are some exercises.

Exercise 10.5 Nash equilibrium in pure strategies Find all Nash equilibria in the games in Table 10.4, where Player I chooses between Up (U), Middle (M), and Down (D) and Player II chooses between Left (L), Middle (M), and Right (R).

Table 10.4 Nash equilibrium exercises

	L	**R**
U	2,2	0,0
D	0,0	1,1

(a)

	L	**R**
U	5,1	2,0
D	5,1	1,2

(b)

	L	**M**	**R**
U	6,2	5,1	4,3
M	3,6	8,4	2,1
D	2,8	9,6	3,0

(c)

Figure 10.1 The suspects. Illustration by Cody Taylor

Notice that in Exercise 10.5(a), there are two Nash equilibria in pure strategies, though one is clearly inferior to the other from the point of view of both players. In Exercise 10.5(b), $\langle U, L\rangle$ and $\langle D, L\rangle$ are not both Nash equilibria although they are "just as good" in the sense that they lead to the same payoffs. And in Exercise 10.5(c), there are outcomes that are better for both players than the Nash equilibrium.

As these games illustrate, there is no straightforward connection between Nash equilibria and "best" outcomes for the players. As a result, it would be a mistake to try to identify the former by searching for the latter. An even more striking example of the general phenomenon is the **prisoners' dilemma** (Figure 10.1).

Example 10.6 Prisoners' dilemma Two criminals are arrested on suspicion of two separate crimes. The prosecutor has sufficient evidence to convict the two on the minor charge, but not on the major one. If the two criminals *cooperate* (C) with each other and stay mum, they will be convicted on the minor charge and serve two years in jail. After separating the prisoners, the prosecutor offers each of them a reduced sentence if they *defect* (D), that is, testify against the other. If one prisoner defects but the other one cooperates, the defector goes free whereas the cooperator serves 20 years in jail. If both defect, both get convicted on the major charge but (as a reward for testifying) only serve ten years. Assume that each prisoner cares about nothing but the number of years he himself spends in jail. What is the payoff matrix? What is the Nash equilibrium?

The payoff matrix in terms of jail sentences is Table 10.5(a); in terms of utilities, the payoff matrix can be represented as Table 10.5(b). Let us consider Player I first. If Player II cooperates, Player I has the choice between cooperating and defecting; by defecting, he can go free instead of serving two years in jail. If Player II defects, Player I still has the choice between cooperating and defecting; by defecting, he can serve 10 instead of 20 years in jail. In brief, Player I is better off defecting no matter what Player II does. But the same thing is true for Player II. Thus, there is only one Nash equilibrium. Both defect and serve 10 years in jail.

Table 10.5 The prisoners' dilemma

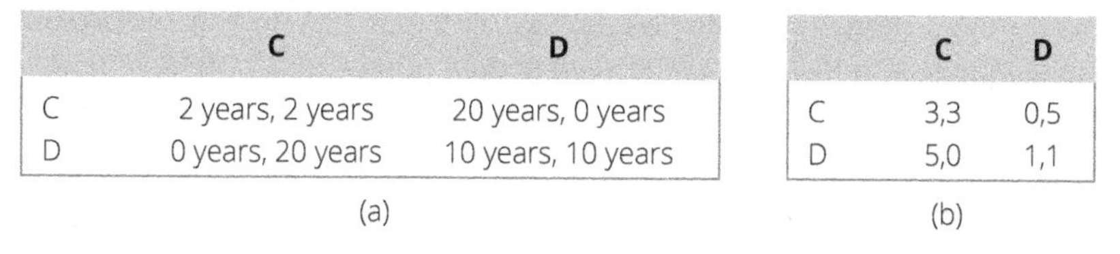

	C	D
C	2 years, 2 years	20 years, 0 years
D	0 years, 20 years	10 years, 10 years

(a)

	C	D
C	3,3	0,5
D	5,0	1,1

(b)

One way to identify the unique Nash equilibrium in the prisoners' dilemma is to eliminate all strictly dominated strategies. A strategy X is said to strictly dominate another strategy Y if choosing X is better than choosing Y no matter what the other player does. Because no rational agent will play a strictly dominated strategy, such strategies can be eliminated from consideration when searching for Nash equilibria. In the prisoners' dilemma, defection strictly dominates cooperation, so cooperation can be eliminated. No rational player will cooperate, and both will defect.

Notice that the result holds even though both prisoners agree that it would have been better if they had both cooperated. An outcome X is said to **Pareto dominate** another Y if all players weakly prefer X to Y and at least one player strictly prefers X to Y. An outcome is **Pareto optimal** if it is not Pareto dominated by any other outcome. In the prisoners' dilemma, the cooperative outcome $\langle C, C \rangle$ Pareto dominates the Nash equilibrium $\langle D, D \rangle$. In fact, both players strictly prefer the former to the latter. Still, rational players will jointly choose an outcome which is not Pareto optimal. For this reason, the prisoners' dilemma is sometimes presented – for example, in the film *A Beautiful Mind*, about game theory inventor and Nobel laureate John Nash – as refuting Adam Smith's insight that the rational pursuit of individual self-interest leads to socially desirable outcomes.

Many real-world interactions have features that are reminiscent of the prisoners' dilemma. Arms races are classic examples. Consider the nuclear buildup in India and Pakistan. Whether or not India has nuclear arms, Pakistan wants them. If India has them, Pakistan needs them to preserve the balance of power; if India does not have them, Pakistan wants them to get the upper hand. For the same reason, India wants nuclear arms whether or not Pakistan has them. Thus, both countries acquire nuclear arms, neither country has the upper hand, and both countries are worse off than if neither had them. Overfishing, deforestation, pollution, and many other phenomena are other classic examples. The idea is that no matter what other players do, each player has an incentive to fish, cut down forests, and pollute, but, if everyone does, everyone is worse off than if nobody had acted.

A number of different solutions might occur to you. What if the two prisoners, before committing the crime, got together and promised to cooperate in the event that they are caught? Surely, you might think, an informal agreement and a handshake would do the trick. The solution fails, however, because whatever verbal agreement the prisoners might have entered into before getting caught will not be binding. At the end of the day, each has to make a choice in isolation, defection strictly dominates cooperation, and a rational agent has no choice but to defect. "Talk is cheap," the saying goes, which is why game theorists refer to non-binding verbal agreements as **cheap talk**.

What if the game could be repeated? Repetition, you might think, should afford a prisoner the opportunity to punish defection by defecting. But suppose the two prisoners play ten prisoners'-dilemma games against each other. To find the equilibrium in the repeated game, we start at the end and use a procedure called **backward induction**. In the last round, no rational prisoner will cooperate, because his opponent has no way to retaliate against defection; so in round ten, both prisoners will defect. In the next to last round, a rational prisoner already knows that his opponent will defect in round ten, which means that it does not matter whether he cooperates or defects; so in round nine, both prisoners will defect. The same thing is true for round eight, round seven, and so on. In this way, the prospect of rational cooperation in the repeated prisoners' dilemma unravels from the end. Repetition does not necessarily solve the problem.

Cooperation can be sustained if there is no last round, that is, if the game is repeated indefinitely. In the indefinitely repeated prisoners' dilemma, there is a Nash equilibrium in which both prisoners cooperate throughout but are prepared to punish defection by defecting. The cooperative solution presupposes that the players do not discount the future too much: if they do, no amount of repetition will save the prisoners. And there is no guarantee that rational individuals will play that particular equilibrium. In fact, there is an *infinite* number of equilibria in the indefinitely repeated prisoners' dilemma, and in one of those equilibria prisoners always defect. In brief, indefinite repetition upholds the prospect of rational cooperation in the prisoners' dilemma, but does not guarantee it.

There is only one sure-fire way for rational agents to avoid defection, and it is to make sure they are not playing a prisoners' dilemma at all. Suppose that the two criminals, before committing their crimes, go to the local contract killer and instruct him to kill anyone who defects in the event that he is caught. If death at the hands of a contract killer is associated with a utility of $-\infty$, the payoff matrix of the two prisoners will now look like Table 10.6. Here, cooperation strictly dominates defection for both players and ⟨C, C⟩ is the unique Nash equilibrium. You might think that it would never be in a person's interest to ask to be killed by a contract killer, no matter what the conditions; yet, by proceeding in this way, the prisoners can guarantee themselves a much better payoff than if they had not. But notice that cooperation is the uniquely rational strategy in this game because it's not a prisoners' dilemma at all.

Table 10.6 The modified prisoners' dilemma

	C	D
C	3,3	$0,-\infty$
D	$-\infty,0$	$-\infty,-\infty$

Example 10.7 The Leviathan The seventeenth-century political philosopher Thomas Hobbes offered a justification of political authority by imagining what life would be like without it. In one of the most famous lines in the history of Western philosophy, Hobbes described this "state of nature" as follows:

> [During] the time men live without a common power to keep them all in awe, they are in that condition which is called war, and such a war as is of

> every man against every man ... In such condition there is ... continual fear and danger of violent death, and the life of man, solitary, poor, nasty, brutish, and short.

The solution, according to Hobbes, is a covenant according to which people give up their right to kill and maim other people in exchange for the right not to be killed and maimed, and which at the same time establishes an overwhelming power – a *Leviathan* – to ensure that people adhere to the terms of the covenant (see Figure 10.2).

Game theory offers a new way to interpret the nature of this "war of all against all." These days, many people think of Hobbes's story as a vivid description of a scenario in which people are forced to play prisoners' dilemmas against each other, and in which the pursuit of rational self-interest therefore leads to the worst possible outcome for all involved. The Leviathan in Hobbes's story serves the same function as the contract killer in the scenario above: by holding people to their promises, he changes the nature of the interaction so as to ensure that rational self-interest coincides with social desirability.

Figure 10.2 The Leviathan. Detail of the frontispiece from the 1651 edition

10.3 Nash equilibrium in mixed strategies

Some games have no Nash equilibria in pure strategies. But that does not mean that they do not have Nash equilibria.

Example 10.8 Coffee shops, cont. Suppose that you still have to go to one of the two coffee shops in Example 10.3 and that your ex has to also. You do not want to run into your ex, but your ex wants to run into you. What kind of game would you be playing against each other?

If a player gets a utility of 1 whenever his or her goal is attained and 0 otherwise, the payoff matrix is Table 10.7.

Table 10.7 A pure coordination game

	Lucy's	Crestwood
Lucy's	1,0	0,1
Crestwood	0,1	1,0

This game has no Nash equilibria in pure strategies. If you go to Lucy's, your ex will want to go there too, but then you want to go to Crestwood, in which case your ex wants to do so too. This coffee-shop game, by the way, has the same payoff structure as a game called **matching pennies**. When two people play matching pennies, each flips a penny. If both coins come up heads, or if both coins come up tails, Player I wins; otherwise, Player II wins. This also happens to be an example of a **zero-sum game**, a game in which whenever one player wins, another player loses.

The game does, however, have a **Nash equilibrium in mixed strategies**. Suppose that you figure out where to go by flipping a coin, and that your ex does the same. Given that you have a 50 percent chance of ending up at Lucy's and a 50 percent chance of ending up at Crestwood, your ex is indifferent between Lucy's and Crestwood and can do no better than flipping a coin. And given that your ex has a 50 percent chance of ending up at Lucy's and a 50 percent chance of ending up at Crestwood, you are indifferent between Lucy's and Crestwood and can do no better than flipping a coin. The two of you are in a Nash equilibrium, though you are playing mixed rather than pure strategies.

In a game like this, the mixed-strategy equilibrium is easy to find. In other games it can be more tricky. Consider the battle of the sexes (Example 10.4). In order to find a mixed-strategy equilibrium in a game like this, there is one critical insight: in order for players to rationally play a mixed strategy, they must be indifferent between the pure strategies they are mixing. Why? If a player strictly preferred one over the other, the only rational thing to do would be to play the preferred strategy with probability one. Thus, you can find the mixed-strategy equilibrium in a game by setting up equations and solving for the probabilities with which the players play different strategies.

Example 10.9 Battle of the sexes, cont. In order to find the mixed-strategy equilibrium in the battle of the sexes (Table 10.8), let us assume that Player I plays U with probability p and D with probability $(1-p)$ and that Player II plays L with probability q and R with probability $(1-q)$.

Consider Player I first. In order to play a mixed strategy, he must be indifferent between U and D, meaning that $u(\text{U}) = u(\text{D})$. The utility of playing U will depend on what Player II does, that is, on what q is. When playing U, Player I has a probability of q of getting 2 utiles and a probability of $(1 - q)$ of getting 0. Consequently, $u(\text{U}) = q * 2 + (1 - q) * 0 = 2q$. When playing D, Player I has a probability of q of getting 0 utiles and a probability of $(1 - q)$ of getting 1. Thus, $u(\text{D}) = q * 0 + (1 - q) * 1 = 1 - q$. So $u(\text{U}) = u(\text{D})$ entails that $2q = 1 - q$, meaning that $q = 1/3$.

Next, consider Player II. In order to play a mixed strategy, she needs to be indifferent between L and R, meaning that $u(\text{L}) = u(\text{R})$. Now $u(\text{L}) = p * 1 + (1 - p) * 0 = p$ and $u(\text{R}) = p * 0 + (1 - p) * 2 = 2 - 2p$. So $u(\text{L}) = u(\text{R})$ entails that $p = 2 - 2p$, meaning that $p = 2/3$.

Hence, there is a Nash equilibrium in mixed strategies in which Player I plays U with probability 2/3 and Player II plays L with probability 1/3. In the mixed-strategy equilibrium, Player I gets payoff $u(\text{U}) = u(\text{D}) = 2q = 2/3$ and Player II gets payoff $u(\text{L}) = u(\text{R}) = p = 2/3$.

Table 10.8 An impure coordination game

	L	R
U	2,1	0,0
D	0,0	1,2

As this example shows, games with pure-strategy equilibria may also have mixed equilibria.

Exercise 10.10 Mixed-strategy equilibrium Find the mixed-strategy Nash equilibria in Tables 10.4(a) and (b).

In the mixed-strategy equilibrium in (a), notice that Player I is more likely to play D than U and that Player II is more likely to play R than L. This might seem strange, since you would perhaps expect the players to be more likely to play the strategy associated with the more desirable equilibrium ⟨U, L⟩. But assume for a proof by contradiction that the two players are in a mixed-strategy equilibrium in which Player I plays U and Player II plays L with some high probability. If so, Player I would strictly prefer U to D and Player II would strictly prefer L to R. Thus, the two players would not be in equilibrium at all, contrary to the initial assumption. For the two players to want to mix, they must be indifferent between the two pure strategies, and this can happen only when Player I is more likely to play D than U and when Player II is more likely to play R than L.

Notice also that the probability p with which Player I plays U is a function not of Player I's own payoffs, but of Player II's payoffs. This might seem equally counterintuitive. Yet, it follows from the fact that p must be selected in such a manner as to make Player II indifferent between her pure strategies. Similarly, the probability q with which Player II plays L is determined not by her payoffs, but by her opponent Player I's payoffs. This is a fascinating feature of Nash equilibria in mixed strategies.

Exercise 10.11 Pure vs. mixed equilibria Find all Nash equilibria (in pure and mixed strategies) in the games depicted in Table 10.9.

Although a mixed-strategy equilibrium may at first blush seem like an artificial construct of mainly academic interest, mixed strategies are important and common in a wide variety of strategic interactions. Even if you are a tennis player with a killer cross-court shot, it would be unwise to hit the cross-court shot every time, or your opponent will learn to expect it. Every so often, you must hit the ball down the line. In games like these, in order to keep your opponent guessing, you must mix it up a bit. This analysis shows that it is not a mistake, but *necessary*, every so often to hit your weaker shot. In games, the fact that a person chooses the inferior option from time to time does not mean that he or she is not playing an equilibrium strategy.

Table 10.9 Mixed Nash equilibrium exercises

	L	R
U	5,2	1,1
D	1,1	2,5

(a)

	L	R
U	4,1	2,0
D	5,1	1,2

(b)

	L	R
U	1,1	0,0
D	0,0	0,0

(c)

Example 10.12 Spousonomics According to the authors of the book *Spousonomics*, "Economics is the surest route to marital bliss" because "it offers dispassionate, logical solutions to what can often seem like thorny, illogical, and highly emotional domestic disputes." Suppose that you are stuck in an equilibrium where you do the dishes, make the bed, and empty the cat litter while your spouse sits back and relaxes. Spousonomics, apparently, teaches that you can turn your spouse into an acceptable (if not ideal) partner by playing a mixed strategy, by sometimes doing the laundry, sometimes not, sometimes making the bed, sometimes not, and so on.

Exercise 10.13 Rock-paper-scissors

(a) Draw the payoff matrix for the game rock-paper-scissors. Suppose that a win gives you 1 utile, a tie 0, and a loss -1.

(b) What is the unique Nash equilibrium in this game?

We already know that not all games have Nash equilibria in pure strategies. But now that we have access to the concept of a Nash equilibrium in mixed strategies it is possible to prove a famous theorem originally due to John Nash. Simplified and expressed in words:

Theorem 10.14 Nash's theorem *Every finite game – that is, every game in which all players have a finite number of pure strategies – has a Nash equilibrium.*

Proof.
Omitted. □

Given this theorem, the search for Nash equilibria is not futile. As long as the number of pure strategies available to each player is finite – and whether this condition is satisfied is fairly easy to determine – we know that the game has at least one Nash equilibrium in pure or mixed strategies. This is neat.

Example 10.15 Chess Chess is a finite game. We know this because every player has a finite number of moves to choose from at any point in the game and because every game ends after a finite number of moves. Because it is a finite game, Nash's theorem establishes that it has an equilibrium.

This suggests that chess should be uninteresting, at least when played by experienced players. Assuming Player I plays the equilibrium strategy, Player II can do no better than playing the equilibrium strategy, and vice versa. Thus, we should expect experienced players to implement the equilibrium strategies every time, and the outcome to be familiar and predictable.

Yet Nash's theorem only establishes the existence of an equilibrium; it does not reveal what the equilibrium strategies are. As of yet, no computer is powerful enough to figure out what they are. And even if we knew what the strategies were, they might be too complex for human beings to implement. Thus, chess is likely to remain interesting for a good long time.

Before we move on, two more exercises.

Exercise 10.16 Chicken The game of **chicken** was popularized in the 1955 film *Rebel Without a Cause*, starring James Dean. The game is typically played by two people who drive cars straight at each other at high speed; the person who swerves first is called "chicken" and becomes an object of contempt. In this game, the British philosopher Bertrand Russell saw an analogy with Cold War policy:

> Since the nuclear stalemate became apparent, the Governments of East and West have adopted the policy which [US Secretary of State] Mr Dulles calls "brinkmanship". This is a policy adapted from a sport which, I am told, is practised by some youthful degenerates. This sport is called "Chicken!" ... As played by irresponsible boys, this game is considered decadent and immoral, though only the lives of the players are risked. But when the game is played by eminent statesmen, who risk not only their own lives but those of many hundreds of millions of human beings, it is thought on both sides that the statesmen on one side are displaying a high degree of wisdom and courage, and only the statesmen on the other side are reprehensible. This, of course, is absurd.

Imagine that each player has the choice between swerving (S) and not swerving (¬S), and that the payoff structure is that of Table 10.10. Find all Nash equilibria in this game.

Table 10.10 Chicken

	S	¬S
S	3,3	2,5
¬S	5,2	1,1

In a branch of game theory called **evolutionary game theory**, this game figures prominently under the heading of **hawk & dove**. Hawks are willing to fight to the death whereas doves easily give up. The best possible outcome for you is when you are a hawk and your opponent is a dove, the second best outcome is when both of you are doves, the third best is when you are a dove and your opponent is a hawk, and the worst possible outcome is when both of you are hawks. If doves "swerve" and hawks do not, the payoff structure of hawk & dove is the same as that of chicken. In evolutionary game theory, the mixed-strategy equilibrium is interpreted as describing a population in which hawks and doves coexist in given proportions – just as they do in the real world.

Exercise 10.17 The stag hunt This game is due to Jean-Jacques Rousseau, the eighteenth-century French philosopher. Rousseau describes a scenario in which two individuals go hunting. The two can hunt hare or deer but not both. Anyone can catch a hare by himself, but the only way to bag a deer is for both hunters to pursue the deer. A deer is much more valuable than a hare. The **stag hunt**, which is thought to provide an important parable for social cooperation, is usually represented as in Table 10.11. What are the Nash equilibria (in pure and mixed strategies) of this game?

Table 10.11 The stag hunt

	D	H
D	3,3	0,1
H	1,0	1,1

Notice how superficially subtle differences in the payoff structure between the prisoners' dilemma (Table 10.5(b)), chicken (Table 10.10), and the stag hunt (Table 10.11) lead to radically different analytical results.

10.4 Equilibrium refinements

The concept of a Nash equilibrium is associated with a number of controversial results. In this section, we consider two alternative equilibrium concepts, designed to deal with supposedly problematic cases.

Example 10.18 Trembling-hand perfection Let us return to Table 10.9(c). As you know, ⟨U, L⟩ is a Nash equilibrium. ⟨D, L⟩ is *not* an equilibrium, since Player I can improve his payoff by playing U instead of D, and neither is ⟨U, R⟩. But consider ⟨D, R⟩. If Player II plays R, Player I can do no better than playing D; if Player I plays D, Player II can do no better than playing R. Thus, ⟨D, R⟩ is a Nash equilibrium. There are no mixed equilibria. No matter what Player II does, Player I will never be indifferent between U and D, and no matter what Player I does, Player II will never be indifferent between L and R.

There is nothing wrong with the analysis here, but there is something odd about the second equilibrium ⟨D, R⟩. A strategy *X* is said to weakly dominate another strategy *Y* if choosing *X* is no worse than choosing *Y* no matter what the other player does, and choosing X is better than choosing Y for at least one strategy available to the other player. In Example 10.18, U weakly dominates D and L weakly dominates R. Thus, there seems to be no reason why rational individuals would ever play the second equilibrium ⟨D, R⟩. And the problem is not that (1,1) Pareto dominates (0,0) (see Section 10.2).

The concept of a **trembling-hand-perfect equilibrium** was designed to handle this kind of situation.

Definition 10.19 Trembling-hand-perfect equilibrium *A trembling-hand-perfect equilibrium is a Nash equilibrium that remains a best response for each player even when others have some minuscule probability of* ***trembling****, that is, accidentally playing an out-of-equilibrium strategy.*

In Table 10.9(c), ⟨U, L⟩ is a trembling-hand-perfect equilibrium: even if there is a minuscule probability $\varepsilon > 0$ that Player II plays R, she still plays L with probability $(1 - \varepsilon)$ and U remains a best response for Player I. (And similarly for the other player.) By contrast, ⟨D, R⟩ is not trembling-hand perfect. If there is a minuscule probability $\varepsilon > 0$ that Player II plays L, no matter how small, U is a strictly preferred strategy for Player I.

Exercise 10.20 Battle of the sexes, cont. Are the two pure-strategy equilibria in the battle of the sexes (Table 10.8) trembling-hand perfect?

Trembling-hand-perfect equilibrium is a **refinement** of Nash equilibrium. This means that every trembling-hand-perfect equilibrium is a Nash equilibrium, but not every Nash equilibrium is trembling-hand perfect.

Exercise 10.21 Trembling-hand perfection Find (**a**) all Nash equilibria in pure strategies in Table 10.12 and (**b**) identify which of them are trembling-hand perfect.

Table 10.12 Trembling-hand perfection, cont.

	L	M	R
U	1,4	0,0	0,0
M	0,0	4,1	0,0
D	0,0	0,0	0,0

Substituting the concept of trembling-hand-perfect equilibrium for the concept of Nash equilibrium would eliminate some problematic implications of the Nash-equilibrium concept. The concept of trembling-hand equilibrium is, however, insufficient to deal with all problematic cases.

Example 10.22 Credible versus non-credible threats Consider a game with two stages. In the first stage, Player I plays U or D. If Player I plays D, both players get a payoff of 2. If Player I plays U, it is Player II's turn. In the second stage, Player II plays L or R; if Player II plays L, Player I gets 5 and Player II gets 1. If Player II plays R, both get 0. What are the Nash equilibria of this game?

The game can be represented as in Table 10.13. There are two Nash equilibria: ⟨U, L⟩ and ⟨D, R⟩.

Table 10.13 Subgame perfection

	L	R
U	5,1	0,0
D	2,2	2,2

Yet there is something odd about the second of the two equilibria. The only thing preventing Player I from playing U is the threat of Player II playing R. But suppose that Player I did play U. Then, Player II would have the choice of playing L (for a payoff of 1) and R (for a payoff of 0). In the second stage, it is not in Player II's interest to play R. So while it is perfectly possible for Player II to threaten to play R if Player I plays U, she would have no interest in carrying out the threat. Knowing this, it appears that Player I should just go ahead and play U. Game theorists say that the problem is that Player II's threat is not **credible**. Many people think that it is problematic that a Nash equilibrium might involve a non-credible threat. And the problem is not that the Nash equilibrium is not trembling-hand perfect.

Games with multiple stages are called **sequential**. To analyze such games, it is often useful to use a tree-like representation called the **extensive form**. The game from Example 10.22 can, for example, be represented as in Figure 10.3. This representation affords another way to spell out the problem. Consider that part of the game which starts at the node where Player II moves (see the shaded area in the figure). We refer to it as a **subgame** of the original game. In the subgame, Player II has two strategies (L and R) and there is only one Nash equilibrium: to play L (for a payoff of 1) rather than R (for a payoff of 0). Yet the Nash equilibrium in the entire game requires Player II to play R in the subgame. One way to spell out the problem, then, is to say that the Nash equilibrium of the *game* requires Player II to play a strategy that is not a Nash equilibrium in the *subgame*.

Consistent with this analysis, game theorists have proposed another equilibrium concept: **subgame-perfect equilibrium**. As suggested in the previous paragraph, a subgame of a game is any part of that game which in itself constitutes a game. A game is always its own subgame, but in this case there is a proper subgame starting at the node where Player II moves.

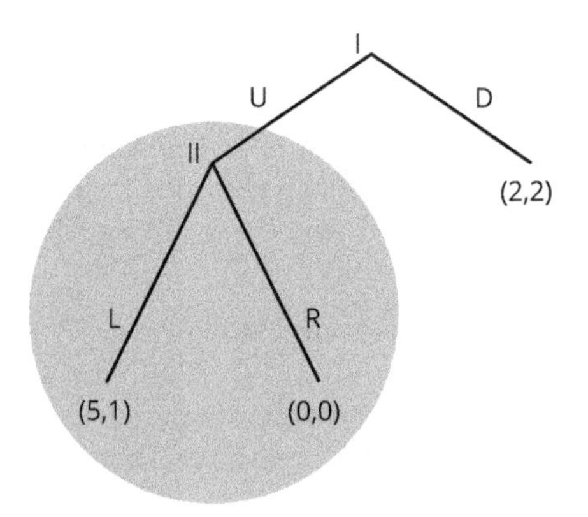

Figure 10.3 Subgame perfection

Definition 10.23 Subgame-perfect equilibrium *A subgame-perfect equilibrium is a strategy profile that constitutes a Nash equilibrium in each subgame.*

Like trembling-hand-perfect equilibrium, subgame-perfect equilibrium is a refinement of Nash equilibrium: all subgame-perfect equilibria are Nash equilibria, but not all Nash equilibria are subgame perfect.

One way to find subgame-perfect equilibria is to start at the end and use backward induction. Backward induction would tell you to start with the last subgame, that is, at the node where Player II moves (the shaded area of the figure). Since L would lead to a payoff of 1 and R would lead to a payoff of 0, L is the unique Nash equilibrium strategy. So, in subgame-perfect equilibrium, Player II will play L. Given that Player II will play L, what will Player I do at the first node? Player I has the choice between playing U for a payoff of 3, and playing D for a payoff of 2. Thus, Player I will play U. In brief, there is only one subgame-perfect equilibrium in this game, and it is $\langle U, L \rangle$.

Example 10.24 MAD Mutually assured destruction (MAD) is a military doctrine according to which two superpowers (such as the US and the USSR) can maintain peace by threatening to annihilate the human race in the event of an enemy attack. Suppose that the US moves first in a game like that in Figure 10.3. The US can launch an attack (U) or not launch an attack (D). If it launches an attack, the USSR can refrain from retaliating (L) or annihilate the human race (R). Given the payoff structure of the game in the figure, $\langle D, R \rangle$ is a Nash equilibrium. The doctrine is flawed, however, in that the threat is not credible: the MAD Nash equilibrium presupposes that USSR forces are willing to annihilate the human race in the event of a US attack, which would obviously not be in their interest. Thus, the MAD Nash equilibrium is not subgame perfect.

In Stanley Kubrick's 1963 film *Dr Strangelove*, the USSR tries to circumvent the problem by building a **doomsday machine**: a machine that in the event of an enemy attack (or when tampered with) automatically launches an attack powerful enough to annihilate the human race. Such a machine would solve the strategic problem, because it guarantees retaliation to enemy attack and therefore makes the threat credible. As the film illustrates, however, such machines are associated with other problems. To begin with, you must not forget to tell your enemy that you built one.

Exercise 10.25 Subgame perfection Use backward induction to find the unique subgame-perfect equilibrium in the game in Figure 10.4. Recall that a strategy is a

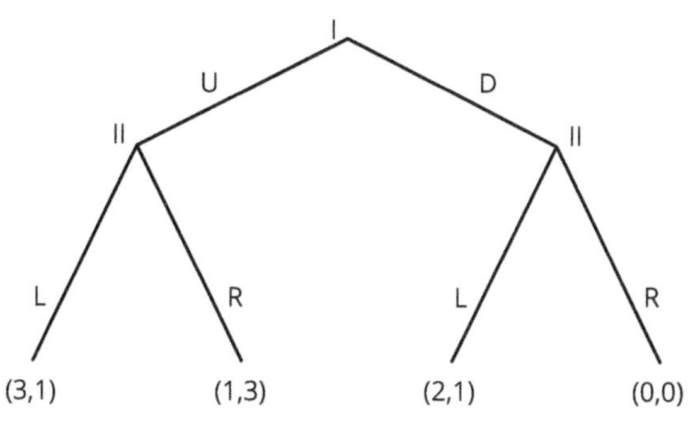

Figure 10.4 Subgame perfection exercise

complete plan of action, which means that a strategy for Player II will have the form "L at the first node and L at the second (LL)," "R at the first node and L at the second (RL)," and the like. In this game, then, whereas Player I only has two strategies to choose from, Player II has four.

Finally, one more exercise:

Exercise 10.26 The centipede game The centipede game has four stages (see Figure 10.5). At each stage, a player can Take, thereby ending the game, or Pass, thereby increasing the total payoff and allowing the other player to move.
(**a**) Use backward induction to find the unique subgame-perfect equilibrium.
(**b**) Would the outcome of the game differ if it had 1000 stages instead of four?

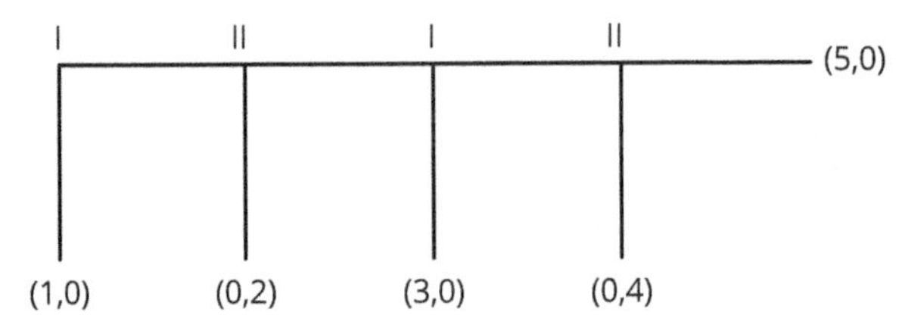

Figure 10.5 The centipede game

10.5 Discussion

Like the theories we came across earlier in this book, analytical game theory admits of descriptive and normative interpretations. According to the descriptive interpretation, game theory captures the manner in which people behave when they engage in strategic interactions. In this view, game theory predicts that people will jointly choose an equilibrium strategy profile. Specific predictions, of course, will depend not only on the game played, but on the equilibrium concept that is employed. According to the normative interpretation, game theory describes how rational agents should behave when they engage in strategic interaction. In this view, game theory says that players should jointly choose an equilibrium strategy profile. Again, the specific advice offered by the theory will depend on the game played and the equilibrium concept employed.

Either way, living up to the demands of game theory is extremely challenging. To employ backwards induction, for one thing, you need to think all the way ahead until the end of the game – the very last round you might find yourself playing – and then work your way back in time to the present moment. Political and military leaders gearing up for some confrontation, in order to be rational, need to map out the entire structure of the game and think through what they would do at each node. Because rationality in games is so very demanding, we should expect many real people to fail at it, even when they try. Unfortunately, people who fail to act rationally in games include actual political and military leaders pushing their countries to the brink of war. And the people who have to suffer the consequences of war rarely if ever can be satisfied that the people in charge have fully thought through the consequences.

One thing to notice is that the nature of the game is defined by the distribution of utilities in the matrix. What makes one game a prisoners' dilemma and another game a stag hunt is the manner in which the utilities are distributed. A strategic interaction defined in terms of dollars won, or the like, therefore is not strictly speaking a game. Hence, if you do not know how much utility each player derives from each outcome, you do not know what game they are playing. This is a conceptual point, but it can lead to practical problems. As long as you do not know what game people are playing, you cannot (without making additional assumptions) even begin to use the tools of game theory – Nash equilibrium, and all that – to analyze the interaction.

Another thing to notice is that a game does not necessarily have a unique equilibrium. And analytical game theory in itself does not contain the resources required to identify *which* equilibrium people will or should play. This fact causes another practical problem. While the theory can be interpreted as predicting that the outcome of strategic interaction will or should be a Nash equilibrium, this is only to say that some Nash equilibrium will or should obtain. In this sense, then, the theory is indeterminate. And because some games (like the indefinitely repeated prisoners' dilemma) have an infinite number of equilibria, the theory is radically indeterminate.

If we want determinate predictions, we must augment the theory with additional resources. The most famous such effort is the theory of **focal points**, due to 2005 Nobel laureate Thomas C. Schelling. According to this theory, some equilibria tend to stand out in the minds of the players. Schelling predicts that people will frequently succeed in selecting such equilibria. The precise feature of an equilibrium that makes it stand out in the minds of the players is far from obvious.

> Finding the key ... may depend on imagination more than on logic; it may depend on analogy, precedent, accidental arrangement, symmetry, aesthetic or geometric configuration, casuistic reasoning, and who the parties are and what they know about each other.

This theory can explain why people favor ⟨U, L⟩ over ⟨D, R⟩ in Table 10.9(c). When there is a unique Pareto-optimal outcome that also happens to be a Nash equilibrium, it seems plausible to assume that people will use Pareto optimality as a focal point. If so, we might be able to explain observed behavior without making the transition to trembling-hand-perfect equilibrium.

In the next chapter, we will explore behavioral economists' challenge to analytical game theory.

Problem 10.27 Lunch date *Suppose you have to meet a friend at noon tomorrow somewhere in your town, but you forgot to specify the exact location. Where would you go? Ask your friends where they would go. Would you succeed in coordinating?*

Additional exercises

Exercise 10.28 For each of the games in Table 10.14, identify the Nash equilibria in pure strategies (if any).

Table 10.14 More Nash equilibrium exercises

(a)

	L	R
U	2,2	0,0
D	0,0	3,3

(b)

	L	R
U	2,2	7,6
D	6,7	2,2

(c)

	L	R
U	3,0	2,1
D	7,5	1,6

(d)

	L	R
U	1,1	1,1
D	0,0	1,1

(e)

	L	R
U	1,1	2,0
D	0,2	2,2

Exercise 10.29 Which of the Nash equilibria in pure strategies are trembling-hand perfect (if any)?

Exercise 10.30 Which of the Nash equilibria in pure strategies are Pareto-optimal outcomes in the relevant game?

Exercise 10.31 For each of the games in Table 10.14, identify the Nash equilibria in mixed strategies (if any). Assume Player I plays U with probability p and Player II plays L with probability q.

Exercise 10.32 Find all subgame-perfect equilibria in the game in Figure 10.6. The game has two players and three stages. Player I moves at the first and third stages, and Player II moves at the second stage.

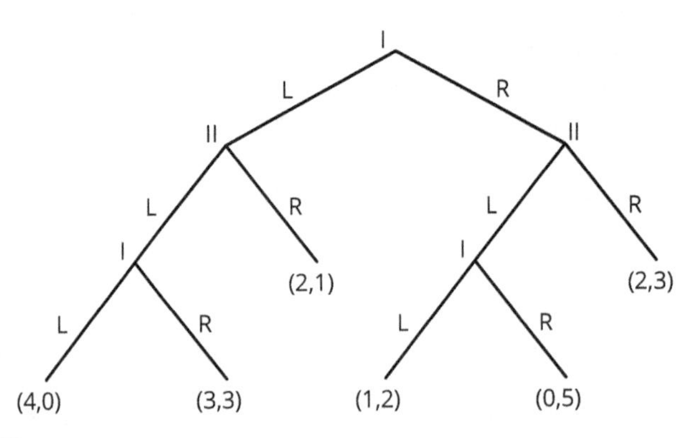

Figure 10.6 Three-stage game

Exercise 10.33 Paradoxes of rationality Experimental economists have invited students with different majors to play prisoners'-dilemma games against each other. In an experiment pitching economics majors against economics majors, and non-majors against non-majors, who would you expect to do better?

Chapter 11 contains more game-theoretic exercises.

Further reading

There are many fine introductions to game theory, including Binmore (2007), Dixit et al. (2009), and Osborne and Rubinstein (1994). *Spousonomics* is Szuchman and Anderson (2011, pp. xii–xv, 294–8). Life in the state of nature is described in Hobbes (1994 [1651], xiii, 8–9, p. 76). Skyrms (1996) discusses the doctrine of mutually assured destruction (pp. 22–5) and the games of chicken and hawk & dove (pp. 65–7); Russell (1959, p. 30) examines the game of chicken. The theory of focal points is due to Schelling (1960, p. 57). Evidence about economics majors' performance in prisoners'-dilemma games can be found in Frank et al. (1993).

11 BEHAVIORAL GAME THEORY

Learning objectives

After studying this chapter you will:

- Know how to incorporate social preferences into the theory of games
- Understand how considerations of fairness, justice, reciprocity, trust, and limited strategic thinking matter to people's performance in games
- Apply game theory in the real world – but also understand when it would be unwise to do so

11.1 Introduction

Analytical game theory is in many ways a huge success story: it is increasingly becoming the foundation of other subdisciplines of economics (including microeconomics), and it has migrated to philosophy, biology, political science, government, public policy, and elsewhere. But, as we will see in this chapter, its descriptive adequacy and normative correctness are controversial. **Behavioral game theory** examines the degree to which analytical game theory succeeds in capturing the behavior of real people engaging in strategic interaction, and proposes extensions of analytical game theory in the interest of capturing that behavior. Some of the proposed extensions to analytical game theory do not in fact constitute deviations from neoclassical orthodoxy. Thus, there is nothing distinctively behavioral about some of the models discussed under the heading "behavioral game theory." Other models, however, constitute real deviations from neoclassical orthodoxy – and as such deserve to be called behavioral.

11.2 Social preferences: Altruism, envy, fairness, and justice

Much of the literature on social preferences is driven by data from two games: the **ultimatum game** and the **dictator game**. Both are played by two agents: a **proposer** (Player I) and a **responder** (Player II). Here, these games are outlined as they are presented to participants in laboratory experiments, where outcomes are described in terms of dollars and cents rather than in terms of the utilities that players derive from them. In order to analyze the interaction, we need to transform the dollars and cents into utilities. Strictly speaking, as you know from the previous chapter, you do not even know what game the players are playing until you have identified payoffs in utility terms.

The ultimatum game has two stages. At the outset, the proposer is given a fixed amount of money; for purposes of this discussion, let us suppose it is $10. In the first

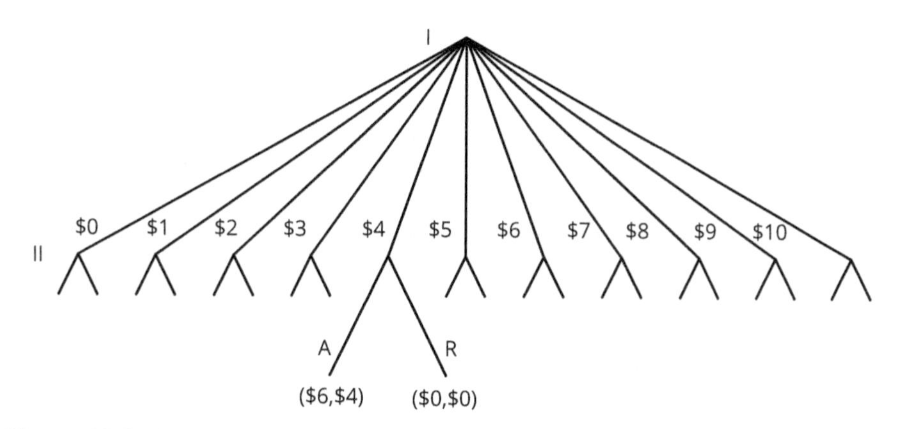

Figure 11.1 The ultimatum game (in dollar terms)

stage, Player I proposes a division of the dollar amount; that is, the proposer offers some share of the $10 to the other player. The proposer might propose to give it all away (leaving nothing for himself), to give none of it away (leaving all of it for himself), or to offer some fraction of the $10 to the other player (leaving the balance for himself). For example, the proposer might offer $4, leaving $6 for himself. In the second stage, the responder accepts or rejects the proposed division. If she accepts, both players receive their proposed share; if she rejects, neither player receives anything. The ultimatum game can be represented as in Figure 11.1. In this figure, I have omitted all branches corresponding to fractional amounts, and I have truncated all but one set of branches representing Player II's decision in the second stage.

Example 11.1 Dividing the cake When two children have to divide a cake, they sometimes follow a procedure in which the first child splits the cake in two and the second chooses first. Given that Kid II will choose the largest piece, Kid I will want to divide the cake as evenly as possible, thereby guaranteeing a 50–50 split. We can easily imagine a variation of this procedure, in which Kid I proposes a division of the cake, and Kid II gets to approve (in which case each child gets the proposed piece) or disapprove (in which case the parents give the entire cake to the dog). The new procedure would constitute an example of the ultimatum game.

The ultimatum game has been extensively studied by experimental economists. According to one survey of the results:

> The results ... are very regular. Modal and median ultimatum offers are usually 40–50 percent and means are 30–40 percent. There are hardly any offers in the outlying categories 0, 1–10, and the hyper-fair category 51–100. Offers of 40–50 are rarely rejected. Offers below 20 percent or so are rejected about half the time.

Based on these results, we should expect responders to reject offers below $2 when playing the (one-shot, anonymous) game in Figure 11.1. But such low offers would be rare. By and large, we should expect offers in the \$3–\$5 range. Many people have drawn the conclusion that these results are inconsistent with analytical game theory.

The observed outcomes are quite consistent with Nash equilibrium predictions, however, even when players care about nothing but their own dollar payoffs. Let us assume that each individual is simply trying to maximize his or her dollar payoffs, and that $u(x) = x$. Player I must choose an amount to offer the other player; suppose that he offers \$4. Player II's strategy is a little more convoluted, because a strategy must specify what a player will do under all possible circumstances (see Section 10.2). Player II's strategy, therefore, must specify what she will do at each of the nodes where she might find herself. Suppose that Player II rejects all proposed divisions in which Player I offers less than \$4 and accepts all others. If so, the two players are in equilibrium. If Player I decreased his offer, it would be rejected, and both would receive nothing; if he increased his offer, it would be accepted, but he would receive less. Given that Player I offers \$4, Player II can do no better than accepting. In brief, data on the ultimatum game do not represent a puzzle from the point of view of the theory we learned in Section 10.2, since observed outcomes are consistent with Nash equilibrium predictions. (Given the many equilibria of this game, though, this is not saying much.)

Nevertheless, many people think the Nash equilibrium prediction is problematic, since it requires players to reject positive offers off the equilibrium path. One way to articulate the problem is to say that the Nash equilibrium in the game requires players to reject what is in effect a dominant strategy (namely, accepting) in subgames that start at nodes where Player II moves. Another way to articulate the problem is to say that the equilibrium is not subgame perfect and that Player II's threat to reject a low offer is not credible (see Section 10.4). We might, therefore, restrict our analysis to subgame-perfect equilibria. There is only one subgame-perfect equilibrium in this game. In this equilibrium, Player I offers nothing and Player II accepts all offers. This might be counterintuitive. But it is a Nash equilibrium because (a) given Player I's offer, Player II would be no better off if she rejected it, and (b) given that Player II accepts all offers, Player I can do no better than to keep all the money for himself. It is a subgame-perfect equilibrium, because Player II's strategy is also a Nash equilibrium strategy in all subgames: no matter what she has been offered, she cannot improve her payoff by rejecting the offer. A prediction based on the idea of subgame-perfect equilibrium, given our assumption about the two players' utility function, is in fact inconsistent with the experimental results.

The dictator game resembles the ultimatum game, except for the fact that the second stage has been eliminated. In dollar terms and assuming that the proposer starts out with \$10, the dictator game can be represented as in Figure 11.2; again, I have left out all the branches representing fractional amounts. On the assumption that the players' utility function remains $u(x) = x$, there is only one Nash equilibrium and therefore only one subgame-perfect equilibrium: the case in which Player I offers nothing to the responder and keeps all the money for himself.

Example 11.2 Charitable donations One example of a dictator game played in the real world involves a person's decision about whether to give money to charity. Whenever you walk by a beggar, for example, you are in effect playing a dictator game

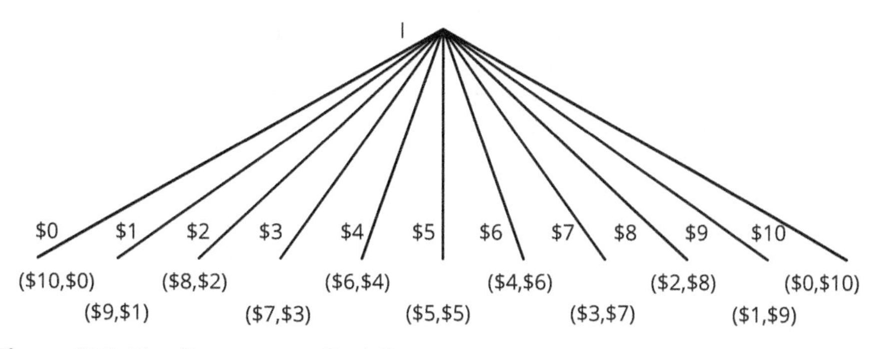

Figure 11.2 The dictator game (in dollar terms)

in which you must allocate some share of the money in your pocket to the panhandler. If you walk on, you have in effect picked the maximally selfish, Nash-equilibrium allocation.

Experimental evidence suggests that proposers in the (one-shot, anonymous) dictator game typically offer less than proposers in the (one-shot, anonymous) ultimatum game. That said, many proposers are nevertheless willing to share a substantial amount (10–30 percent) of their initial allocation. In the version of the game in Figure 11.2, this means that the proposer would be willing to share $1–$3 with the responder, even though the latter has no way to penalize proposers who offer nothing.

The literature on **social preference** grapples with these phenomena. This literature is based on the assumption that people sometimes care not only about their own attainment but about other people's attainment too. We can model this by assuming that a person P's utility function $u_P(\cdot)$ has two or more arguments. Thus, P's utility function may be given by $u_P(x, y)$, where x is P's attainment and y is the other person Q's attainment.

It is quite possible for P to derive positive utility from Q's attainment, so that $u_P(x, y)$ is an increasing function of y. For example, P's utility function may be $u_P(x, y) = 3/5\sqrt{x} + 2/5\sqrt{y}$. If so, P is said to be **altruistic** and to have **altruistic preferences**. Some parents, relatives, friends, and admirers are willing to make real sacrifices in order to improve another person's situation. This is easily explained if we assume that their utility is (in part) a function of the other person's attainment. Altruism in this sense might be what Adam Smith had in mind when he said: "[There] are evidently some principles in [man's] nature, which interest him in the fortune of others, and render their happiness necessary to him" (quoted in Section 1.2).

There is no requirement that P derive *positive* utility from Q's attainment. In fact, $u_P(x, y)$ may be a decreasing function of y. For example, P's utility function may be $u_P(x, y) = \sqrt{x} - \sqrt{y}$. This specification entails that P's utility goes up when Q's attainment goes down and *vice versa*. If so, P is said to be **envious**. Some gas–electric hybrid-car owners derive deep satisfaction from rising gasoline prices. This cannot be explained by reference to the financial effects of gasoline prices on the hybrid owner: rising gasoline prices will hurt hybrid owners too. But it can be explained if we assume that the disutility hybrid owners derive from getting less gasoline for their dollar is outweighed by the utility they derive from knowing that SUV owners suffer even more.

There is no reason to restrict our analysis to these functional forms. According to a common interpretation of John Rawls's theory of justice (see Section 6.2), societies should be ordered with respect to justice based on the welfare of the least fortunate in each society. Thus, a person with **Rawlsian preferences** might try to maximize the minimum utility associated with the allocation. If each individual derives $\sqrt{x}$ utiles from his or her private consumption x, the Rawlsian P might maximize $u_P(x, y) = \min(\sqrt{x}, \sqrt{y})$. Rawls uses the term "justice as fairness" to describe his theory, so Rawlsian preferences could also be described as preferences for **fairness**.

Another agent might care about the degree of inequality among the relevant agents, so as to rank allocations based on the absolute difference between the best and worst off. Such an agent is said to be **inequality averse** and to have inequality-averse preferences. If each individual derives $\sqrt{x}$ utiles from his or her private consumption x, the inequality-averse P might wish to minimize the absolute difference between each person's utility. This amounts to maximizing $u_P(x, y) = -|\sqrt{x} - \sqrt{y}|$. Such agents care about equality for its own sake, unlike the Rawlsians who (given the definition above) care about equality only insofar as it benefits the least well off. Because the inequality-averse agent ends up assessing the outcomes of ultimatum and dictator games so similarly to the Rawlsian agent, I will not discuss this case further.

Utilitarians such as Bentham, whom we came across in Section 1.2, believe that we should pursue the greatest good for the greatest number. Thus, a utilitarian agent might try to maximize the total amount of utility derived from private consumption. If each individual derives $\sqrt{x}$ utiles from his or her private consumption x, the utilitarian P might maximize $u_P(x, y) = \sqrt{x} + \sqrt{y}$. So understood, **utilitarian preferences** constitute a special case of altruistic preferences. Obviously, this list is far from exhaustive: any agent who derives utility from another agent's private consumption counts as having social preferences.

To see how the shape of the proposer's utility function affects his assessment of the various outcomes in the ultimatum and dictator games, see Table 11.1. When the payoff is ($0, $0), all kinds of agents receive zero utility.

Table 11.1 Dictator game utility payoffs (maxima in boldface)

Payoffs	Player *P*'s utility function $u_P(x, y)$				
(x, y)	$\sqrt{x}$	$\sqrt{x}+\sqrt{y}$	$\sqrt{x}-\sqrt{y}$	$\min(\sqrt{x}, \sqrt{y})$	$3/5\sqrt{x}+2/5\sqrt{y}$
($10, $0)	**3.16**	3.16	**3.16**	0.00	1.90
($9, $1)	3.00	4.00	2.00	1.00	2.20
($8, $2)	2.83	4.24	1.41	1.41	2.26
($7, $3)	2.65	4.38	0.91	1.73	**2.28**
($6, $4)	2.45	4.45	0.45	2.00	2.27
($5, $5)	2.24	**4.47**	0.00	**2.24**	2.24
($4, $6)	2.00	4.45	−0.45	2.00	2.18
($3, $7)	1.73	4.38	−0.91	1.73	2.10
($2, $8)	1.41	4.24	−1.41	1.41	1.98
($1, $9)	1.00	4.00	−2.00	1.00	1.80
($0, $10)	0.00	3.16	−3.16	0.00	1.26

Egoists and enviers prefer the outcome where they get all the money. Utilitarians and Rawlsians prefer outcomes where the dollar amount is split evenly. Finally, an altruist who gives a little more weight to her own private utility than does a utilitarian might prefer the outcome ($7, $3) to all others.

It goes without saying that certain kinds of social preference would go a long way toward explaining proposers' behavior in the ultimatum and dictator games. Altruists, Rawlsians, and utilitarians actually *prefer* more equal outcomes. For such agents, there is nothing mysterious about the fact that they voluntarily offer non-zero amounts to responders.

Exercise 11.3 Altruism and the ultimatum game Imagine the ultimatum game from Figure 11.1 played by two utilitarians with $u(x, y) = \sqrt{x} + \sqrt{y}$. Find the unique subgame-perfect equilibrium in this game.

The literature on social preferences underscores a point from the previous chapter, namely that the game people are playing depends on their utility functions. The following exercise illustrates how agents with different utility functions end up playing very different games, even when their interactions superficially might look identical.

Exercise 11.4 Social preferences and the prisoners' dilemma Find the Nash equilibria in pure strategies in Table 11.2, when played by:
(**a**) Two egoists, for whom $u(x, y) = \sqrt{x}$.
(**b**) Two utilitarians, for whom $u(x, y) = \sqrt{x} + \sqrt{y}$.
(**c**) Two enviers, for whom $u(x, y) = \sqrt{x} - \sqrt{y}$.
(**d**) Two Rawlsians, for which $u(x, y) = \min(\sqrt{x}, \sqrt{y})$.

Notice that this game superficially (in dollar terms) has the payoff structure of the prisoners' dilemma (Table 10.5 on page 223).

Table 11.2 Prisoners' dilemma (in dollar terms)

	C	D
C	$16, $16	$0, $25
D	$25, $0	$9, $9

Social preferences are fascinating and important. Economists who do not allow for the possibility of social preferences run the risk of committing terrible mistakes, whether they are trying to explain or predict behavior or to design optimal incentives. If we are mistaken about the players' utility function, we will not even know what game they are playing. Consequently, our analysis of their interaction is likely to fail. Consider the game in Table 11.2. Superficially, when expressed in dollar terms, this looks like a prisoners' dilemma. But as Exercise 11.4 showed, the players may in fact be playing a very different game.

Notice, however, that the entire analysis in this section can be completed without departing from neoclassical orthodoxy. As we know from Sections 1.1 and 2.6, the standard approach makes no assumptions about the nature of people's preferences. Consequently, it makes no assumptions about what can enter as an argument in people's utility function. It is sometimes argued that the results from the dictator and ultimatum games refute the "selfishness axiom" of neoclassical economics. But this charge is misguided: not only is there no such axiom in the calculus, but selfishness is not even entailed by the theory. There is nothing specifically behavioral about models of social preferences; if anything, the analysis shows the strength and power of the neoclassical framework.

Exercise 11.5 The Stora Rör Swimming Association The Stora Rör Swimming Association is a non-profit organization teaching open-water swimming skills and water safety to kids in the Baltic. The Association is funded in large part by means of an annual auction, where members donate home-baked goods, and the like, to be auctioned off to other members. It is an open-bid auction, meaning that all bids immediately become common knowledge. The event generates a lot of revenue because members compete in bidding the most absurd amounts: a modest, although delicious, home-baked cake can easily fetch a hundred dollars or more. Is this kind of behavior consistent with standard rationality? How so?

11.3 Intentions, reciprocity, and trust

There is, however, something awkward about the account offered in the previous section. In order to accommodate the behavior of the proposer in the dictator game, we postulate that proposers are largely altruistic. But in order to accommodate the behavior of responders in the ultimatum game, this approach is inadequate. As you can tell from the third column of Table 11.1, an altruist with utility function $u(x, y) = \sqrt{x} + \sqrt{y}$ would prefer any outcome to (\$0, \$0). In subgame-perfect equilibrium, therefore, an altruistic responder would accept all offers (see Exercise 11.3). But this is inconsistent with the observation that low offers frequently are rejected. Of all the agents described in the table, only the enviers prefer (\$0, \$0) to a sharply unfavorable division like (\$8, \$2). And it would be inconsistent to postulate that people are simultaneously altruistic (to explain their behavior in the dictator game) and envious (to accommodate their behavior in the ultimatum game).

There are other awkward results. In a variation of the ultimatum game, responders were found to reject the uneven division (\$8, \$2) if the proposer had the choice between (\$8, \$2) and the even division (\$5, \$5), but accept it if the proposer had the choice between (\$8, \$2) and the maximally uneven division (\$10, \$0). This makes no sense from the point of view of a responder who evaluates final outcomes in accordance with either one of the social preference functions in the previous section. According to each of those models, either (\$8, \$2) is better than (\$0, \$0) or it is not; the choices available to the proposer do not matter at all.

To some analysts, these results suggest that responders do not base their decisions on the final outcome of the game alone, but are (at least in part) responsive to what they see as the proposer's **intentions**. In this view, people are willing to reward people who are perceived as having good intentions and to punish people who are perceived as having bad intentions. A proposer who offers \$2 rather than \$0 is interpreted as having good intentions, even if the resulting allocation is uneven, whereas a proposer who offers \$2 rather than \$5 is interpreted as having bad intentions. Sometimes, these results are discussed in terms of **reciprocity** or **reciprocal altruism**. Respondents are said to exhibit **positive reciprocity** when they reward players with good intentions and **negative reciprocity** when they punish proposers with bad intentions. Thus, a responder in the ultimatum game who rejects a small positive offer from the proposer is said to exhibit negative reciprocity.

Reciprocity is often invoked in discussions of the **trust game**. This game is played by two players: a **sender** (Player I) and a **receiver** (Player II). At the outset, both are awarded some initial amount, let us suppose \$10. In the first stage, the sender sends some share $\$x$ of his \$10 to the receiver. The amount is sometimes called an "investment." Before the investment is received by the receiver, it is multiplied by some factor, let us say three. Thus, the receiver receives $\$3x$. In the second stage, the receiver returns to the sender some share $\$y$ of her total allocation $\$10 + \$3x$. The final outcome, then, is $(\$10 - \$x + \$y, \$10 + \$3x - \$y)$. The game is called the trust game for the obvious reason that the sender might **trust** the receiver to return some of her gains to him. When both agents maximize $u(x) = x$, there is only one subgame-perfect equilibrium in this game. Since the receiver maximizes by keeping all her money, y will equal zero. Notice that the receiver is in effect playing a dictator game with the sender as a beneficiary. Given that none of the share $\$x$ will be returned to him, the sender will keep all his money and x will equal zero. Notice that the resulting allocation (\$10, \$10) is Pareto inferior to many other attainable allocations. If $x = 10$ and $y = 20$, for example, the final outcome is (\$20, \$20).

Example 11.6 Investment decisions Suppose that you have the opportunity to invest in a promising business venture, but that you have no way to recover your expenses in case your business partner turns out to be unreliable. In order to capture the gains from trade, you simply must trust the partner to do the job. If so, you and your business partner are playing a trust game against each other. If you play the subgame-perfect equilibrium strategy, you will never invest. But if you never invest, you will never capture any of the available surplus.

Experimental economists have found that senders in the (one-shot, anonymous) trust game on the mean send about half of their initial allocation, and that receivers return a little less than what was invested. Given the figures from the previous paragraph, we should expect senders to send about \$5 and receivers to return somewhere between \$4 and \$5. Thus, Player II succeeds in capturing some, but not all, of the available surplus. (There is a great deal of variability across studies, however.)

Why would a responder care to return some of the sender's investment when the latter has no way to penalize a receiver who returns nothing? According to one frequent answer, the receiver feels like she must reciprocate the sender's investment. Thus, a receiver who returns some of the sender's investment is said to exhibit positive reciprocity. The receiver's behavior is also consistent with altruism and inequality aversion. Meanwhile, the sender's behavior is thought to reflect the expectation that his investment will be repaid in combination with some degree of altruism.

A similar analysis obtains in the case of prisoners'-dilemma and **public-goods games**. We know the prisoners' dilemma from Section 10.2. In a typical public-goods game, there are n players. For purposes of this discussion, let us assume that there are three players. Each is given an initial allocation, say $10. The game has only one stage, in which all players move simultaneously. Each player has the option of transferring a share of their initial allocation to a public account. The money in the public account is multiplied by some factor between one and three, say two, and split evenly between the players.

Given the nature of the game, the Pareto-optimal outcome results when all players transfer all their money to the public account. In this case, the payoff is ($20, $20, $20). But the Pareto-optimal outcome is not a Nash equilibrium, for each player can improve his outcome by transferring less. Indeed, there is only one Nash equilibrium in this game, and it is when nobody transfers any money to the public account and the outcome is ($10, $10, $10). Public-goods games therefore have certain structural similarities to the prisoners' dilemma.

The nature of this interaction should be familiar. Perhaps a set of roommates all prefer the state in which all assist in washing the dishes to the state in which nobody washes the dishes, but no matter what the others do, each roommate prefers not to wash any dishes. Thus, nobody washes the dishes and all suffer the Pareto-inferior outcome. Or, all members in a neighborhood association prefer the state in which all members spend one day a year cleaning up the neighborhood to the state in which nobody spends any time cleaning up the neighborhood, but no matter what the others do, each member prefers not to clean the neighborhood. Thus, nobody cleans the neighborhood and all suffer the Pareto-inferior outcome.

Yet, in experimental studies, cooperation remains a remarkably stubborn phenomenon. In the prisoners' dilemma, the fraction of people playing the cooperative strategy is not 100 percent. But neither is it zero, even when people are playing the one-shot game anonymously. And in anonymous, one-shot public-goods games, Robyn M. Dawes and Thaler report:

> While not everyone contributes, there is a substantial number of contributors, and the public good is typically provided at 40–60 percent of the optimal quantity. That is, on average, the subjects contribute 40–60 percent of their stake to the public good. In [one study], these results held in many conditions: for subjects playing the game for the first time, or after a previous experience; for subjects who believed they were playing in groups of 4 or 80; and for subjects playing for a range of monetary stakes.

Why do people cooperate in one-shot prisoners'-dilemma and public-goods games? The experimental results are consistent with a high level of trust in the other players and with a desire to reciprocate what they expect to be generous contributions from the others. Contrary to predictions in our discussion about cheap talk (see Section 10.2), pre-play communication actually increases cooperation in prisoners' dilemmas and contributions in public-goods games. It may promote group identity – a sense of belonging – which is conducive to cooperation. In this sense, talk – even when cheap – might serve to promote reciprocity. Other explanations are consistent with the experimental data. Players might, for example, be altruistic as well. It should be noted that when the game is repeated, the level of contributions tends to decline. Thus, repetition appears to bring the players in closer accord with subgame-perfect-equilibrium predictions.

Predictions based on egoistic utility functions in combination with game-theoretic equilibrium concepts suggest that people will be unable to coordinate their actions even when it is in their interest to do so. Yet, this is a needlessly pessimistic vision of human nature. The economist Elinor Ostrom won the 2009 Nobel Prize for exploring ways in which people develop sophisticated mechanisms that allow them to reach beneficial outcomes in trust and public-goods-style games. There is plenty of evidence from the field and from the lab suggesting that people do succeed in coordinating their behavior under a wide range of conditions. Some roommates do succeed in developing mutually acceptable arrangements to make sure the dishes get washed, and some neighborhood associations do succeed in getting their members to participate in neighborhood-cleanup operations. Bad social and political philosophy, and bad social and political institutions, might result from the false assumption that cooperation cannot emerge spontaneously.

Unsurprisingly, there appear to be framing effects in strategic contexts as well. One well-studied phenomenon, for example, is that it is possible to influence cooperation rates in prisoners'-dilemma games by changing the manner in which the game is described. One study on such **social framing effects** had participants play standard prisoners'-dilemma games against each other, although the games were not described as such. The study found that only 26 percent of participants were willing to cooperate when the game was described as the "Stock Market Game," whereas almost 45 percent of participants were willing to cooperate when it was described as the "Community Game." Similarly, people are more likely to cooperate in a "social-exchange study" than in a "business-transaction study," and in a "community game" than in a "Wall Street game."

How do we explain such behavior? According to one prominent hypothesis, labels operate by triggering **social norms**. A label such as "community" triggers norms of cooperation, engendering a desire to cooperate, whereas a label such as "stock market" triggers norms of competition, engendering a desire to compete.

The role of social norms in human behavior has recently attracted a great deal of attention from philosophers and social scientists alike. These thinkers in various ways emphasize that humans are not only rational animals, but norm-following animals as well. Suppose you find yourself having a meal alone in an unfamiliar part of the world. Chances are you will be looking around for cues about proper behavior in the context, and do your best to fit in – even if you are among strangers you will never meet again. The point is that you *want* to follow the relevant social norms. That said, norm following is *conditional* in the sense that you only want to follow a given norm if you believe that others will follow it too. If everyone else is lining up neatly, waiting politely for their turn to buy a drink, chances are you will want to line up neatly too. But if everyone else is bellying up to the bar, trying to get the attention of the bartender, chances are you would rather join the scrum than wait politely for your turn. It is clear, then, that the game in question has two equilibria: one in which people form an orderly queue, and one in which they belly up to the bar. Which equilibrium gets selected may depend on what norm gets triggered in the moment, and what norm gets triggered may in turn depend on superficially insignificant factors such as frames, labels, and descriptions.

As you know by now, participants may not be playing prisoners' dilemmas at all, but rather something like the stag hunt game (see Exercise 10.17 on page 230). The game people are actually playing has two equilibria: $\langle$cooperate, cooperate$\rangle$ and $\langle$defect, defect$\rangle$. The same thing may be true in the event that people have social preferences. In this setting, social framing effects can occur because the label of the game is used as a coordination device, allowing participants acting independently to settle on equilibrium strategies.

This analysis suggests that a story about intentions, trust, and reciprocity can be incorporated into the traditional neoclassical model just like a story about social preferences can. There appears to be no principled reason why it cannot be done, and some game theorists have tried. Because the specifications are a little more complicated than those in the previous section, they have been left out of this discussion. That said, it might well be possible to fit the analysis of intentions, reciprocity, and trust into the traditional neoclassical framework. Or it might not.

11.4 Limited strategic thinking

John Maynard Keynes, one of the most influential economists of the twentieth century, drew an analogy between investing in the stock market and participating in a certain kind of newspaper contest. In *The General Theory of Employment, Interest and Money*, Keynes wrote:

> [Professional] investment may be likened to those newspaper competitions in which the competitors have to pick out the six prettiest faces from a hundred photographs, the prize being awarded to the competitor whose choice most nearly corresponds to the average preferences of the competitors as a whole; so that each competitor has to pick, not those faces which he himself finds prettiest, but those which he thinks likeliest to catch the fancy of the other competitors, all of whom are looking at the problem from the same point of view. It is not a case of choosing those which, to the best of one's judgment, are really the prettiest, nor even those which average opinion genuinely thinks the prettiest.

The basic structure of this strategic interaction is captured by the **beauty-contest game**. Here, n players simultaneously pick a number between zero and 100 inclusive. The person whose number is closer to seven-tenths of the average number wins a fixed prize. (The fraction does not have to be seven-tenths, but it must be common knowledge among the players.)

There is only one Nash equilibrium in this game. In this equilibrium, every player picks the number zero and everyone ties for first place. Suppose everyone picked 100, the highest number available. If so, the winning number would be 70. Thus, no rational person would ever pick a number greater than 70. But if no one picks a number greater than 70, the winning number cannot be greater than 49. Yet, if nobody picks a number greater than 49, the winning number cannot be greater than about 34. And so on, all the way down to zero.

Real people, however, do not play the Nash equilibrium strategy in the one-shot game. When played for the first time, answers might fall in the 20–40 range. Interesting things happen when the game is repeated with feedback about the average number in the previous round. During subsequent rounds, the average number will decrease and eventually approach zero. This suggests that, over time, real people converge to the Nash equilibrium prediction.

The favored explanation of the result in the one-shot game is based on the idea that people have different degrees of cognitive sophistication. "Level-0" players just pick a number between zero and 100 randomly. "Level-1" players believe that all the other players are level-0 players. Level-1 players, therefore, predict that the mean number will be 50, and therefore pick $0.7 * 50 = 35$. Level-2 players believe that all other players are level-1 players and that the average will be 35, and therefore pick $0.7 * 35 \approx 25$, and so on. Using statistical techniques, behavioral game theorists can estimate what proportion of each sample is level 0, level 1, and so on. The results suggest that most humans are level-1 or level-2 players. This kind of analysis is often referred to as **level-*k*** thinking, where *k* represents a person's position in a hierarchy of cognitive sophistication.

One fascinating feature of this game is that, even if you know the unique Nash equilibrium strategy, you may not want to play it. As long as you expect other players to play out-of-equilibrium strategies and choose a positive number, you will want to do so too. And as long as other players expect you to expect them to pick a positive number, they will want to pick a positive number. And so on. Thus, everyone plays a positive number and an out-of-equilibrium strategy. But, although you want to pick a number greater than zero, the number must not be too high: the aim is to stay one step ahead of the other players.

In keeping with Keynes's analogy, it has been suggested that this kind of game captures the dynamics of real markets and can explain bubbles in stock and real-estate markets. Even if all investors know that the market will ultimately crash, and that the unique Nash equilibrium strategy is to exit the market, they might assume that others will continue to buy for just a little while longer. As long as individual investors think that they can stay one step ahead of the competition and exit the market just before everybody else does, they will want to continue to buy. In the process, of course, they will drive prices even higher – before it all comes crashing down.

Example 11.7 Rock-paper-scissors, cont. The game rock-paper-scissors has a unique Nash equilibrium, in which both players randomize with probability 1/3, 1/3, and 1/3, and in which both players have an equal probability of winning (see Exercise 10.13 on page 228). So it might surprise you to hear that there is a World Rock Paper Scissors Society and a World Championship competition. According to the Society website, rock-paper-scissors is a game of skill, not chance: "Humans, try as they might, are terrible at trying to be random [and] in trying to approximate randomness become quite predictable."

Here is one piece of advice. The pros will tell you that "rock is for rookies," because inexperienced males tend to open with rock. Thus, if your opponent is one, open with paper. If your opponent is slightly more sophisticated and may be thinking you will open with rock and therefore opens with paper, open with scissors. If you are playing against an even more sophisticated agent, who will for the reason just identified open with scissors, open with rock.

Here is another piece of advice. Inexperienced players will not expect you to call your throw ahead of time. Thus, if you announce that you are throwing rock next, an inexperienced opponent will assume that you will not and instead will choose something other than paper. So you will want to throw rock. If you play against a slightly more sophisticated agent, though, they will expect you to throw rock after announcing that you will throw rock; so what you need to do is to throw scissors.

As in the beauty-contest game, the goal is to stay exactly one step ahead of your opponent. A similar analysis might apply to people's performance in the centipede game (see Exercise 10.26 on page 234). The typical finding is that people Pass until a few stages from the end, when they Take. This outcome would be expected if neither player thinks the other will play the unique subgame-perfect equilibrium, and that both attempt to Take one stage before the other one does.

Unlike models of social preferences, in this case there is little hope of capturing observed behavior in the one-shot game within the traditional neoclassical model. Things are quite different in the repeated version of the game. As the same group plays the game again and again, they approximate equilibrium predictions.

11.5 Discussion

According to the great Austrian economist Friedrich A. Hayek, the existence of spontaneous coordination constitutes the central problem of economic science. Hayek wrote:

> From the time of Hume and Adam Smith, the effect of every attempt to understand economic phenomena – that is to say, of every theoretical analysis – has been to show that, in large part, the co-ordination of individual efforts in society is not the product of deliberate planning, but has been brought about, and in many cases could only have been brought about, by means which nobody wanted or understood.

In this view, economists have never seriously doubted *that* coordination takes place; the question is *how* it emerges and is sustained. Much of the behavioral game theory literature on social preferences, trust, and reciprocity was developed in large part to answer this question. Obviously, this chapter does not contain a complete account of strategic interaction that fails to fit the picture painted by analytical game theory, or of the models behavioral game theorists have offered to capture the way in which people really interact with each other.

To what extent is the work presented under the heading "behavioral game theory" compatible with the traditional neoclassical framework? As we have seen, much of what goes under the heading is in fact consistent with analytical game theory. Models of social preferences are clearly consistent, since they proceed by allowing a person *P*'s utility function to reflect another person *Q*'s attainment. The degree to which ideas of intentions, reciprocity, and trust can be incorporated into neoclassical theory remains unclear, though game theorists have tried. Models that try to capture people's limited ability to think strategically, by contrast, are more clearly inconsistent with any model that relies on Nash or subgame-perfect equilibrium concepts.

In defense of analytical game theory, it has been argued that neoclassical theory is only intended to apply under sharply circumscribed conditions. Thus, the prominent game theorist Ken Binmore writes:

> [Neoclassical] economic theory should only be expected to predict in the laboratory if the following three criteria are satisfied:
>
> - The problem the subjects face is not only "reasonably" simple in itself, but is framed so it seems simple to the subjects;
> - The incentives provided are "adequate";
> - The time allowed for trial-and-error adjustment is "sufficient".

Binmore recognizes that he is also denying the predictive power of neoclassical economics in the field, and adds: "But have we not got ourselves into enough trouble already by claiming vastly more than we can deliver?" Binmore's view is nicely illustrated by the beauty-contest game (see Section 11.4). Here, real people's behavior differs dramatically from Nash equilibrium predictions during the first round, but converges to it as the game is repeated and players learn the outcome of previous rounds. The same thing might be true for public-goods and other games. Binmore's defense of analytical game theory does not constitute an argument against behavioral game theory, however, provided behavioral game theory is intended to apply when one or more of the three conditions are not satisfied.

Additional exercises

Exercise 11.8 Explain how positive offers in the ultimatum game are consistent with Nash-equilibrium predictions. (There are many misconceptions about this, so it is good to make sure you understand.)

Exercise 11.9 Gandhi A leader of India's independence movement and advocate of non-violent social change, Mahatma Gandhi is supposed to have said: "The true measure of any society can be found in how it treats its most vulnerable members." If we interpret this line as an expression of social preferences, what kind of preferences would that be?

Exercise 11.10 Egoists, utilitarians, and enviers Consider the game in Table 11.3, expressed in dollar terms.

Table 11.3 Egoists and others

	L	R
U	$3, $2	$0, $1
D	$1, 0	$2, $1

(a) Suppose, first, that this game is played by two egoists, for whom $u(x, y) = x$. Compute all Nash equilibria.
(b) Suppose, next, that this game is played by two utilitarians, for whom $u(x, y) = x + y$. Compute all Nash equilibria.
(c) Suppose, finally, that this game is played by two enviers, for whom $u(x, y) = x - y$. Compute all Nash equilibria.

Exercise 11.11 In the **volunteer's dilemma**, at least one person has to volunteer to provide some good in order for everyone to enjoy it. For example, in order for you and your friends to have snacks while watching a movie, somebody has to bring snacks. We will assume for simplicity that snacks are shared, and that their quantity and quality do not matter. Throughout, we will assume that watching a movie without snacks gives each person 0 units of utility. A volunteer's dilemma can involve any number of people, but here we will limit ourselves to the case when there are two.
(a) Let us suppose, first, that players have purely self-interested preferences. Having snacks to eat while watching a movie gives a player 2 units of utility, independently of who brought them, but bringing snacks costs 1 unit of utility. What is the payoff matrix of the game? What are the Nash equilibria?
(b) Suppose instead that players have social preferences. Having snacks to eat still gives a player 2 units of utility, but they get 2 additional units of utility whenever the other player has snacks to eat. Bringing snacks still costs 1 unit of utility. What is the payoff matrix of the game? What are the Nash equilibria?
(c) Suppose now that players get utility not from the other's consumption, but from the act of bringing the snacks. Having snacks to eat again gives each player 2 units of utility, but now bringing the snacks generates 1 additional unit of utility. What is the payoff matrix of the game? What are the Nash equilibria?

Exercise 11.12 That fridge There's a fridge in the office that needs to be cleaned out once a week. Nobody does, though, and you hate it. You do not hate it enough to clean it every week yourself. But you would gladly clean it out once in a while, if only your colleagues took turns with you. Anyway, you decide to give it a shot. One random Friday you clean it out well, in the hope that your efforts will be appreciated and returned in kind. What sort of game are you playing with your colleagues?

Exercise 11.13 The watch collection In the interest of acquiring yet another hobby you cannot afford, you have decided to start collecting fine Swiss watches. You find that there are some watches that you like and some that you do not. Now, you only want to buy watches that will appreciate over time, so you cannot just buy what you like – you need to buy watches that will be good investments. Watches are good investments only to the extent that people will want to buy them in the future for even more than you paid when you bought them. But people will want to buy the watches in the future only if those people think that the watches will continue to appreciate, which happens only if yet other people will want to buy them later. And so on… What sort of game is this?

Exercise 11.14 Kong Ming *The Game Theorist's Guide to Parenting*, by Paul Raeburn and Kevin Zollman, may be the most useful parenting book of all time. In passing, it tells the story of famed Chinese military strategist Kong Ming, also known as Zhuge Liang. In 149 BC, Kong Ming and his forces were under siege and ridiculously outnumbered in the town of Yangping. They could not escape, and if they did nothing, the town would quickly be overrun by the enemy. In a stroke of genius, Kong Ming and his forces went into hiding and left the gates to the city wide open, leaving nobody but a solitary lute player to greet the besieging army. The enemy could not understand why Kong Ming would do this, concluded it must be a trap, and withdrew. Use level-*k* thinking to explain how Kong Ming outsmarted the enemy.

Problem 11.15 Equilibrium concepts *We know from Section 10.4 that some game theorists think the Nash equilibrium concept is problematic, and that the concept of subgame-perfect equilibrium better captures the behavior of rational agents. But as the single subgame-perfect equilibrium in the ultimatum game suggests (see Section 11.2), there is something funny about subgame-perfect equilibrium too. For one thing, the subgame-perfect equilibrium requires the responder to accept an offer of $0. And in most Nash equilibria, the responder does better than this. In your view, which equilibrium concept best captures the behavior of rational agents: Nash or subgame-perfect equilibrium?*

Further reading

Holt (2019) contains an excellent introduction to behavioral game theory. The most thorough treatment is Camerer (2003), which includes the quotation summarizing results from the ultimatum game (p. 49). Kagel and Roth (1995) explore experimental methods and results; Durlauf and Blume (2010) offer a more concise and up-to-date treatment. A good review of cooperation in laboratory experiments is Dawes and Thaler (1988), who among other things summarize the results of public-goods experiments (p. 189). Social framing effects are explored in Ellingsen et al. (2012), and the nature and dynamics of norms in Bicchieri (2005). The World RPS Society (2011) will tell you how to beat anyone at rock-paper-scissors. The two historical quotations are from Keynes (1936, p. 156) and Hayek (1933, p. 129); for more on Hayek's take on information and cooperation, see Angner (2007). The Binmore quotation is from Binmore (1999, p. F17).

PART 6

CONCLUDING REMARKS

12 BEHAVIORAL POLICY

Learning objectives

After studying this chapter you will:

- Know what behavioral welfare economics, libertarian paternalism, and the nudge agenda are
- Know how the economics of happiness can contribute to behavioral policy
- Be able to assess competing policy proposals in terms of different normative foundations

12.1 Introduction

Although behavioral economics was already firmly established as a subdiscipline of economics by the first decade of the twenty-first century, the enterprise appears to have received a boost from the economic crisis that struck around then. As David Brooks put it in the *New York Times:* "My sense is that this financial crisis is going to amount to a coming-out party for behavioral economists and others who are bringing sophisticated psychology to the realm of public policy." Brooks is frequently described as a conservative, but commentators across the political spectrum have blamed the crisis in part on inadequate economic models. The former chairman of the Federal Reserve Alan Greenspan is known as a follower of Ayn Rand's objectivism, which celebrates the value of rational self-interest. Yet, in 2008 Congressional testimony, Greenspan said: "I made a mistake in presuming that the self-interests of organizations, specifically banks and others, were such as that they were best capable of protecting their own shareholders and their equity in the firms." Similarly, the Nobel laureate and liberal economic commentator Paul Krugman argues:

> [Economists] need to abandon the neat but wrong solution of assuming that everyone is rational and markets work perfectly. The vision that emerges as the profession rethinks its foundations may not be all that clear; it certainly won't be neat; but we can hope that it will have the virtue of being at least partly right.

Brooks, Greenspan, and Krugman all seem to agree that economic theory can have a substantial impact on policy and therefore on the quality of people's lives. They also seem to agree that this impact can be better or worse depending on the appropriateness of the theory for the task.

Armed with a set of economic theories and a desire to influence policy and improve lives, behavioral economists have become increasingly confident working on

behavioral policy: economic policy informed by behavioral theory. There is nothing unusual about the fact that behavioral economists wish to change, and not only to understand, the world in which they live. Neoclassical and other economists do too. As Hayek put it in 1933:

> It is probably true that economic analysis has never been the product of detached intellectual curiosity about the why of social phenomena, but of an intense urge to reconstruct a world which gives rise to profound dissatisfaction. This is as true of the phylogenesis of economics as the ontogenesis of probably every economist.

Since behavioral (like neoclassical) economists take their central normative concern to be welfare, and perhaps its distribution, this work is often discussed under the heading of **behavioral welfare economics**. Behavioral welfare economics subsumes behavioral law and economics, which incorporates behavioral economic theories into law and economics, as well as behavioral public economics, which relies on behavioral economics to describe and evaluate the effects of public policies.

As part of their exploration of behavioral policy, behavioral economists have developed a doctrine variously referred to as **libertarian, light/soft**, or **asymmetric paternalism**, and a series of policy proposals collectively referred to as **the nudge agenda**. Advocates claim that the nudge agenda allows us to improve people's choices and thereby their well-being on their own terms at minimal cost and without interfering with their liberty or autonomy. Critics warn that the nudge agenda represents an ineffective and dangerous intrusion into the sphere of personal decision-making by bureaucrats who may be no better at making decisions than the people whose choices they are trying to improve.

Other behavioral economists have argued that empirical findings – especially the result that choices do not in general mirror a complete, transitive preference ordering – mean that the traditional preference-based welfare measures must be abandoned. Instead, these economists argue, welfare measures should be based on "what really matters," which they often take to be the happiness of the people involved. Advocates maintain the **economics of happiness** permits radical reforms that would make economic policy more responsive to what ultimately counts. Critics respond that the whole project is misinformed, and that those policy proposals, if consistently implemented, would have unintended and unfortunate consequences indeed.

In this chapter, we will have a closer look at libertarian paternalism and the nudge agenda, first, and the economics of happiness, second. The goal is to explore what these proposals are, how they differ (or not) from standard approaches, and what their promises and limitations might be.

12.2 Behavioral welfare economics, libertarian paternalism, and the nudge agenda

The policies proposed by orthodox economists tend to fall in a small number of categories. Some such policies remove barriers to mutually beneficial interactions, as when free-trade agreements permit small-scale organic coffee growers on one continent to sell coffee beans to discerning coffee drinkers on another. Some policies ban harmful actions, as when it is illegal to steal other people's property and to sell your children to the highest bidder; some mandate beneficial actions, as when it is required to wear a helmet when riding a motorcycle or to vaccinate your children against preventable diseases. Other policies impose incentives that encourage people to engage in more beneficial actions or fewer harmful ones. Thus, tax incentives for the installation of solar panels serve to encourage the production of clean energy and carbon taxes serve to discourage the use of fossil fuels.

The distinction between bans/mandates on the one hand and incentives/disincentives on the other matters legally and psychologically. But economists like to point out that there is a sense in which all these policy proposals are economically equivalent: they all work by shifting the incentives associated with different courses of action. So, for example, a ban can be analyzed as just another kind of incentive, where the disincentive takes the form of fines, jail time, loss of privileges, or the like. It is no coincidence that traditional economists' policy proposals tend to have this form. A theory according to which behavior is determined by a system of consistent and stable preferences suggests that effectively the only way to change behavior, other than expanding menus, is to shift incentives.

Without denying that modifying incentives is a powerful way to influence behavior, behavioral economists' policy proposals tend to have a different character. Based on their research, behavioral economists have proposed a whole other form of welfare-enhancing policy intervention. To the extent that people's actual choices deviate in significant and systematic ways from what they would have chosen if they had been fully rational, informed, etc., it is at least in theory possible to *improve* people's choices in the sense of making their actual choices conform more closely to the rational ones. On the assumption that people are made better off by whatever they would choose if they were fully rational, etc., improving people's choices in this manner means making them better off. And they are better off *by their own lights*, that is, as defined by their own rational and informed preferences. Making people better off in this manner is, in a nutshell, the goal of libertarian paternalism and the nudge agenda.

The project is illustrated in Figure 12.1. The person in Figure 12.1(a) is rational and informed. Her preferences tell her she wants to go from A to B, and so she does. The person in Figure 12.1(b), if he were rational and informed, would also want to go from A to B. Unfortunately, he is not. Doing as well as he can under the circumstances, he goes from A to B' instead. The libertarian paternalist, noticing that the person on the right is not as well off as he could be in terms of his own rational, informed preferences, aspires to change the conditions under which he makes his

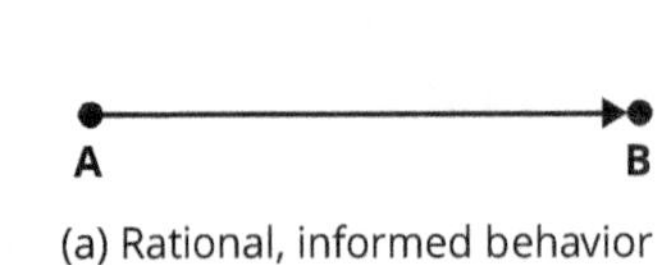

(a) Rational, informed behavior

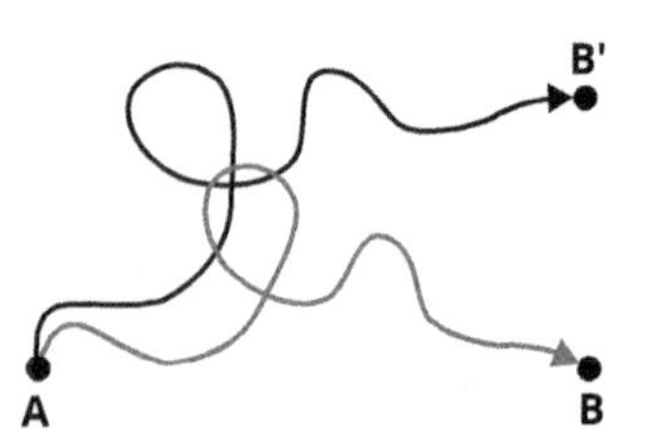

(b) Irrational, uninformed behavior

Figure 12.1 Two kinds of behavior

decision so that his actions take him along the shaded line to B instead of B'. Note that the behavioral economist can explore policy alternatives that make no sense within the traditional framework. In this sense, (descriptive) behavioral economics provides the policy maker with entirely new levers that can be used for a variety of policy purposes.

The fact that we sometimes make suboptimal decisions might suggest that it would be best if people's choices were made *for* them by a third party: a disinterested bureaucrat, perhaps, or an enlightened despot or Leviathan. Yet, no behavioral economist advocates a dictatorship as a solution. There are many obvious reasons: bureaucrats and despots will be ignorant about people's preferences and the circumstances of their lives, less than fully benevolent, and on the average no more rational than the people whose choices they are trying to improve. Moreover, the very act of making decisions ourselves might enhance our welfare; even making mistakes, if the consequences are not too severe, can be edifying.

Libertarian paternalism says that it is legitimate to help people make better decisions themselves, by their own lights, if it is possible to do so without interfering with their liberty or autonomy. The approach is described as "paternalistic" because it aims to make people better off, but "libertarian" because it tries to make people better off in a manner that respects their liberty and autonomy. Others call it soft or light paternalism, to distinguish it from more heavy-handed kinds of paternalism. The term asymmetric paternalism is sometimes used to underscore that soft paternalistic interventions are expected to impose little to no cost on individuals who are fully rational and informed, while having potentially large benefits for those who are not. Interventions that target the environment in which choices are made – the **choice architecture**, as we will call it – and that are designed to have this effect are called **nudges**. A paradigmatic nudge has several properties. (i) It aims to help people make better decisions themselves, rather than making decisions for them. (ii) It imposes little to no cost on those who are exposed to it. (iii) It has little to no effect on the choices of those who are already rational and well informed. (iv) The effect on the choices of those who are not already rational and well informed is potentially beneficial for them, by their own lights.

Figure 12.2 illustrates the manner in which nudges differ from other forms of policy. What the three traffic signs have in common is that they try to reduce the

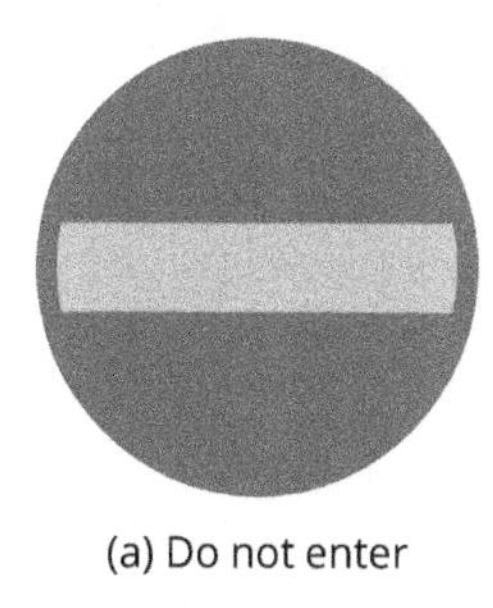

(a) Do not enter

(b) Congestion charging zone

(c) Right turn ahead

Figure 12.2 Three kinds of traffic sign. © The Swedish Transport Agency. Used with permission number of people who drive straight ahead. But they do so in very different ways. The do-not-enter sign in Figure 12.2(a) represents a ban. It signals that it is illegal to drive straight ahead, and that doing so is punishable by fines or jail time. The congestion-charging-zone sign in Figure 12.2(b) represents a disincentive. It signals that it is legal to continue on, but that anyone who does so will incur an additional charge. By contrast, the right-turn-ahead sign in Figure 12.2(c) represents a nudge. The sign does not say that it is illegal to drive straight ahead; nor does it say that there is a charge associated with doing so.

The right-turn-ahead sign works differently from the other two. (i) The sign aims to help people make better driving decisions themselves, rather than making decisions for them. (ii) The sign imposes no cost on people exposed to it beyond the trivial amount of time it takes for experienced drivers to scan it as part of the environment, and this is true whether they intend to stay on the road or not. (iii) The sign has no effect on the decisions of drivers who are rational and informed: with or without the sign, a rational, informed driver who intends to stay on the road would bear right at the curve, and a rational, informed driver who intends to drive straight ahead and off the cliff would do that. (iv) The sign can, however, have an asymmetric and potentially large beneficial effect (by their own lights) on drivers who are irrational or ignorant of the upcoming turn. To the extent that the sign changes the behavior of drivers it is because it "nudges" them into adjusting the speed and turning the wheel as appropriate.

Exercise 12.1 Traffic signals What kind of policy is represented by (**a**) a stop sign; (**b**) a lane-ending sign; (**c**) a red light; (**d**) a yellow light; (**e**) a parking meter labeled "¢25 per hour"; and (**f**) a no-parking sign?

Behavioral welfare economists have proposed a number of specific interventions that they believe are welfare enhancing, and which are motivated by theoretical developments in behavioral economics. Here is a small sample:

- **Default options** are options that will be selected in case the decision-maker fails to make an active choice. Insofar as people are prone to status quo bias, they will exhibit a tendency to stick with the default even when it would be virtually costless to make an active decision. By having a choice architect carefully determine what option should be the default, in light of people's own interests, behavioral economists believe that more people will end up with the option that is best for them.
- The **Save More Tomorrow (SMarT) Program** encourages workers to save more for retirement by giving them the option of committing in advance to allocating a portion of their future raises toward savings. Because committing a future raise feels like a forgone gain whereas committing money in the pocket feels like a loss, prospect theory predicts that workers will find it easier to save future salary increases than money in the pocket. The SMarT program is designed to increase savings rates by leveraging this effect.
- **Cooling-off periods** are periods of time following a decision during which decision-makers have the option to reverse their choices. Cooling-off periods are based on the idea that people in a transient "hot" emotional state sometimes make suboptimal decisions. Behavioral economists argue that cooling-off periods offer

people the opportunity to reevaluate their decisions from the perspective of a "cool" state, which is likely to lead to a better decision.

As always, the devil is in the details. If the burden imposed on rational, informed decision-makers by cooling-off periods, etc., is non-trivial, the intervention would no longer constitute a nudge – whether or not it succeeds in enhancing total welfare.

Exercise 12.2 Apples or fries? US hamburger chains have started serving apple slices instead of French fries with their "meal" options. Fries are still available, but the customer has to ask to have fries substituted for apples. Preliminary reports suggest that as a result, customers end up eating more apples and less fries. What kind of intervention is this?

Since the definition of a nudge has some vagueness built into it, we should expect there to be a grey area where reasonable people can disagree. Yet, it is easy to think of examples that do not constitute nudges. The weather is not a nudge, for example, since it is not an intervention. Subliminal advertising is not a nudge, since if it is effective it may interfere with the choices of the rational and irrational alike. And as we saw above, stop signs and parking bans are not nudges.

Exercise 12.3 The Bloomberg ban As part of a 2013 public-health initiative spearheaded by then-mayor Michael Bloomberg, New York City banned the sale of sodas larger than 16 ounces (or about 0.5 liters) in restaurants, movie theaters, and the like. (The law was later struck down by the state's highest court.) Explain why the "Bloomberg ban" does not constitute a nudge.

Any argument to the effect that some intervention is or is not welfare enhancing presupposes a welfare criterion: a rule that (at least in principle) allows us to determine who is well off and who is not. Neoclassical economics relies on a preference-based criterion, according to which people are well off to the extent that their preferences are satisfied. Given the neoclassical understanding of "utility," this is equivalent to saying that people are well off to the extent that their utility is high. It has been argued that behavioral economics entails that this criterion is inadequate and that we must instead accept a happiness-based criterion, according to which people are well off to the extent that they are happy. Notice that these criteria are substantially different, since it is possible to have satisfied preferences without being happy and vice versa (cf. Section 12.4 below). Others maintain that it is possible to maintain the preference-based criterion by assuming (as some neoclassical economists already do) that the preferences that count are not the preferences actually revealed in people's choices, but the preferences that they would have if they were perfectly rational, ideally informed, and so on. To my mind, behavioral welfare economics is most helpfully understood as presupposing such a criterion.

The nudge agenda has already proven enormously influential. Among other things, it has inspired the development of **behavioral-insights (BI) teams**, colloquially known as **nudge units**, across the world. A BI team is an organization, inside or outside of government, that is tasked with translating insights from the behavioral sciences into practical policy proposals. The UK BI Team, set up inside government in 2010 and awarded some independence in 2014, is often described as the first one. Less than a decade later, the OECD has identified more than 200 institutions around

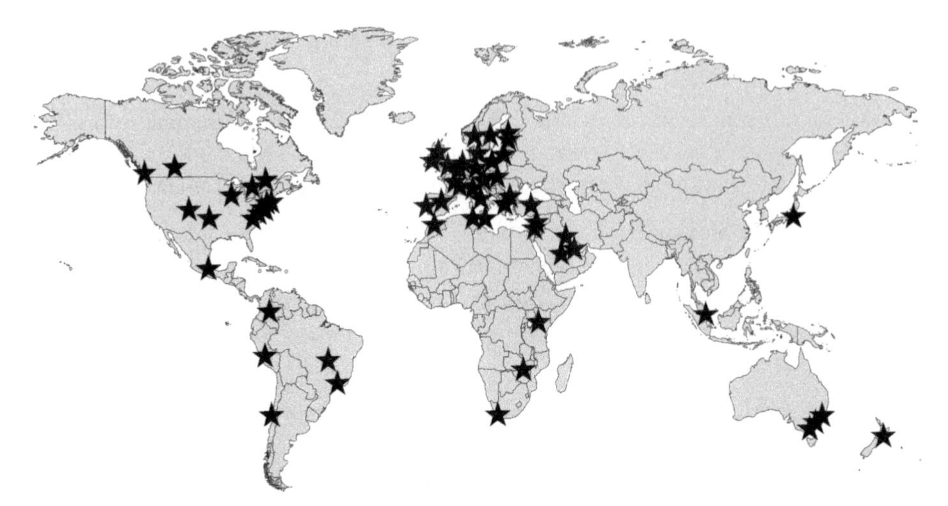

Figure 12.3 Behavioral-insight teams across the world. Data from OECD (2017)

the globe applying behavioral insights to policy (see Figure 12.3). If there is one near you, feel free to check it out! A 2017 OECD report titled *Behavioural Insights and Public Policy* reviews 100 case studies drawn from the domains of "consumer protection, education, energy, environment, finance, health and safety, labour market policies, public service delivery, taxes and telecommunications." Although challenges remain, the report argues, behavioral insights permit governments and other organizations to develop strategies that enhance efficiency without resorting to costly regulation or sanctions. Importantly, BI teams are no mere consumers of behavioral science; they are producers as well, running, e.g., large-scale randomized controlled trials in the field to secure the data required for policy solutions to urgent problems.

12.3 Assessing the nudge agenda

The appeal of libertarian paternalism and the nudge agenda is obvious: Would it not be wonderful if people could be helped to make better decisions themselves, even by their own lights, without interfering with their liberty and autonomy? Proponents are enthusiastic indeed. They point out that nudges can be cheap to administer, that results can be immediate, and that the effects can be large. The right-turn-ahead sign in Figure 12.2(c) is a case in point: presumably these signs are so common because the cost of putting them up and maintaining them is low, they begin to operate immediately, and expected benefits in terms of reduced car crashes are high. Harvard Law professor Cass R. Sunstein, the author (with Thaler) of *Nudge*, writes:

> In the context of retirement planning, automatic enrollment has proved exceedingly effective in promoting and increasing savings. In the context of consumer behavior, disclosure requirements and default rules have protected consumers against serious economic harm, saving many millions of

> dollars. Simplification of financial aid forms can have the same beneficial effect … as thousands of dollars in additional aid (per student) … In some cases, nudges have a larger impact than more expensive and more coercive tools.

It is worth pointing out that nudging is not limited to governments. Corporations, charities, religious organizations, managers, teachers, and parents can all use nudges to help people make better decisions, by their own lights. And many of them already do.

Example 12.4 Google Google wants to encourage its employees to eat healthily. *Fast Company* reports: "In pursuit of that healthiness, happiness, and innovation, the software giant has turned to 'nudges': simple, subtle cues that prompt people to make better decisions." Borrowing ideas from behavioral economics, Google is now: (1) putting candy in opaque bins rather than clear dispensers; (2) placing salad in full view to people entering the cafeteria and dessert much further down; (3) encouraging people to use smaller plates by pointing out that people with bigger plates tend to eat more; (4) color-coding foods in accordance with how healthy they are; and more. All these interventions can plausibly be characterized as nudges as we understand the term here.

Problem 12.5 *Identify three features of your car, computer, or any other device that constitute nudges.*

So when it comes to libertarian paternalism and the nudge agenda, what is not to love? Of course, every tool can be used badly. A hammer is a wonderful thing if you have to drive nails, but it can cause a lot of damage if swung in the vicinity of crystal glass or human heads. Similarly, nudges can presumably be inelegant, inappropriate, misguided, mean-spirited, and/or outright harmful. But just like it would be unhelpful to criticize the use of hammers based on the fact that people get hurt if they are whacked over the head, it would be odd to criticize nudges based on the fact that they have harmful uses. For a general critique of nudging to get traction, the critique needs to make the case that nudges are harmful to human welfare (or autonomy, or whatever) even when used correctly.

Critics point out that the people doing the nudging – the "choice architects," in Sunstein's lingo – will themselves be lacking in rationality and information, not to mention benevolence. Yet, the objection largely misses the point. Behavioral economists are well aware of the fact that policy-makers suffer the same limitations as the rest of us do. Not only is this the reason why behavioral economists rejected the dictatorship solution, but it is part of the reason why they developed the nudge agenda in the first place. And choice architects need not be superhuman. Consider the right-turn-ahead sign again. Posting such a sign does not mean that you pretend to know what is good for people, that you want to impose your vision of the good life on them, or even that you are telling them what is good for them; you can post the sign even if you recognize that some people have reason to drive straight ahead. And posting such a sign does not mean that you are deluded into thinking that you are more rational, better informed, or more benevolent than anybody else. It just means that you think there is a chance that the sign will help some people who want to stay on the road to do so.

Some critics argue that all forms of paternalism are objectionable (offensive, illegitimate) and that, therefore, libertarian paternalism is too. But this criticism equivocates on the meaning of the term "paternalism." If by "paternalism" we mean the disposition or desire to make others better off, there is nothing objectionable about it; paternalism in this sense is what is otherwise known as "benevolence" and it is widely considered a virtue. In this sense of "paternalism," increasing total welfare by removing trade barriers is a paternalistic intervention and virtually every welfare economist is a paternalist. The sense in which paternalism is objectionable is the sense in which it entails a violation of people's liberty or autonomy, and libertarian paternalism by definition does not. In fact, when Sunstein notes that nudging may be more effective than more coercive forms of policy, he is raising the possibility that the same or better results can be attained by substituting the former for the latter. If so, nudging would actually *increase* people's freedom of choice. Libertarian economist Bryan Caplan has endorsed the nudge agenda to the extent that it substitutes softer for harder forms of paternalism. He writes:

> "Nudging" is a great idea. We should start by ending existing hard paternalism in favor of gentle (or even subliminal) persuasion. Instead of prohibiting drugs, we should allow anyone who wants to use currently illegal drugs to go to a government website to request an Authorized Narcotics User Card ... Analogous opt-out rules should be devised for government health care programs, worker protection laws, consumer protection laws, and so on.

At any rate, to the extent that we are concerned with objectionable forms of paternalism, it would be more motivated to target policies that actually do interfere with people's liberty and autonomy – such as the bans and mandates represented by Figure 12.2(a) and the (dis)incentives represented by Figure 12.2(b) – and there are plenty of them already.

Other critics allege that the nudge agenda is dangerous because there is a slippery slope from perfectly innocuous nudges to more coercive forms of policy. If we allow policy-makers to nudge us at all, even in ways that do not violate our liberty and autonomy, the argument goes, policy-makers will soon engage in more coercive forms of policy. It is quite true that all policy interventions (even those that aim to remove barriers to mutually beneficial interaction and to enhance freedom of choice) can have unintended and unanticipated adverse consequences. It is also true that the wolf of hard paternalism may appear in the clothing of soft-paternalist sheep. But there is something odd about the slippery-slope argument. Note that it does not say that there is anything wrong with libertarian paternalism or nudging *per se*, but rather with the harder forms of paternalistic intervention that will likely follow. But why should we think that they will? Consider traffic signs of the third type. The critic alleges that allowing the use of signs of the third type will necessarily or likely lead to a proliferation of signs of the first two kinds. But there seems little reason to think this is so. Indeed, to the extent that signs of the third kind work, they may obviate the need for signs of the first two kinds. This is the thought underlying Caplan and Sunstein's hope that soft paternalism will lead to an expansion of personal freedom. Again, given that signs of the first two kinds are already in widespread use, to the extent that we are concerned about limitations of liberty and autonomy, our time would be better spent arguing against them instead. Considerations like this have encouraged philosophers to reject all slippery-slope arguments. Moral philosopher Simon Blackburn, for example, writes: "'Slippery slope' reasoning needs to be resisted, not just here but everywhere."

A very different kind of criticism comes from within behavioral economics itself. Some behavioral economists have cautioned that nudging has sharp limitations, and that other forms of policy may be necessary to solve many policy challenges. Thus, George Loewenstein and Peter Ubel have argued that a wide variety of problems – including the obesity epidemic, conflicts of interest in medicine, and the challenge of energy preservation – are best addressed using more traditional forms of policy. They continue:

> As policymakers use it to devise programs, it's becoming clear that behavioral economics is being asked to solve problems it wasn't meant to address. Indeed, it seems in some cases that behavioral economics is being used as a political expedient, allowing policymakers to avoid painful but more effective solutions rooted in traditional economics.

In terms of traffic signs, the point here is that there are conditions under which signs of the first two kinds in Figure 12.2 are called for, and that signs of the third kind (though potentially effective under a limited range of conditions) cannot be expected to replace them. Loewenstein and Ubel could bolster their claim by pointing out that more heavy-handed policies often serve not to benefit the decision-maker, but to prevent him or her from harming others. To deal with things like theft, assault, and murder, outright bans – not nudges – are particularly appealing. Either way, unlike Caplan and perhaps Sunstein, then, Loewenstein and Ubel think the nudge agenda should be seen as a complement to, rather than a substitute for, traditional forms of policy.

12.4 The economics of happiness

The economics of happiness offers a very different approach to behavioral policy. This approach often takes as its starting point the fact that people's choices do not in general reflect a transitive, complete preference ordering (as we know from Chapter 3). This fact is thought to undermine not just the standard approach to welfare economics, but any approach that defines welfare in terms of preference satisfaction, in at least two ways. First, it is argued, it does not even make sense to think about welfare in preference-satisfaction terms when the preferences revealed in a person's choices are inconsistent. Second, and for sure, people who act inconsistently do not necessarily choose so as to maximize their happiness. Either way, the conclusion is that we need an approach to policy that does not presuppose that welfare can be understood in terms of preference satisfaction.

The concept of happiness offers a solution to these problems. The idea that happiness is what ultimately matters in life – what we have reason to pursue in our own lives, and to promote in the lives of others – has had remarkable staying power in philosophy, social science, medicine, religion, and beyond. Aristotle's concept of *eudaimonia*, which is sometimes translated as flourishing but often as happiness, was picked up by the Church Fathers during the Middle Ages, and ended up having a huge influence on the Western intellectual tradition. Bentham, whom we came across already in Section 1.2, famously argued that every private action and every public policy should be assessed by reference to their contribution to the total happiness – and nothing else. Behavioral scientists who are skeptical of the idea of welfare as preference satisfaction often feel compelled to go "back to Bentham," as they say. For one example among many, famous happiness economist P. Richard G. Layard says: "I believe that Bentham's idea was right and that we should fearlessly adopt it and apply it to our lives."

The economics of happiness is often described as a new development. The subtitle of Layard's book, for instance, describes it as "a new science." But in reality the systematic empirical study of happiness and related mental states goes back almost a century. In the decades following World War I, psychologists turned the tools of the nascent subdiscipline of personality psychology to the study of happiness and satisfaction, and there is in fact an uninterrupted research stream all the way up to the present day. Happiness research was brought to the attention of mainstream economists in the 1970's, when Richard A. Easterlin published a paper on what he called the "paradox of happiness." That is the idea that happiness is increasing with income in any cross-section but does not appear to be rising over time, even during a period of rapid economic growth. More recently, the economics of happiness has benefitted greatly from the rise of **positive psychology**, the psychological study of positive and desirable mental states, as well as from its association with Kahneman, Layard, and other high-profile economists.

People's happiness levels cannot be measured simply by observing their choices – and to proponents, this is a good thing too. Instead, welfare assessments within the economics of happiness are typically based on questionnaires with one or more fairly straightforward questions. A question that has been in use for decades reads: "Taking things all together, how would you say things are these days – would you say you're *very happy*, *pretty happy*, or *not too happy* these days?" A more recent questionnaire offers four prompts of the form "In general, I consider myself… " and invites participants to respond on a seven-point scale, where 1 represents "… not a very happy person" and 7 "… a very happy person." Occasionally, researchers use prompts such as horizontal lines, ladders and mountains, or happy and sad faces. The faces scale is not unlike the ones used for customer satisfaction surveys on the go in airports and elsewhere, where you are invited to smack a brightly colored button with a happy or sad face on it as you speed by.

What have economists found, using such measures? Here are a couple of stylized facts from a recent survey by happiness economist Andrew Clark:

- Happiness is positively associated with income, both in cross-sectional studies and in panel data, although the marginal happiness of income is sharply diminishing.
- Happiness is negatively associated with unemployment, both in cross-sectional studies and in panel data.

- Happiness levels over the course of the life cycle age is U-shaped, meaning that young and old people are happier than middle-aged people.
- Happiness is positively associated with marriage, although the explanation seems to be that happier people are more likely to get married, not that marriage makes people happier.
- Happiness levels are determined in part by social comparisons, that is, by how people's accomplishments compare with those of other people.
- People have a remarkable ability to adapt to life changes, meaning that the effect of those changes is attenuated over time (cf. Section 9.5). But adaptation does not always occur, and when it does it is not always complete.

What sort of behavioral policy can be defended on the basis of these results? Recall that the entire enterprise was motivated in large part by the promise that it would give us a principled way to assess policy proposals, even in the presence of irrational behavior. As it happens, answers are not always clear, often controversial, and sometimes distinctly unpalatable. It is uncontroversial that unemployment is bad, and that policies which all things equal reduce unemployment are worth pursuing. More controversially, a number of happiness economists have defended Robin Hood-style redistribution of resources. Since the marginal happiness of money is sharply diminishing, they maintain, total happiness can be increased by taking from the rich and giving to the poor. Some have argued that high-end conspicuous consumption ought to be taxed, because it imposes sharply negative externalities on others engaged in social comparison. Finally, the fact that people adapt to disability, and the like, might suggest that governments should allocate resources away from things like traffic safety. This might strike you not only as unpalatable, but misconceived too, since injury and disability are arguably bad for people even if they adapt to their new condition.

Policy implications aside, the economics of happiness has had to overcome a great deal of skepticism but is now a relatively uncontroversial subdiscipline of economics. There is no doubt that economists, with their sophisticated econometrics and enormous datasets, have contributed a great deal to our understanding about who is happy and why. And it is hard to deny that this kind of information can be a useful input into our private lives, in organizations, and indeed in public policy. People want to be happy, and they want to make others happy, and the economics of happiness gives us clues about how.

That said, there is a sense in which the entire enterprise is based upon a misconception. The fact that people's choices do not in general reflect a transitive, complete preference ordering does not mean that we have to abandon the idea that welfare can be understood in terms of preference satisfaction. Recall from Section 12.3 that it is possible to think about a person's welfare in terms of what she would want if she were ideally rational, perfectly informed, and so on. According to an idealized preference-satisfaction criterion, you are well off to the extent that you have what you would prefer under some idealized (counterfactual) conditions. The fact that people's actual preferences, as revealed in their choices, are intransitive and/or inconsistent is not a threat to such a welfare criterion. As noted above, behavioral welfare economics can helpfully be understood as presupposing such a criterion. On this view, the fact that people do not always choose so as to maximize their own well-being is a feature, not a bug, since there are many thing that people may have reason to value beyond their own happiness. There may, of course, be other reasons to adopt a happiness-based welfare criterion: maybe Bentham really was right and happiness is the only thing that ultimately matters. But even then, it may be good to know that philosophers, who have thought about the nature of welfare and wellbeing for 2,500 years straight in the Western tradition alone, overwhelmingly reject the idea.

12.5 Discussion

The notion that economic theory can have a real impact on worldly events was wholeheartedly endorsed by Keynes, who wrote:

> [The] ideas of economists and political philosophers, both when they are right and when they are wrong, are more powerful than is commonly understood. Indeed the world is ruled by little else. Practical men, who believe themselves to be quite exempt from any intellectual influences, are usually the slaves of some defunct economist. Madmen in authority, who hear voices in the air, are distilling their frenzy from some academic scribbler of a few years back. I am sure that the power of vested interests is vastly exaggerated compared with the gradual encroachment of ideas ... it is ideas, not vested interests, which are dangerous for good or evil.

The ideas of behavioral economists are no exception. How exactly behavioral economics will affect "practical" women and men remains to be seen. But there is no doubt that behavioral economists have already had a sizable impact on the world of policy – and on the people affected by it. There is also no doubt that policies informed by the best available behavioral science – policies developed, perhaps, by you, the reader of this book – might help make the world a better place.

Behavioral policy and behavioral welfare economics remain relatively young, and are inspiring a rapidly developing set of policy proposals. Yet, it is not too early to draw some interim conclusions.

First off, the terms behavioral economists chose to discuss their policy proposals could probably have been chosen more wisely. The appropriation of the term "paternalism" in particular has invited many misguided responses. It would appear that Thaler and Sunstein had hoped to appeal to the union of libertarians and paternalists, but instead ended up appealing to their intersection – which is a small subset indeed. The discussion might have been less heated if they instead had used an unsexy term such as **ergonomics** or **human-factors engineering**. They could have done so without great loss, since ergonomics is defined as "the scientific discipline concerned with the understanding of interactions among humans and other elements of a system, and the profession that applies theory, principles, data and methods to design in order to optimize human well-being and overall system performance." Economist Raj Chetty has suggested we think of the nudge agenda in terms of model uncertainty: as a solution to a problem that arises when we do not know what model best captures the behaviors of the people involved. Assuming that neoclassical (optimizing) agents are insensitive to nudges while no behavioral agent is harmed and at least one of them is made better off, nudging becomes a weakly dominant strategy. I like to think of the nudge agenda as an effort to make social institutions, retirement plans, health-care systems, etc., robust under predictable variations in rationality. Thus, it is possible to think of the agenda as an effort to make social and other institutions trembling-hand perfect, as it were (see Section 10.4).

Once the conceptual confusion has been cleared up, it seems that the debate has been needlessly amped up. When used correctly and carefully, there is little reason to think that a nudge (such as the right-turn-ahead sign) needs to be dangerous, sneaky, or intrusive, and the fact that choice architects are mere mortals is no show-stopper. We have no reason to fear a slippery slope to harder kinds of paternalism; indeed, the nudge agenda may even expand the sphere of personal liberty. (We do need to be alert to unintended and unanticipated adverse consequences and to wolves in sheep's clothing, but that is true for all policy proposals.) That said, the enthusiasm among the biggest proponents of the nudge agenda is probably unwarranted. When even fellow behavioral economists ask proponents to hold their horses, we should at most be cautiously optimistic.

Most importantly, the debate about libertarian paternalism and the nudge agenda has been unhelpfully framed. The question is not whether people who design cars, phones, retirement plans, and health-care systems should nudge or not. Your car is full of nudges – such as the fasten-the-seat-belt indicator – and it would make little sense to demand that car manufacturers or anybody else cease to use them when they work. The real question, instead, is whether nudges should be used as complements to or substitutes for other forms of policy in the interest of attaining legitimate policy goals. But the answer to this question depends both on what the legitimate policy goals are – which can only be established by means of philosophical analysis – as well as empirical facts about the efficacy of nudges – which can only be established by means of systematic, empirical research. And on these two questions, we still have much to learn.

Further reading

The quotations in the introductory section are from Brooks (2008), who cites Greenspan, and Krugman (2009, p. 43). Hayek (1933, pp. 122–2) discusses the phylogenesis and ontogenesis of economists. Thaler and Sunstein (2008) is the go-to source about the nudge agenda; Sunstein (2014) helpfully catalogs ten nudges (pp. 585–7) and supplies the block quotation at the beginning of Section 12.4 (p. 584). OECD (2017) explores BI teams and similar efforts across the world. Caplan (2013) defends nudges from a libertarian perspective; Blackburn (2001, p. 64) dismisses slippery-slope reasoning; and Loewenstein and Ubel (2010) caution against the over-reliance on nudges. The connection to human-factors engineering was suggested by Robin Hogarth (personal communication); the term "ergonomics" is defined by the International Ergonomics Association (2015). Chetty (2015) suggests we think of the nudge agenda in terms of model uncertainty. Layard (2005) defends Bentham and calls the economics of happiness "a new science"; Clark (2018) surveys four decades of happiness research. Angner (2015b) discusses conceptions of welfare or well-being in economics. The passage from Keynes appears in *The General Theory of Employment, Interest and Money* (Keynes, 1936, pp. 383–4).

13 GENERAL DISCUSSION

Learning objective

After studying this chapter you will:

- Be able to assess the relative virtues (and perhaps vices) of behavioral economics

If the upshot of the previous chapter is correct, there may be little difference in the normative foundations of neoclassical and behavioral economics; the difference when it comes to policy proposals by and large reflects differences in descriptive theory. This, then, brings us to the question of how to assess the relative merits of neoclassical and behavioral economics. It is not my intention to try to settle the argument here. A proper assessment would require a thorough discussion of experimental and other methods, statistical methodology, and interpretation of a wide range of empirical results. All this is beyond the scope of this book, which (as explained in the Preface) is primarily an exercise in exposition. Yet, in the preceding chapters, I have aspired to offer some indication of what is at stake in the debate between behavioral and neoclassical economists, as well as what a proper assessment would look like.

One important insight is that neoclassical economics is not as silly as some of its critics make it out to be, and that many of the objections against the enterprise are misguided. As we saw in Section 11.2, for example, observed behavior in the ultimatum game is perfectly consistent with Nash equilibrium predictions. And Sections 2.6 and 11.2 have shown that neoclassical economics does not say that people are selfish, materialistic, greedy, or anything of the sort. Thus, attacks on what some critics have called the "selfishness axiom" of neoclassical economics are misguided not just in the sense that selfishness is not an axiom in the calculus, but in the sense that selfishness is not implied by the theory. Relatedly, standard theory does not say that people relentlessly pursue their own happiness or pleasure, meaning that any criticism based on the assumption that it does is inevitably flawed. Moreover, as Sections 4.7 and 6.6 indicate, the standard approach does not say that people (consciously or not) perform any sort of calculations in their head. Thus, any criticism premised on the notion that most people are unable, for example, to apply Bayes's theorem in their head is misguided. The quality of the conversation would be greatly enhanced if critics would abandon these straw-man attacks.

For practical purposes, economists have no choice but to use additional assumptions, often called auxiliary assumptions, in conjunction with their theories. In order to make substantive predictions, for example, the theorist might need to make more specific assumptions about what, exactly, people have preferences over, and how, exactly, these preferences order the available options. Depending on context, auxiliary assumptions might say that people only care about their own payoffs in dollar terms. The auxiliary assumptions need to be justified on independent grounds, and the justification may or may not be convincing. Such auxiliary assumptions, though, form no essential part of the neoclassical enterprise and can easily be replaced by others.

A no less important insight, however, is that anecdotal, experimental, and field evidence all suggest that people's observed behavior deviates from neoclassical theory in predictable fashion. There is little doubt that people (sometimes) do things such as honoring sunk costs, relying on adaptive but imperfect heuristics, violating the sure-thing principle, acting impulsively, and exhibiting limited strategic thinking. If deviations were random and unsystematic, they would (for many purposes) be theoretically uninteresting. In fact, however, deviations are frequently substantial and systematic, which means that they are predictable and can be captured by using a descriptive, scientific theory. At a fundamental level, behavioral economics is the result of this insight. The fact that people's behavior is irrational does not mean that it is unpredictable, nor that it cannot be described using scientific means.

The models developed by behavioral economists can be challenged on various grounds. For one thing, the evidence supporting them can be questioned. As the discussion of the replication crisis (in Section 1.3) underscored, it is likely that at least some behavioral results will turn out to be mere experimental artifacts, but it is extremely unlikely that all (or even most) will. A challenger can also point out that it is sometimes possible, when the results hold up under scrutiny, to accommodate these results within the standard framework, either by redescribing the choices available to the agent or by admitting additional arguments into the utility function. And it is important not to attribute irrationality to people when their behavior is better described as consistent with standard theory. That said, neoclassical economists frequently bend over backwards to accommodate empirical results in a manner that is both artificial and *ad hoc*. In the interest of defending standard theory, orthodox economists sometimes cook up a contorted – so-called "exotic" – utility function that after the fact makes observed behavior consistent with the neoclassical utility-maximization narrative, though it is not supported by any other independent evidence. It is often both simpler and more plausible to infer that people sometimes are in violation of standard theory.

Some neoclassical economists, as we have seen, are happy to admit that this is so. In defense of analytical game theory, as we know from Section 11.5, it has been argued that neoclassical theory is only intended to apply under sharply circumscribed conditions. Of course, this defense of neoclassical economics does not constitute an argument against behavioral economics. Instead, this response might offer a way to reconcile neoclassical and behavioral economics. Many behavioral economists are happy to admit that observed behavior sometimes approaches or coincides with neoclassical predictions under certain conditions. But if those conditions do not hold, neoclassical economists should be able to agree that a non-neoclassical theory is required to explain and predict behavior. This is the domain of behavioral economics.

Note that nothing prevents behavioral economists from continuing to use neoclassical models for various purposes. The fact that neoclassical models sometimes do a great job of capturing observed choices is part of the reason why those models often survive as special cases of a more general behavioral model. Thus, the exponential discounting function (from Section 8.3) survives as a special case of the hyperbolic one (from Section 9.2): just set beta to one and the latter reduces to the former. There is nothing inconsistent about this. But there is an interesting asymmetry between neoclassical and behavioral economics. Postwar neoclassical orthodoxy insists that economics must rid itself of all ties to psychology – hedonic and otherwise (see Section 1.2). An economist committed to this view cannot consistently help himself or herself to theories making references to things "in the head" whenever it is convenient to do so. In this respect too, behavioral economics has a distinct advantage over its neoclassical counterpart.

Science progresses in fits and starts. Rather than a steady progression from darkness to light, science tends to offer a series of increasingly complex models that capture to a greater or a lesser extent empirical phenomena that for whatever reason attract scientists' interest. The "final" theory is likely to remain out of reach. The same is true of economics. In *Worstward Ho*, Samuel Beckett wrote: "Ever tried. Ever failed. No matter. Try again. Fail again. Fail better." To use Beckett's phrase, progress in science in general and economics in particular can be thought of as a matter of failing better. Incidentally, Beckett might just as well have been describing the study of science, which is never finished, or the writing of textbooks, which can always be improved.

To what extent do behavioral economists fail better than neoclassical economists? I do not pretend to have the answer. But I do hope to have shed some light on the nature of both neoclassical and behavioral economics, and to have underscored the power and promise of economic analysis of social phenomena. Moreover, I hope to have demonstrated how behavioral economics can advance our understanding of people's decisions under conditions of scarcity – and of the results of those decisions for society at large.

Further reading

Worstward Ho, first published in 1983, is included in *Nohow On* (Beckett, 1989); the quotation appears on p. 101. Behavioral economists' use of neoclassical theory is discussed at length in Angner (2015a, 2019). For a deeper discussion about the nature of behavioral economics, its strengths and weaknesses, see Davis (2011) and Ross (2005). For a more advanced textbook, see Wilkinson and Klaes (2017).

APPENDIX: ANSWER KEY

Chapter 1

Exercise 1.1 (**a**) Descriptive. (**b**) Normative. (**c**) Descriptive.

Exercise 1.3 (**a**) 100. (**b**) $400,000. (**c**) $242 (see Section 8.2).

Chapter 2

Exercise 2.1 (**a**) fBn. (**b**) nBf. (**c**) nBn.

Exercise 2.2 {Afghanistan, Albania, Algeria, Andorra, Angola, . . . , Zimbabwe}. Any order is fine, but the curly brackets are part of a correct answer. Notice that if you were to spell it out, this would be a pretty long list.

Exercise 2.3 (**a**) In all likelihood: d $\succsim$ r. (**b**) In all likelihood: r $\succsim$ d.

Exercise 2.7 (**a**) Intransitive, incomplete. (**b**) Transitive, incomplete. (**c**) Transitivity depends on whether we consider half-siblings; either way, it is incomplete. (**d**) Intransitive, incomplete. (**e**) Transitive, incomplete. (**f**) Transitive, incomplete. (**g**) Transitive, incomplete.

Exercise 2.8 If the enemy of your enemy is not your enemy, then "is the enemy of" is intransitive.

Exercise 2.9 (**a**) Transitive, complete. (**b**) Transitive, incomplete. (**c**) Transitive, incomplete. (**d**) Transitive, incomplete.

Exercise 2.10 (**a**) Transitivity implies that if apples are at least as good as bananas, and bananas are at least as good as starvation, then apples are at least as good as starvation; that if starvation is at least as good as bananas, and bananas are at least as good as apples, then starvation is at least as good as apples; and so on. (**b**) Completeness implies that either apples are at least as good as bananas or bananas are at least as good as apples, but also that apples are at least as good as apples, and that bananas are at least as good as bananas.

Exercise 2.13 Assume $x \succsim y$ & $y \sim z$. The fact that $y \sim z$ implies that $y \succsim z$. Given that $x \succsim y$ & $y \succsim z$, it follows that $x \succsim z$.

Exercise 2.14 Here are the complete proofs.

(a)	1.	$x \sim y \,\&\, y \sim z \,\&\, z \sim p$	by assumption
	2.	$x \sim z$	from (1), by transitivity of $\sim$
	3.	$x \sim p$	from (1) & (2), by transitivity of $\sim$
	$\therefore$	$x \sim y \,\&\, y \sim z \,\&\, z \sim p \rightarrow x \sim p$	QED □
(b)	1.	$x \sim y \,\&\, y \sim z \,\&\, z \sim p$	by assumption
	2.	$x \succsim y \,\&\, y \succsim x$	from (1), by definition of $\sim$
	3.	$y \succsim z \,\&\, z \succsim y$	from (1), by definition of $\sim$
	4.	$z \succsim p \,\&\, p \succsim z$	from (1), by definition of $\sim$
	5.	$x \succsim z$	from (2) & (3), by transitivity of $\succsim$
	6.	$x \succsim p$	from (4) & (5), by transitivity of $\succsim$
	7.	$z \succsim x$	from (2) & (3), by transitivity of $\succsim$
	8.	$p \succsim x$	from (3) & (7), by transitivity of $\succsim$
	9.	$x \sim p$	from (6) & (8), by definition of $\sim$
	$\therefore$	$x \sim y \,\&\, y \sim z \,\&\, z \sim p \rightarrow x \sim p$	QED □

Exercise 2.17 Here is the complete proof of Proposition 2.16(i). What this exercise calls the first part is lines (2)–(4).

1.	$x \succ y \,\&\, y \succ z$	by assumption
2.	$x \succsim y \,\&\, \neg y \succsim x$	from (1), by definition of $\succ$
3.	$y \succsim z \,\&\, \neg z \succsim y$	from (1), by definition of $\succ$
4.	$x \succsim z$	from (1) and (2), by transitivity of $\succsim$
5.	$z \succsim x$	by assumption, for proof by contradiction
6.	$y \succsim x$	from (3) and (5), by transitivity of $\succsim$
7.	$\bot$	from (2) and (6)
8.	$\neg z \succsim x$	from (5)–(7), by contradiction
9.	$x \succ z$	from (4) and (8), by definition of $\succ$
$\therefore$	$x \succ y \,\&\, y \succ z \rightarrow x \succ z$	QED □

Exercise 2.18 Assume, for proof by contradiction, that "is the enemy of" is in fact transitive. That means that whenever A is the enemy of B, and B is the enemy of C, A must be the enemy of C. But assuming that the enemy of the enemy of a person is his or her friend, that is impossible. Therefore, "is the enemy of" is not transitive.

Exercise 2.19 Here is the proof:

1.	$x \succ x$	by assumption
2.	$x \succsim x \,\&\, \neg x \succsim x$	from (1), by definition of $\succ$
3.	$\bot$	from (2)
$\therefore$	$\neg x \succ x$	QED □

Exercise 2.20 Here is the complete proof:

1. $x \succcurlyeq y \,\&\, y \succcurlyeq z$	by assumption
2. $x \succcurlyeq y \,\&\, \neg y \succcurlyeq x$	from (1), by definition of $\succ$
3. $x \succcurlyeq z$	from (1) and (2), by transitivity of $\succcurlyeq$
4. $z \succcurlyeq x$	by assumption, for proof by contradiction
5. $y \succcurlyeq x$	from (1) and (4), by transitivity of $\succcurlyeq$
6. $\bot$	from (2) and (5)
7. $\neg z \succcurlyeq x$	from (4)–(6), by contradiction
8. $x \succ z$	from (3) and (7), by definition of $\succ$
$\therefore\ x \succ y \,\&\, y \succcurlyeq z \rightarrow x \succ z$	QED □

Exercise 2.21 See below:

(a)	1. $x \succ y$	by assumption
	2. $x \succcurlyeq y \,\&\, \neg y \succcurlyeq x$	from (1), by definition of $\succ$
	3. $x \succcurlyeq y$	from (2), by logic
	$\therefore\ x \succ y \rightarrow x \succcurlyeq y$	QED □
(b)	1. $x \succ y$	by assumption
	2. $x \succcurlyeq y \,\&\, \neg y \succcurlyeq x$	from (1), by definition of $\succ$
	3. $\neg y \succcurlyeq x$	from (2), by logic
	$\therefore\ x \succ y \rightarrow \neg y \succcurlyeq x$	QED □
(c)	1. $x \succcurlyeq y$	by assumption
	2. $y \succ x$	by assumption, for proof by contradiction
	3. $y \succcurlyeq x \,\&\, \neg x \succcurlyeq y$	from (2), by definition of $\succ$
	4. $\bot$	from (1) & (3)
	5. $\neg y \succ x$	from (2)–(4), by contradiction
	$\therefore\ x \succcurlyeq y \rightarrow \neg y \succ x$	QED □
(d)	1. $x \succ y$	by assumption
	2. $x \sim y$	by assumption, for proof by contradiction
	3. $x \succcurlyeq y \,\&\, \neg y \succcurlyeq x$	from (1), by definition of $\succ$
	4. $x \succcurlyeq y \,\&\, y \succcurlyeq x$	from (2), by definition of $\sim$
	5. $\bot$	from (3) & (4)
	6. $\neg x \sim y$	from (2)–(5), by contradiction
	$\therefore\ x \succ y \rightarrow \neg x \sim y$	QED □
(e)	1. $x \sim y$	by assumption
	2. $x \succ y$	by assumption, for proof by contradiction
	3. $x \succcurlyeq y \,\&\, y \succcurlyeq x$	from (1), by definition of $\sim$
	4. $x \succcurlyeq y \,\&\, \neg y \succcurlyeq x$	from (2), by definition of $\succ$
	5. $\bot$	from (3) & (4)
	6. $\neg x \succ y$	from (2)–(5), by contradiction
	$\therefore\ x \sim y \rightarrow \neg x \succ y$	QED □

(f)	1. $\neg x \succsim y$	by assumption	
	2. $x \succsim y \vee y \succsim x$	by completeness	
	3. $y \succsim x$	from (1) & (2), by logic	
	$\therefore\ \neg x \succsim y \rightarrow y \succsim x$	QED	□
(g)	1. $\neg x \succsim y$	by assumption	
	2. $x \succsim y \vee y \succsim x$	by completeness	
	3. $y \succsim x$	from (1) & (2), by logic	
	4. $y \succ x$	from (1) & (3), by definition of $\succ$	
	$\therefore\ \neg x \succsim y \rightarrow y \succ x$	QED	□
(h)	1. $\neg x \succ y$	by assumption	
	2. $\neg(x \succsim y \ \&\ \neg y \succsim x)$	from (1), by definition of $\succ$	
	3. $\neg x \succsim y \vee y \succsim x$	from (2), by logic	
	4. $y \succsim x$	from (3), by part (f) above	
	$\therefore\ \neg x \succ y \rightarrow y \succsim x$	QED	□

Exercise 2.22 Begin by assuming what is to the left of the arrow, in this case $x \sim y$ & $y \sim z$. Then, on a separate line, assume (for a proof by contradiction) the opposite of what you are trying to prove; that is, assume that $x \succ z$. Finally, derive a contradiction.

Exercise 2.23 (**a**) Begin by assuming that $\neg x \succsim y$ and $\neg y \succsim z$. Apply Proposition 2.21(g) twice to get $y \succ x$ and $z \succ y$. Transitivity will yield $z \succ x$, which by Proposition 2.21(b) gives you the result $\neg x \succsim z$. (**b**) Begin by assuming that $\neg x \succ y$ and $\neg y \succ z$. Proposition 2.21(h) applied twice, transitivity, and 2.21(c) will give you the result.

Exercise 2.24 The answer is, of course, that $f^+ \succ c$. Begin by assuming that $f \sim c$ & $f^+ \succ f$. You need to prove two things: that $f^+ \succsim c$ and that $\neg c \succsim f^+$. After using the definitions of indifference and strict preference, the first part follows by transitivity of weak preference. For the second part, assume (for a proof by contradiction) that $c \succsim f^+$.

Exercise 2.25 A rational person is indifferent. Because $c_1 \sim c_2$ & $c_2 \sim c_3$, Proposition 2.14 implies that $c_1 \sim c_3$. Because, in addition, $c_3 \sim c_4$, the same proposition implies that $c_1 \sim c_4$ and so on. Ultimately, you will find that $c_1 \sim c_{1000}$. QED.

Exercise 2.26 Assume that $x \succ y$ & $y \succ z$ & $z \succ x$. Apply the definition of strict preference a few times, and a contradiction is immediate.

Exercise 2.27 Assume that $x \succsim y$ & $y \succsim z$ & $z \succsim x$. Use the transitivity of weak preference and the definition of indifference to establish that $x \sim y$ & $y \sim z$ & $z \sim x$.

Exercise 2.28 See Figure A.1.

Exercise 2.30 The new menu would be {nothing at all, soup, salad, chicken, beef, soup-and-chicken, soup-and-beef, salad-and-chicken, salad-and-beef}.

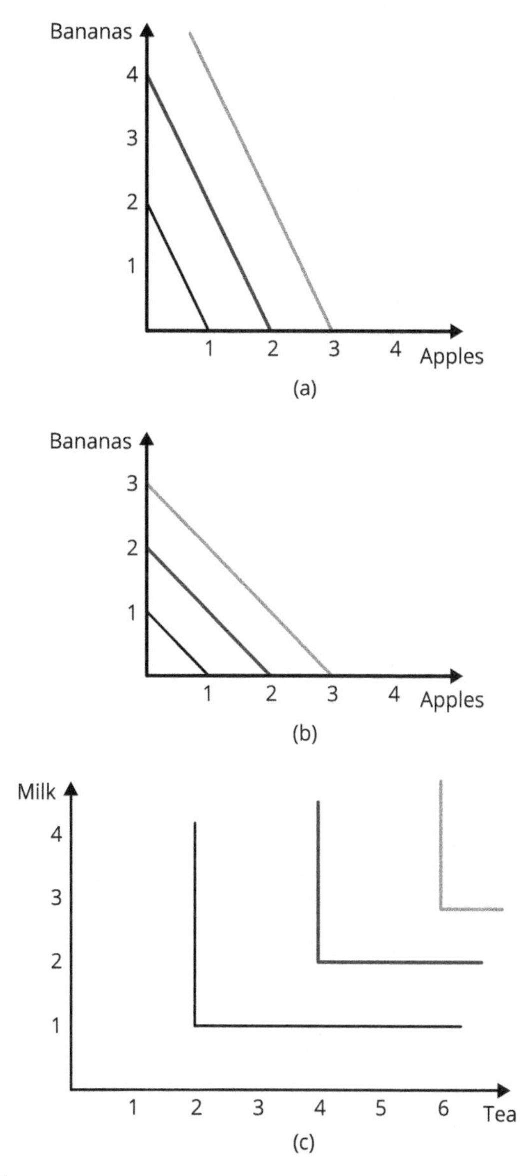

Figure A.1 Indifference curves

Exercise 2.31 See Figure A.2.

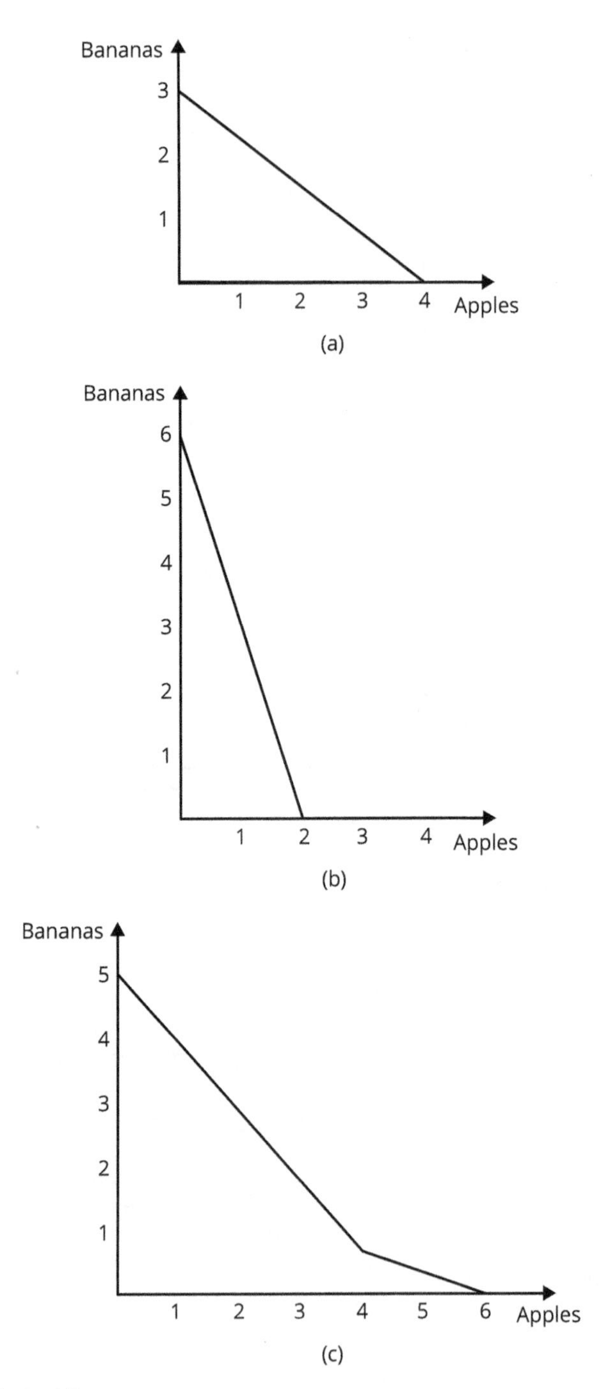

Figure A.2 Budget line

Exercise 2.35 In order to prove this proposition, you need to do two things. First, assume that $x \sim y$ and prove that $u(x) = u(y)$. Second, assume that $u(x) = u(y)$ and prove that $x \sim y$. Here is the complete proof:

1.	$x \sim y$	by assumption
2.	$x \succsim y \;\&\; y \succsim x$	from (1), by definition of ~
3.	$u(x) \geq u(y) \;\&\; u(y) \geq u(x)$	from (2), by definition of $u(\cdot)$ (twice)
4.	$u(x) = u(y)$	from (3), by math
5.	$u(x) = u(y)$	by assumption
6.	$u(x) \geq u(y) \;\&\; u(y) \geq u(x)$	from (5), by math
7.	$x \succsim y \;\&\; y \succsim x$	from (6), by definition of $u(\cdot)$ (twice)
8.	$x \sim y$	from (4) and (8), by definition of ~
∴	$x \sim y \Leftrightarrow u(x) = u(y)$	QED □

Exercise 2.36 (**a**) Neoclassical economic agents are assumed to be "rational"– which is used in a technical sense quite different from that of "intelligent" and "analytic" in everyday speech – and there is no assumption that they are selfish. Classical economic agents, like those who appear in the work of Adam Smith (see Section 1.2), are not assumed to be rational, intelligent, analytic, or selfish – although they may be on occasion. (b) The notion that money makes people happy is no part of fundamental economic theory. To the extent that people prefer more money to less, we can say that money gives them utility. But the concept of utility has no essential connection to happiness.

Exercise 2.37 See Table A.1.

Table A.1 Properties of weak preference, indifference, and strong preference

	Property	Definition	$\succsim$	$\sim$	$\succ$
(**a**)	Transitivity	$xRy \;\&\; yRz \rightarrow xRz$ (for all x, y, z)	✓	✓	✓
(**b**)	Completeness	$xRy \vee yRx$ (for all x, y)	✓		
(**c**)	Reflexivity	xRx (for all x)	✓	✓	
(**d**)	Irreflexivity	$\neg xRx$ (for all x)			✓
(**e**)	Symmetry	$xRy \rightarrow yRx$ (for all x, y)		✓	
(**f**)	Anti-symmetry	$xRy \rightarrow \neg yRx$ (for all x, y)			✓

Exercise 2.38 (**a**) "is not married to" (**b**) "is married to." (**c**) both. (**d**) neither. (**e**) neither.

Exercise 2.39 Omitted.

Exercise 2.40 (**a**) She violates completeness. (**b**) He violates transitivity. (**c**) It violates completeness.

Exercise 2.41 (**a**) Agrees with convexity. (**b**) Agrees with non-satiation. (**c**) Disagrees with continuity. (**d**) Agrees with non-satiation.

Exercise 2.42 (**a**) Curly (or snake-like) indifference curves are ruled out by convexity. Notice that the line segment between the two X's lies below the indifference curve, which violates convexity. The example shows how indifference curves that violate convexity might lead to a situation where there is no unique best choice, given the budget set and indifference curves. (**b**) Concentric indifference curves are ruled out by non-satiation. The point marked X in the image is a **bliss point**, where the agent is perfectly satiated. Non-satiation says there is no such thing. (**c**) Indifference curves that cross are ruled out by transitivity. Notice that in the image, the agent is indifferent between A and B and between A and C. By transitivity of indifference (Proposition 2.12(iii)), the agent has to be indifferent between B and C, but that is not the case here. (**d**) Thick indifference curves are ruled out by non-satiation. Pick any point on the indifference curve. When the curve is thick, you can draw a tiny circle around the point such that the entire circle fits within the indifference curve. But then non-satiation tells us there is another point within the circle that is preferred to the original point, which violates the assumption that both are on the same indifference curve.

Chapter 3

Exercise 3.1 (**a**) See Figure A.3. (**b**) \$1000. (**c**) \$1000.

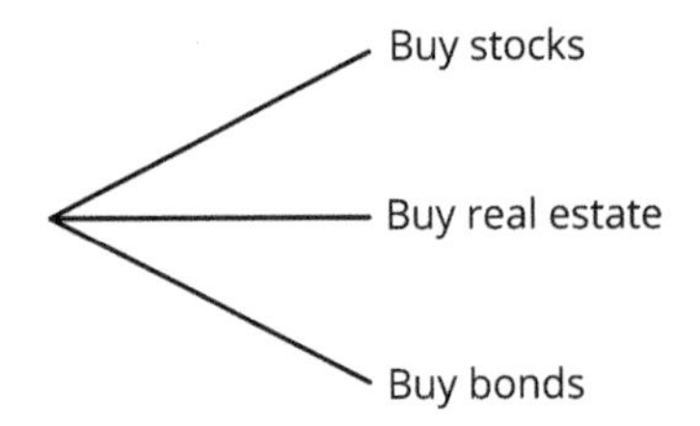

Figure A.3 Investment problem

Exercise 3.3 (**a**) See Figure A.4.

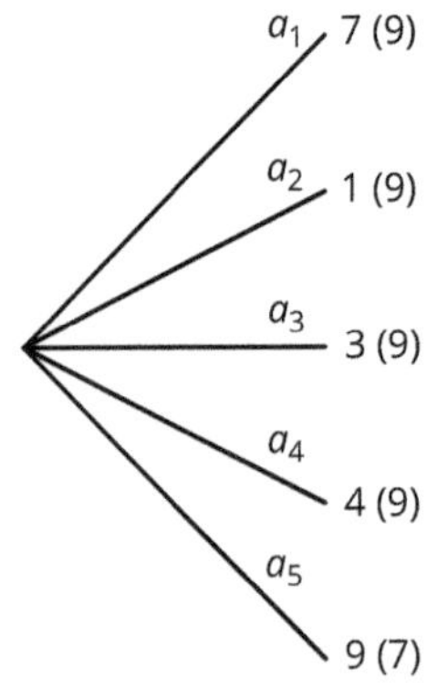

Figure A.4 Opportunity costs

Exercise 3.6 The answers will depend on your preferences, but they could be (**a**) the most fulfilling relationship you could have instead, (**b**) the course of study that excites you most, and (**c**) the most satisfying activity you could engage in instead of sleeping in the morning.

Exercise 3.7 For highly paid people, the opportunity cost of mowing lawns, etc., is greater.

Exercise 3.10 You may be ignoring the fact that there are better things you could spend $60 on, including a $10 meal and 50 dollars' worth of other fun or useful things.

Exercise 3.11 If a person is willing to do "whatever it takes" to attain some goal, that means he or she is willing to ignore all opportunity costs – which is irrational.

Exercise 3.13 Not necessarily: if another, even more successful campaign to boost revenue were available to you at the time, it would have been irrational to invest in the advertising campaign.

Exercise 3.14 No: military action is associated with huge explicit and implicit costs, which are often underestimated by its advocates.

Exercise 3.17 See Figure A.5.

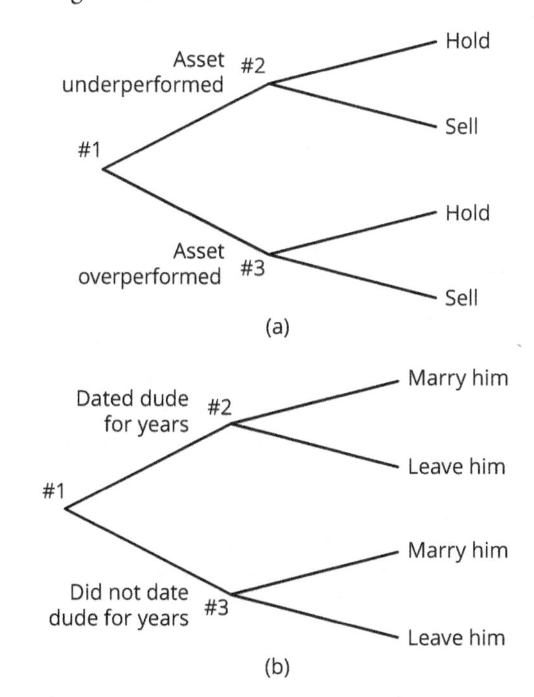

Figure A.5 Sunk costs, again

Exercise 3.19 The difference between the cheese you manage to throw away and the one you do not is that the former is not associated with a sunk cost, whereas the latter is.

Exercise 3.20 Sign up for the course at the public university. The tuition already paid to the liberal arts college is a sunk cost.

Exercise 3.27 (**a**) You would want to put it in area B. (**b**) Before.

Exercise 3.28 (**a**) Show a property that is not in as good shape as and even farther from the office than the first property, but which is still in better shape than the second property. (**b**) Show a property that is in even worse shape and slightly farther from the office than the second property, but which is still closer to the office than the first property.

Exercise 3.29 If the third-party politician C promises neither higher nor lower taxes but massive cuts to public services, he will be asymmetrically dominated by A, as intended.

Exercise 3.30 (**a**) Choose a wingman or wingwoman who is less desirable than you are along all relevant dimensions, but who beats each competitor along at least one dimension. (**b**) You want your wingman or wingwoman to fall in the 8–9 range (exclusive) both with respect to attractiveness and intelligence. (**c**) He or she thinks you are less desirable than he or she is along all relevant dimensions.

Exercise 3.31 You would need to sell a vehicle that is super-fast and super-unsafe: something like a rocket-propelled bicycle.

Exercise 3.32 The third-party politician must promise even lower taxes, and even greater cuts to public services than politician A.

Exercise 3.36 From the point of view of the waitress without a car, the value of getting one is the value of going from 0 to +1. In terms of value, this amounts to going from $v(0) = 0$ to $v(+1) = 0.5$. The change, therefore, is $v(+1) - v(0) = 0.5$. From the point of view of the waitress with a car, the value of losing one is the value of going from 0 to −1 cars. In terms of value, that amounts to going from $v(0) = 0$ to $v(-1) = -2$. The change, then, is $v(-1) - v(0) = -2 - 0 = -2$. The total change for a person experiencing the gain *and* the loss is $0.5 + (-2) = 0.5 - 2 = -1.5$. This amounts to a loss of 1.5 units of value, leaving her worse off than she was before the sequence of events took place.

Exercise 3.37 In value terms, gaining \$6 and losing \$4 amounts to a change in value terms of $v(+6) + v(-4) = 3 - 8 = -5$. In value terms, that is as bad as suffering a loss of \$2.50, since $v(-2.50) = -5$, in spite of the fact that you are left with \$2 more than you had at the outset.

Exercise 3.38 (**a**) In terms of deviations from a reference point of \$0, the drop in price corresponds to a drop from +1 to 0. In value terms, that corresponds to a drop from $v(+1) = 0.5$ to $v(0) = 0$. So the change in value is $v(0) - v(+1) = 0 - 0.5 = -0.5$.

(**b**) In terms of deviations from a reference point of \$1, the drop in price corresponds to a drop from 0 to –1. In value terms, that corresponds to a drop from $v(0) = 0$ to $v(-1) = -2$. So the change in value is $v(-1) - v(0) = -2 - 0 = -2$. (**c**) With a reference point of \$0, you experience a loss of 0.5 units of value; with a reference point of \$1, you experience a loss of 2 units of value. Since a loss of 2 is worse than a loss of 0.5, using a reference point of \$1 makes you feel worse than a reference point of \$0.

Exercise 3.39 (**a**) Given her reference point of \$12, Alicia thinks of the price drop from \$17 to \$12 as a change from +5 to 0. The change in value terms is $v(0) - v(+5) = 0 - 2.5 = -2.5$, meaning a loss of 2.5. (**b**) Given her reference point of \$17, Benice thinks of the price drop as a drop from 0 to –5. The change in value terms is $v(-5) - v(0) = -10 - 0 = -10$, meaning a loss of 10. (**c**) Given her reference point of \$10, Charlie thinks of the price drop as a change from +7 to +2. The change in value terms is $v(2) - v(7) = 1 - 3.5 = -2.5$, meaning a loss of 2.5. (**d**) Benice is most disappointed.

Exercise 3.40 (**a**) 50. (**b**) 200. (**c**) The net effect is –150. (**d**) Bad.

Exercise 3.41 (**a**) Alex thought of the \$2 as a forgone gain, so for her the absolute value of the \$2 was 1. Mathematically, the change in value can be computed as $v(0) - v(+2) = 0 - 2/2 = -1$. (**b**) Bob thought of the \$2 as a loss, so for him the absolute value of the \$2 was 4. Mathematically, the change in value can be computed as $v(-2) - v(0) = -4 - 0 = -4$. (**c**) Bob.

Exercise 3.42 To people who do not bring their own mugs the discount seems like a forgone gain, whereas a penalty would feel like an actual loss. Since forgone gains are easier to stomach, customers are less likely to be alienated this way.

Exercise 3.43 A person who does not expect a raise would experience $v(+5)$. A person who does expect a raise would experience $v(-5)$. See Figure A.6 for a graphical representation of the difference.

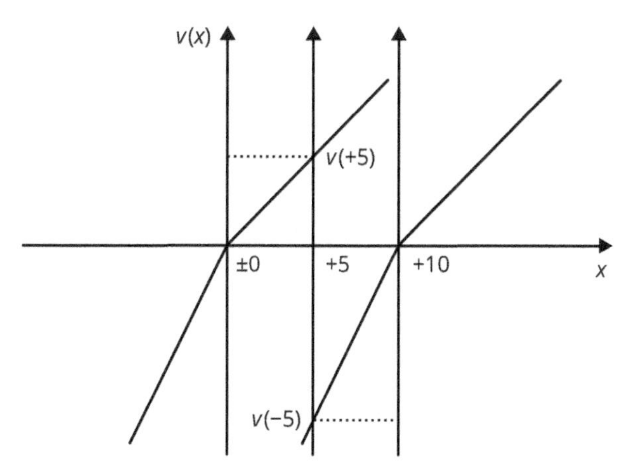

Figure A.6 The raise

Exercise 3.44 (**a**) $v(+93 - 75) = v(+18) = 9$. (**b**) $v(+67 - 75) = v(-8) = -16$. (**c**) The theory would suggest that you should set low expectations and perform well.

Exercise 3.45 The theory would suggest that you should surround yourself with low-paid people and make a lot of money.

Exercise 3.47 See Figure A.7.

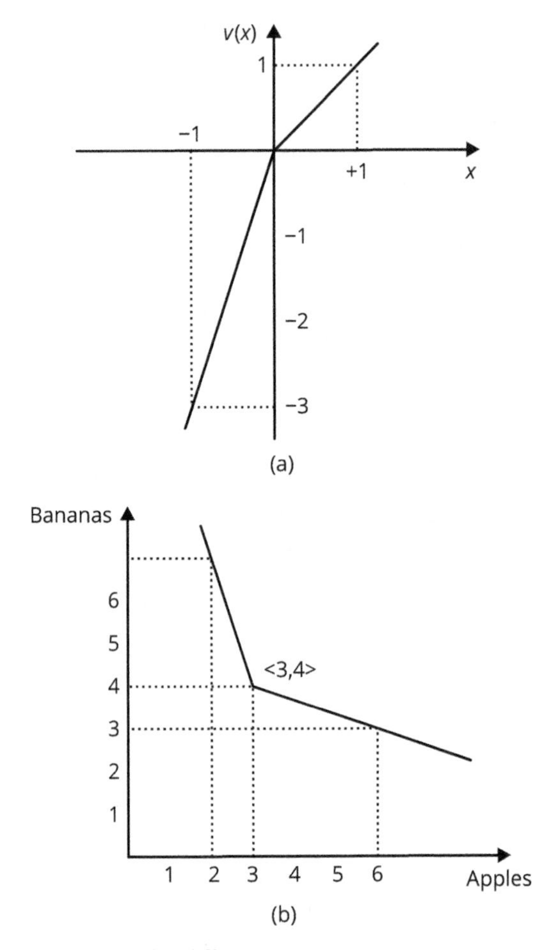

Figure A.7 Value function and indifference curve

Exercise 3.48 (**a**) Status quo bias would entail that Europeans would tend to favor the European system while Americans would tend to favor the American system. The bias is driven by loss aversion. For Europeans, the loss of government-provided health care would not be outweighed by the gain in disposable income; for Americans, the loss of disposable income would not be outweighed by the gain in government-provided health care. (**b**) See Figure A.8. (**c**) Loss aversion suggests that after adaptation

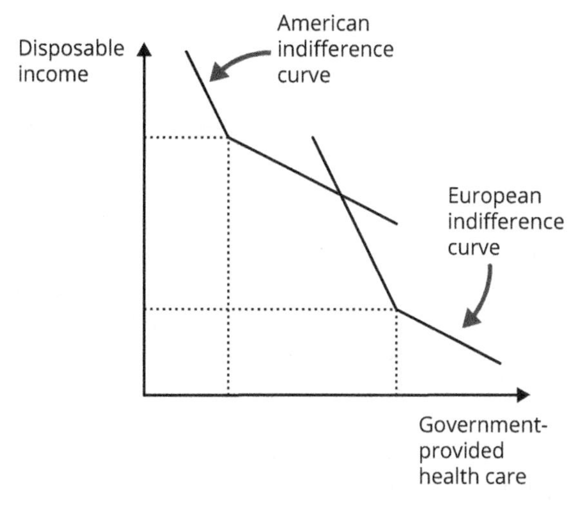

Figure A.8 Health-care systems

has occurred, Americans would be unwilling to give up the new system, and the opposition party would find it difficult to engineer a return to the old one.

Exercise 3.49 Loss aversion/status quo bias makes it very difficult to take money or other benefits away from people.

Exercise 3.50 (**a**) Once a program is enacted, loss aversion will make the beneficiaries of that program extremely averse to canceling it. This example also illustrates the concept of status quo bias. (**b**) The sunset provision is supposed to shorten the lifespan of temporary programs, since simply letting legislation expire is easier than taking positive action to cancel it.

Exercise 3.54 Assume that the emperor computed the number of grains of rice that he would owe the inventor during the first n days, where n is a number considerably less than 64, used that number as an anchor, and insufficiently adjusted the number upwards.

Exercise 3.55 Advertise it as being sharply reduced from an original, even higher price.

Exercise 3.56 When asked to whom custody should be awarded, people naturally search for reasons to award custody to a person, which favors the parent with the biggest strengths (in this case B). When asked to whom custody should be denied, people naturally search for reasons to deny custody to a person, which disfavors the parent with the biggest weaknesses (also B). The outcome, then, depends on whether person making the choice is searching for reasons to accept or reasons to reject.

Exercise 3.57 The answer is (**b**) \$10. The answer is the value to you of going to the Dylan concert (\$50) minus what you would have to pay to go (\$40). Only 21.6 percent of the professional economists in the study got the answer right, which is particularly embarrassing if you reflect on the fact that they could have done better had they simply picked their answers randomly.

Exercise 3.58 The conclusion overlooks the opportunity costs of making money. In real life, all things are not equal: if you decide to do something like work more to increase your income, there is an opportunity cost in terms of forgone leisure, time with family and friends, and so on. As you work and earn more and more, the marginal benefit of money will go down, and you will ultimately hit a point where you are better off switching to leisure. Working more than that is working too much.

Exercise 3.59 The receptionist might be unfamiliar with the sunk-cost fallacy, which is another reason people who just purchased a one-week pass come back for more.

Exercise 3.60 The sunk-cost fallacy, again.

Exercise 3.61 See Figure A.9.

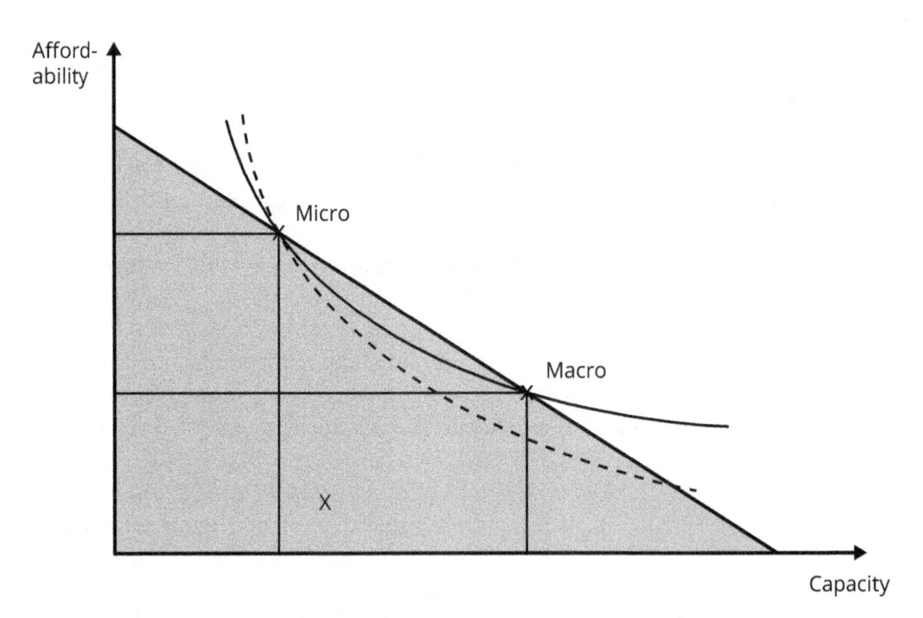

Figure A.9 Pear Corporation hijinks

Exercise 3.62 (**a**) If Tim returned the phone, he would go from $v(0) = 0$ to $v(-1) = -3$, meaning that he would experience a loss of 3 units of value. (**b**) If Bill were to pick up a phone, he would go from $v(0) = 0$ to $v(+1) = 1$, meaning that he would experience a gain of 1 unit of value. Bill's forgone gain when he does not pick up a phone, then, is 1. (**c**) Because Tim's loss in value terms would exceed Bill's forgone gain, Tim is more likely to end up the owner (proud or not) of the new iPhone.

Exercise 3.63 (**a**) When the price went from \$7 to \$4, in terms of deviations from his reference point, Larry went from $\pm$ 0 to –3. In terms of value, therefore, he went from $v(0) = 0$ to $v(-3) = -9$, meaning that he experienced a loss of 9 units of value. (**b**) When the price went from \$7 to \$4, in terms of deviations from her reference point, Janet went from +3 to $\pm$ 0. In terms of value, then, she went from $v(+3) = 3/3 = 1$ to $v(0) = 0$, meaning that she experienced a loss of 1 unit of value. (c) Larry is more disappointed. Note that the difference is that Larry views the decline in stock price as a loss, whereas Janet views it as a forgone gain.

Exercise 3.64 Loss aversion makes the gain of *what you might become* seem small relative to the loss of *what you are* – prompting people to forgo the former in order to avoid the latter, even when they acknowledge that it would be better to act differently.

Exercise 3.65 The aspiration treadmill.

Exercise 3.66 (**a**) Sunk-cost fallacy. (**b**) Anchoring and adjustment. (**c**) Loss aversion. (**d**) Sunk-cost fallacy. (**e**) Failure to consider opportunity costs. (**f**) The compromise effect.

Chapter 4

Exercise 4.8 1/52. Whatever card you pick the first time around, you have a 1/52 chance to pick the same card again the second time.

Exercise 4.9 You can apply the rule only when the outcomes in question are equally likely, and there is no reason to think that is true here.

Exercise 4.10 The outcome space is reduced to {GB, GG}, and the probability is 1/2.

Exercise 4.11 (**a**) {BBB, GGG, BBG, GGB, BGB, GBG, BGG, GBB}. (**b**) {GGG, BBG, GGB, BGB, GBG, BGG, GBB}. (**c**) 1/7. (**d**) {GGG, GGB, GBG, BGG}. (**e**) 1/4.

Exercise 4.12 (**a**) {W/W, W/W, R/R, R/R, W/R, R/W}. (**b**) {W/W, W/W, W/R}. (**c**) 1/3.

Exercise 4.13 (**a**) {W/W, W/W, B/B, B/B, R/R, R/R, R/W, W/R}. (**b**) {B/B, B/B}. (**c**) 1. (**d**) {R/R, R/R, R/W}. (**e**) 1/3.

Exercise 4.14 The analysis of this problem is not completely uncontroversial, but it is fairly widely agreed that the probability is 1/3.

Exercise 4.16 (**c**) and (**d**).

Exercise 4.17 4/52 = 1/13.

Exercise 4.21 It is equally likely: the probability is 1/36 either way.

Exercise 4.22 (**d**).

Exercise 4.23 Not independent.

Exercise 4.24 Because there are two (mutually exclusive) ways for the dots to add up to eleven, the answer is 1/36 + 1/36 = 1/18.

Exercise 4.25 (**a**) 1/52 * 1/52 = 1/2704. (**b**) 1/13 * 1/13 = 1/169.

Exercise 4.26 (**a**) 1/6 * 1/6 = 1/36. (**b**) (1 – 1/6) * (1 – 1/6) = 25/36. (**c**) 1/6 * (1 – 1/6) + (1 – 1/6) * 1/6 = 10/36. (**d**) 1 – (1 – 1/6) * (1 – 1/6) = 11/36.

Exercise 4.27 It would be a mistake because you would be applying the OR rule to two outcomes that are not mutually exclusive.

Exercise 4.28 The answer is:

$$\frac{6}{49} * \frac{5}{48} * \frac{4}{47} * \frac{3}{46} * \frac{2}{45} * \frac{1}{44} = \frac{1}{13{,}983{,}816}$$

This amounts to about 0.000,000,07. So, if you were to play once a year, on the average you would win once every 13,983,816 years. If you played once per day, given that there are 364.25 days in a year, on the average you would win once every 268,920 years.

Exercise 4.30 If people assess the value of the ticket by using the amount that can be won as an anchor and adjusting downwards, insufficient adjustment would imply that they overestimate the value of the ticket. If, in addition, people assess the probability of winning by using the probability of picking the first number correctly as an anchor and adjusting downwards, insufficient adjustment would imply that they overestimate the probability of winning.

Exercise 4.31 (**a**) Pr(H | T) means "The probability that the patient has a headache given that he or she has a tumor," whereas Pr(T | H) means "The probability that the patient has a tumor given that he or she has a headache." (**b**) The probabilities are clearly different. In general, we should expect that Pr(H | T) > Pr(T | H).

Exercise 4.33 You already know the answer is one in four, since there are four aces in a deck of cards and the ace of spades is one of them. But you can compute the answer using Definition 4.31 as follows:

$$\Pr(A\spadesuit|A) = \frac{\Pr(A\spadesuit \& A)}{\Pr(A)} = \frac{\Pr(A\spadesuit)}{\Pr(A)} = \frac{1/52}{4/52} = 1/4$$

Exercise 4.35 $\Pr(\mathbf{A}\spadesuit_1 \& \mathbf{A}\spadesuit_2) = \Pr(\mathbf{A}\spadesuit_1) * \Pr(\mathbf{A}\spadesuit_2 | \mathbf{A}\spadesuit_1) = 1/52 * 0 = 0$.

Exercise 4.38 (**a**) Assume that $\Pr(A | B) = \Pr(A)$, then use Proposition 4.34 to derive $\Pr(B | A) = \Pr(B)$. (**b**) Assume that $\Pr(B | A) = \Pr(B)$, then use Proposition 4.32 to derive $\Pr(A \& B) = \Pr(A) * \Pr(B)$. (**c**) Assume that $\Pr(A \& B) = \Pr(A) * \Pr(B)$, then use Proposition 4.34 to derive $\Pr(A | B) = \Pr(A)$.

Exercise 4.40 (**a**) See Figure A.10. (**b**) The probability that your patient dies within the year is

$$\Pr(D) = \Pr(D | A) * \Pr(A) + \Pr(D | B) * \Pr(B) = 4/5 * 1/3 + 1/5 * 2/3 = 2/5$$

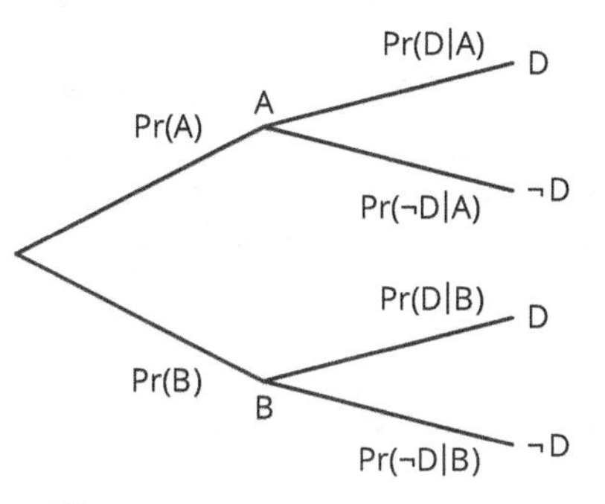

Figure A.10 Cancers A and B

Exercise 4.41 The probability is

$$\begin{aligned}\Pr(P) &= \Pr(P | E) * \Pr(E) + \Pr(P | \neg E) * \Pr(\neg E) \\ &= .90 * .60 + .50 * .40 = .54 + .20 = .74\end{aligned}$$

Exercise 4.43 The answer is

$$\Pr(A|D) = \frac{\Pr(D|A) * \Pr(A)}{\Pr(D|A) * \Pr(A) + \Pr(D|B) * \Pr(B)} = \frac{4/5 * 1/3}{4/5 * 1/3 + 1/5 * 2/3} = 2/3$$

Exercise 4.44 The probability that your test was easy given that you passed is

$$\Pr(E|P) = \frac{\Pr(P|E) * \Pr(E)}{\Pr(D|E) * \Pr(E) + \Pr(P | \neg E) * \Pr(\neg E)} = \frac{.54}{.74} \approx .73$$

Your friend is probably right.

Exercise 4.45 (**a**) 1/4 * 1/6 = 1/24. (**b**) 3/4 * 2/3 = 6/12 = 12/24. (**c**) 1/24 + 12/24 = 13/24. (**d**) (1/24)/(13/24) = 1/13. Good news!

Exercise 4.46 (**a**) The probability assigned to the hypothesis goes from 1/2 to 2/3 after the first trial. (**b**) It goes from 2/3 to 4/5 after the second trial.

Exercise 4.47 Given the way we have defined H and E for the purposes of this exercise, $\Pr(E|H)$ is now zero, since a coin with two heads cannot come up tails. Therefore, the posterior probability will equal 0 no matter the prior:

$$\Pr(H|E) = \frac{0 * \Pr(H)}{0 * \Pr(H) + 0.5 * \Pr(\neg H)} = 0$$

Exercise 4.48 The outcomes are dependent, but not mutually exclusive.

Exercise 4.49 The answer is $(1/25{,}000)^3 = 1/15{,}625{,}000{,}000{,}000$. So if Langford gambles on undoctored machines three times a year, he could expect to win three times straight once every 15,625,000,000,000 years. Notice, though, that the probability that the machines were doctored given that he won does not necessarily equal the probability that he would win given that the machines were not doctored.

Exercise 4.50 The supposed reason is that economists (who tend to be highly numerate) tend not to gamble, which is a big problem in a hospitality industry heavily dependent on gambling revenue.

Exercise 4.51 (**a**) $(1/2)^{20} \approx 0.000{,}001$. (**b**) $(2/3)^{20} \approx 0.0003$. (**c**) $(4/5)^{20} \approx 0.01$. All-male editorial boards of that size are unlikely to result from a purely random process.

Exercise 4.52 (**a**) $0.250^3 \approx 0.016$. (**b**) $(1 - 0.250)^3 \approx 0.42$. (**c**) $3 * 0.250 * (1 - 0.250)^2 \approx 0.42$. (**d**) $1 - (1 - 0.250)^3 \approx 0.58$.

Exercise 4.53 (**a**) Because there are 26 letters in the alphabet, the probability is $(1/26)^8 = 1/208{,}827{,}064{,}576 \approx 0.000{,}000{,}000{,}005$. (**b**) The probability that any one letter will *not* spell out the vulgarity is $1 - 1/208{,}827{,}064{,}576 = 208{,}827{,}064{,}575/208{,}827{,}064{,}576$. So the probability that at least one of the letters will spell out the vulgarity is $1 - (208{,}827{,}064{,}575/208{,}827{,}064{,}576)^{100} \approx 0.000{,}000{,}000{,}5$. That is a little higher, but not much.

Exercise 4.54 Note that there are $2^{10} = 1024$ different ways to answer ten true/false questions. So: (**a**) 1/1024. (**b**) 1/1024. (**c**) 1/1024. (**d**) 10/1024. (**e**) 11/1024.

Exercise 4.55 (**a**) 1/2. (**b**) 1/2. (**c**) $(1/2)^{10} = 1/1024$. (**d**) $1 - (1/1024) = 1023/1024$.

Chapter 5

Exercise 5.1 If you do put all of your eggs in one basket, the events "Egg 1 breaks," "Egg 2 breaks," etc., would not be independent – which is bad if you want to make sure that some eggs remain whole.

Exercise 5.2 (**a**) You want them to be independent. (**b**) They would be dependent.

Exercise 5.3 The sellers want you to think you are more likely to win if you buy a ticket in this particular location. Thus, they hope (and perhaps expect) that you will take the outcomes "A previous ticket sold here was a winner" and "A future ticket sold here will be a winner" not to be independent, when in fact they are.

Exercise 5.4 People assume that triples of bad things – such as the three deaths – are dependent, when in fact they are not.

Exercise 5.5 (**a**) $(1/2)^8 = 1/256$. (**b**) 1/2.

Exercise 5.6 The outcome 4-3-6-2-1 will seem more representative, and therefore more likely.

Exercise 5.7 (**a**) 1/10,000. (**b**) 1/100.

Exercise 5.8 1/25,000.

Exercise 5.13 The answers to all these questions are given by the expression $1 - (7/10)^t$, where t is the number of hours. So: (**a**) 0.51. (**b**) Approximately 0.66. (**c**) Approximately 0.97. Notice that under these circumstances, it is highly likely that you will come across at least one tornado during a 10-hour hike.

Exercise 5.14 The probability of a flood in any given year is 1/10. So: (**a**) 0.81. (**b**) 0.18. (**c**) 0.19. (**d**) Approximately 0.65.

Exercise 5.15 The probability of having no attack on any given day is 1 – 0.000, 1 = 0.999,9. There are 365.25 * 10 = 3652.5 days in ten years. So the probability of at least one attack in ten years is $1 - (0.999{,}9)^{3652.5} \approx 0.306 = 30.6$ percent.

Exercise 5.16 (**a**) Approximately 0.08. (**b**) Approximately 0.15. (**c**) Approximately 0.55. (**d**) Approximately 0.98.

Exercise 5.17 Imagine that you line up the 30 students in a row. The first student can be born on any day of the year, and the probability of this happening is 365/365; the probability that the second student does not share a birthday with the first is 364/365; the probability that the third student does not share a birthday with either of the first two is 363/365; and so on, until you get to the 30th student: the probability that this student will not share a birthday with any of the other 29 students is 336/365. So the probability you are looking for is 365/365 * 364/365 * … 336/365 ≈ 29.4 percent. Thus, in a class this size, the probability that at least two students share a birthday is quite high: about 70.6 percent.

Exercise 5.18 (**a**) Approximately 0.634. (**b**) Approximately 0.999,96.

Exercise 5.19 (**a**) The probability of a catastrophic engine failure is p. (**b**) The probability of a catastrophic engine failure is $1 - (1 - p)^2 = 2p - p^2$. To see why this is so, refer to Figure A.11. (**c**) The single-engine plane. Notice that when p is small, p^2 will be so small as to be negligible. If so, the twin-engine plan is virtually twice as likely to experience a catastrophic engine failure as the single-engine plane! (**d**) Now, the probability of a catastrophic engine failure is p^2.

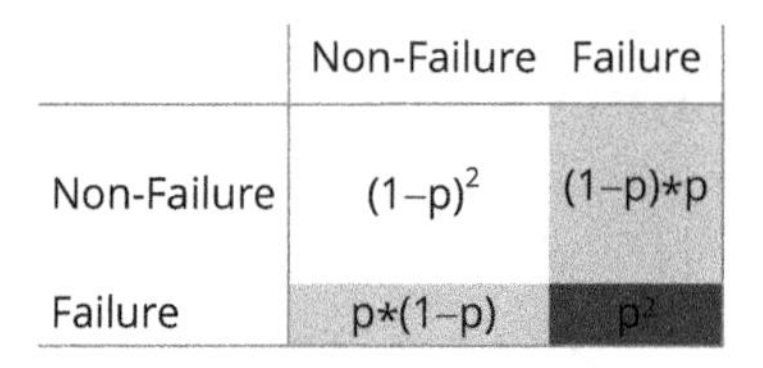

	Non-Failure	Failure
Non-Failure	$(1-p)^2$	$(1-p)*p$
Failure	$p*(1-p)$	p^2

Figure A.11 The private jets

Exercise 5.21 Because the base rate in men is extremely low, the test would not be diagnostic.

Exercise 5.22 Let B mean "The cab is blue," and let P mean "The witness says that the cab is blue." Here is the equation that produces the right answer:

$$\Pr(B|P) = \frac{8/10 * 15/100}{8/10 * 15/100 + 2/10 * 85/100} \approx 41\%$$

Notice that, in spite of the fact that the witness is relatively reliable, the cab that was involved in the accident is more likely to be green than blue.

Exercise 5.23 The answer is given by this equation:

$$\frac{75/100 * 20/100}{75/100 * 20/10 + 25/100 * 80/100} \approx 43\%$$

Exercise 5.24 The answer is given by this equation:

$$\frac{1/1000 * 90/100}{1/1000 * 90/100 + 10/1000 * 10/100} \approx 47\%$$

Exercise 5.25 The probability is:

$$\frac{\frac{10}{10{,}000{,}000} * \frac{999}{1000}}{\frac{10}{10{,}000{,}000} * \frac{999}{1000} + \frac{9{,}999{,}990}{10{,}000{,}000} * \frac{1}{1000}} \approx 0.001 = 0.1\%$$

Exercise 5.26 (**a**) 98/1,000,000,000. (**b**) 19,999,998/1,000,000,000. (**c**) 98/20,000,096 ≈ 0.000005 = 0.0005 percent. (**d**) No.

Exercise 5.27 In Kabul the base rate is likely to be higher, and this might make the test diagnostic.

Exercise 5.28 The correct answer is C.

Exercise 5.30 If, for whatever reason, you manage to acquire a reputation for being smart, honest, diligent, and cool, you can ride that wave for a long time: confirmation bias means that people will continue to think of you in that way even if you do not always act the part. If, by contrast, people start thinking of you as stupid, dishonest, lazy, or uncool, the reputation will be very difficult to get rid of: confirmation bias means that whatever you do is liable to be interpreted as supporting that view.

Exercise 5.32 Books with titles of that kind will largely be read by people who already believe that the liberal mob/the Christian right are destroying America. Then, confirmation bias will set in and further support the readers' preexisting convictions.

Exercise 5.34

$$\text{(a)} \quad \Pr(T|H) = \frac{\Pr(H|T) * \Pr(T)}{\Pr(H|T) * \Pr(T) + \Pr(H|\neg T) * \Pr(\neg T)}$$

$$= \frac{99/100 * 1/10{,}000}{99/100 * 1/10{,}000 + 1/10 * 9999/10{,}000} \approx 0.001$$

$$\text{(b)} \quad \Pr(F|G) = \frac{\Pr(G|F) * \Pr(F)}{\Pr(G|F) * \Pr(F) + \Pr(G|\neg F) * \Pr(\neg F)}$$

$$= \frac{95/100 * 999/1000}{95/100 * 999/1000 + 5/100 * 1/1000} \approx 0.99996$$

Exercise 5.35 Given that the optometrist mainly sees contact users without problems, an image of a healthy user is most available to her. Given that the ophthalmologist mainly sees users with problems, an image of an unhealthy user is most available to him. Insofar as the two are prone to the availability bias, the optometrist is likely to underestimate, and the ophthalmologist to overestimate, the probability of developing serious problems as a result of wearing contacts.

Exercise 5.37 If people are more likely to remember cases when meteorologists' predictions were off, which is likely, the availability heuristic will cause people to think meteorologists are more poorly calibrated than they really are.

Exercise 5.39 Hindsight bias.

Exercise 5.40 The Dunning–Kruger effect.

Exercise 5.41 The heuristics-and-biases program says heuristics are largely functional and only sometimes lead to bias, which is a far cry from saying that people are irredeemably stupid or hopelessly lost.

Exercise 5.42 $Pr(HHH) = (2/3)^3 = 8/27$ whereas $Pr(HHT) = (2/3)^2(1/3) = 4/27$. It is best to bet on HHH.

Exercise 5.43 Since we are talking about overestimating the probability of a conjunction – the fact that the first member is male AND the second member is male AND so on – the person may have committed the conjunction fallacy.

Exercise 5.44 (**a**) 0.04. (**b**) 0.64. (**c**) 0.36. (**d**) Approximately 0.67.

Exercise 5.45 (**a**) 0.072. (**b**) 0.092. (**c**) 0.164. (**d**) Approximately 0.439. (**e**) The base-rate fallacy.

Exercise 5.46 This problem is in effect the same as Exercise 5.22, so the answer is the same: approximately 41 percent.

Exercise 5.47 (**a**) Let T mean that a person is a terrorist and M mean that a person is Muslim. Based on the figures provided, I assume that $Pr(T) = 10/300{,}000{,}000$; that $Pr(M \mid T) = 9/10$; and that $Pr(M \mid \neg T) = 2/300$. If so:

$$Pr(T|M)\frac{\frac{10}{300{,}000{,}000}*\frac{9}{10}}{\frac{10}{300{,}000{,}000}*\frac{9}{10}+\frac{299{,}999{,}990}{300{,}000{,}000}*\frac{2}{300}} \approx 0.000{,}005 = 0.0005\%$$

(**b**) Obviously, there are many more dangerous things for Juan Williams to worry about. But if the image of a Muslim terrorist is particularly available to him, he would be prone to exaggerating the probability that a random Muslim would fall in that category.

Exercise 5.48 … confirmation bias.

Exercise 5.49 Representativeness.

Exercise 5.50 The affect heuristic.

Exercise 5.51 (**a**) Confirmation bias would make Schumpeter more likely to remember and give weight to cases that support his hypothesis than cases that do not. (**b**) Assuming episodes where Schumpeter performed well in the three domains are particularly salient to him, availability bias would cause him to exaggerate the likelihood of a good performance. (**c**) An overconfident Schumpeter would be wrong more often than he thinks when it comes to his abilities in various domains. (**d**) The conjunction fallacy would cause him to overestimate the probability of the conjunction "I am the greatest economist in the world" AND "I am the best horseman in all of Austria" AND "I am the greatest lover in all of Vienna."

Exercise 5.52 (**a**) Confirmation bias. (**b**) Disjunction fallacy. (**c**) Availability bias. (**d**) Base-rate neglect. (**e**) Availability bias. (**f**) Conjunction fallacy. (**g**) Hindsight bias. (**h**) Availability bias.

Chapter 6

Exercise 6.1 (**a**) A maximin reasoner would purchase the warranty. (**b**) A maximax reasoner would not.

Exercise 6.2 (**a**) C. (**b**) A. (**c**) B. The risk-payoff matrix is Table A.2.

Table A.2 Risk-payoff matrix

	S_1	S_2
A	2	0
B	1	1
C	0	4

Exercise 6.5 See Figure A.12.

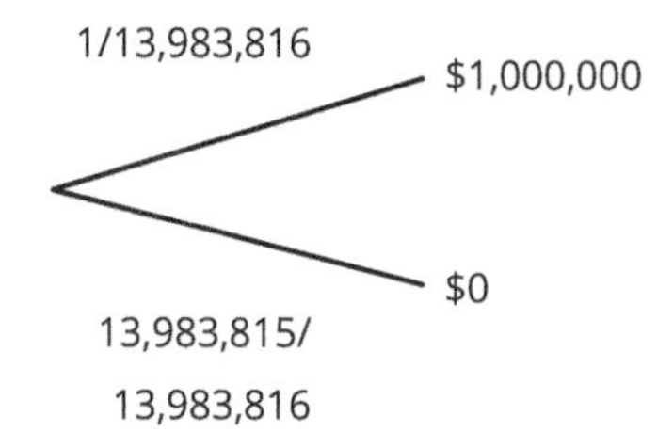

Figure A.12 Lotto 6/49 Tree

Exercise 6.8 (**a**) $EV(\text{A}) = 1/2 * 10 + 1/2 * 0 = 5$. (**b**) $EV(\text{R}) = 4$.

Exercise 6.10 See Table A.3.

Exercise 6.11 (**a**) The expected value is 1/5 ∗ (–\$30) = –\$6. (**b**) Yes.

Exercise 6.13 \$3.50.

Table A.3 Roulette payoffs

Bet	Description	Payout	Pr(win)	Expected value
Straight Up	One number	\$36	1/38	\$36/38
Split	Two numbers	\$18	2/38	\$36/38
Street	Three numbers	\$12	3/38	\$36/38
Corner	Four numbers	\$9	4/38	\$36/38
First Five	0, 00, 1, 2, 3	\$7	5/38	\$35/38
Sixline	Six numbers	\$6	6/38	\$36/38
First 12	1–12	\$3	12/38	\$36/38
Second 12	13–24	\$3	12/38	\$36/38
Third 12	25–36	\$3	12/38	\$36/38
Red		\$2	18/38	\$36/38
Black		\$2	18/38	\$36/38
Even		\$2	18/38	\$36/38
Odd		\$2	18/38	\$36/38
Low	1–18	\$2	18/38	\$36/38
High	19–36	\$2	18/38	\$36/38

Exercise 6.14 (**a**) \$400,020. (**b**) Open the boxes. (**c**) \$150,030. (**d**) Take the sure amount.

Exercise 6.16 (**a**) $-5/-100 = 1/20$. (**b**) $-5/-10 = 1/2$.

Exercise 6.18 The answer is 1/1,000,000.

Exercise 6.19 $p = -79/-325 \approx 0.24$

Exercise 6.23 (**a**) $EU(\text{R}) = u(4) = 4^2 = 16$. (**b**) $EU(\text{A}) = 1/2 * u(10) + 1/2 * u(0) = 1/2 * 10^2 + 1/2 * 0^2 = 50$. (**c**) You should accept the gamble.

Exercise 6.24 (**a**) About 0.000,07. (**b**) 1. (**c**) The dollar.

Exercise 6.25 (**a**) $EU(\text{A}) = 1/3 * \sqrt{9} = 1$. $EU(\text{B}) = 1/4 * \sqrt{16} = 1$. $EU(\text{C}) = 1/5 * \sqrt{25} = 1$. Choose either one. (**b**) $EU(\text{A}) = 1/3 * 9^2 = 27$. $EU(\text{B}) = 1/4 * 16^2 = 64$. $EU(\text{C}) = 1/5 * 25^2 = 125$. Choose C.

Exercise 6.26 (**a**) (i) $EV(\text{G}) = 1/4 * 25 + 3/4 * 1 = 7$. (ii) $EU(\text{G}) = 1/4 * \sqrt{25} + 3/4 * \sqrt{1} = 2$. (**b**) (i) $EV(\text{G}^*) = 2/3 * 7 + 1/3 * 4 = 6$. (ii) $EU(\text{G}^*) = 2/3 * \sqrt{7} + 1/3 * \sqrt{4} \approx 2.43$.

Exercise 6.27 (**a**) See Figure A.13(a). (**b**) See Figure A.13(b). (**c**) $EU(\neg S) = 0$. (**d**) $EU(S) = 0.85 * 10 + 0.10 * (-2) + 0.05 * (-10) = 7.8$. (**e**) Have the operation.

Exercise 6.28 (**a**) See Figure A.14. (**b**) The expected utility of going home is $3/4 * 12 + 1/4 * (-2) = 8.5$. The expected utility of staying put is $2/3 * 9 + 1/3 * 3 = 7$. Thus, you should go home, in spite of the possibility that your aunt might show up.

Exercise 6.29 There are many ways to complete this exercise, but the important result is that B is the rational choice no matter what.

Exercise 6.31 (**a**) $p = 1/2$. (**b**) $p = 3/4$. (**c**) $p = 2/3$.

Exercise 6.33 In this case, $EU(A) = 1/2 * 3^2 + 1/2 * 1^2 = 5$, whereas $EU(R) = 2^2 = 4$. So you should definitely accept the gamble.

Exercise 6.34 (**a**) Risk prone. (**b**) Risk averse. (**c**) Risk prone. (**d**) Risk averse. (**e**) Risk prone. (**f**) Risk neutral. (**g**) Risk prone.

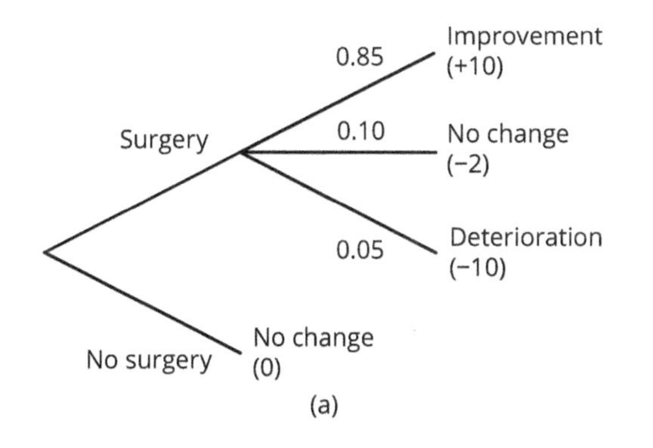

(a)

	Improvement	No change	Deterioration
S	10	−2	−10
¬S	0	0	0

(b)

Figure A.13 Hearing loss

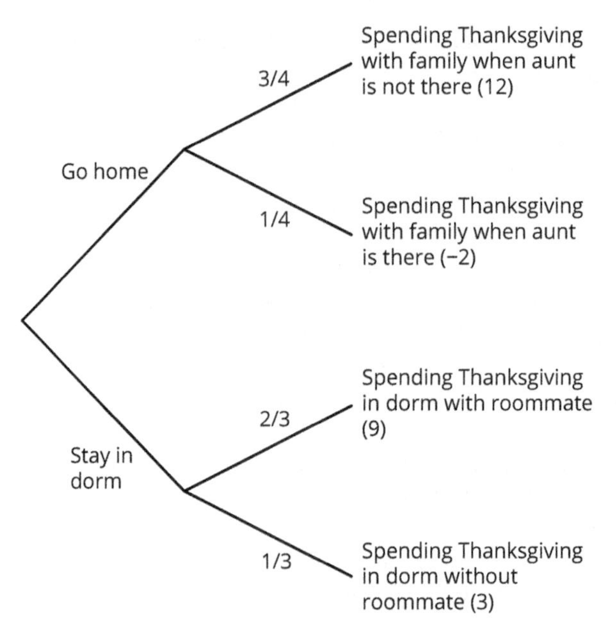

Figure A.14 Thanksgiving indecision

Exercise 6.36 See Figure A.15.

Exercise 6.38 $\sqrt{5}$.

Exercise 6.39 (**a**) The utility of \$4 is 2. The expected utility of G is 3/2. The certainty equivalent is 9/4. Choose \$4. (**b**) The utility of \$4 is 16. The expected utility of G is 21. The certainty equivalent is $\sqrt{21}$. Choose G.

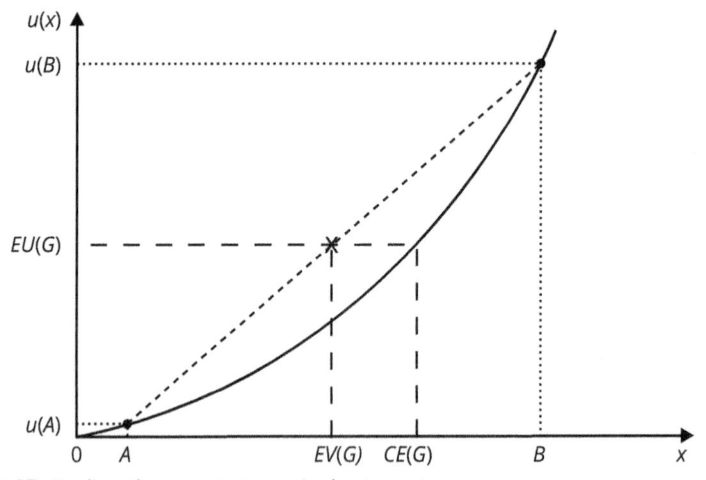

Figure A.15 Finding the certainty equivalent, cont.

Exercise 6.40 (**a**) The expected value of the gamble is 7/8. (**b**) The expected utility is 5/4. (**c**) The certainty equivalent is 25/16. (**d**) $p = 1/2$.

Exercise 6.41 (**a**) The expected utility is 5/2. (**b**) The certainty equivalent is 25/4. (**c**) The probability is 1/8. (**d**) You are risk averse.

Exercise 6.42 Approximately 0.000,000,49 cents.

Exercise 6.44 The maximin criterion.

Exercise 6.45 (**a**) The boxes. (**b**) The fixed amount. (**c**) $118,976.

Exercise 6.46 (**a**) $EU(\text{B}) = 3$. (**b**) $EU(\text{R}) = 4$. (**c**) Press the red button.

Exercise 6.47 Nowhere does the theory say that people perform these calculations in their heads – for the obvious reason that doing so would be impossible or prohibitively slow.

Chapter 7

Exercise 7.4 The value of ± 0 is still $v(\pm 0) = 0$, but the value of -10 is $v(-10) = -2\sqrt{|-10|} = -2\sqrt{10} \approx -6.32$. Thus, the absolute difference is $|v(-10) - v(\pm 0)| \approx |-6.32 - 0| = 6.32$. The value of -1000 is $v(-1000) = -2\sqrt{|-1000|} = -2\sqrt{1000} \approx -63.25$; the value of -1010 is $v(-1010) = -2\sqrt{|-1010|} = -2\sqrt{1010} \approx -63.56$. Hence, the absolute difference is $|v(-1010) - v(-1000)| \approx |-63.56 - (-63.25)| = 0.31$. The absolute difference between $v(\pm 0)$ and $v(-10)$ is much greater than the absolute difference between $v(-1000)$ and $v(-1010)$.

Exercise 7.5 The value of -10 is $v(-10) = -2\sqrt{10} \approx -6.32$. The value of -15 is $v(-15) = -2\sqrt{15} \approx 7.75$. So the absolute difference is 1.43. The value of -120 is $v(-120) = -2\sqrt{120} \approx 21.91$. The value of -125 is $v(-125) = -2\sqrt{125} \approx 22.36$. The absolute difference is 0.45. The difference between $v(-10)$ and $v(-15)$ is much greater than the difference between $v(-120)$ and $v(-125)$. Thus, an *S*-shaped value function can in fact account for the observed behavior.

Exercise 7.6 This problem can be analyzed using Figure 7.3 on page 153. Jen is risk averse because she takes "no animals saved" to be her reference point. Joe is risk prone because he takes "no animals lost" to be his reference point.

Exercise 7.8 (**a**) Omitted. (**b**) $v(\text{A}) = 1/2 * \sqrt{1000/2} \approx 11.18$. (**c**) $v(\text{B}) = \sqrt{500/2} \approx 15.81$. (**d**) $v(\text{C}) = 1/2 * (-2)\sqrt{1000} \approx -31.62$. (**e**) $v(\text{D}) = 2\sqrt{500} \approx -44.72$.

Exercise 7.9 (**a**) The value of the gain is $v(+4) = \sqrt{+4/2} = \sqrt{2} \approx 1.41$, and the value of the loss is $v(-4) = -2\sqrt{|-4|} = -2 * 2 = -4$. In absolute terms, the loss is greater than the gain. (**b**) Relative to a \$0 reference point, \$0 is coded as ± 0, \$2 as +2, and \$4 as +4. The value of the sure thing is $v(+2) = \sqrt{+2/2} = 1$. The value of the gamble is $1/2 * v(\pm 0) + 1/2 * v(+4) = 1/2 * \sqrt{\pm 0/2} + 1/2 * \sqrt{+4/2} \approx 0.71$. The person prefers the sure amount. (**c**) Relative to a \$4 reference point, \$0 is coded as –4, \$2 as –2, and \$4 as ± 0. The value of the sure thing is $v(-2) = -2\sqrt{|-2|} = -2\sqrt{2} \approx -2.83$. The value of the gamble is $1/2 * v(-4) + 1/2 * v(\pm 0) = 1/2 * (-2)\sqrt{|-4|} + 1/2 * (-2)\sqrt{|\pm 0|} = -2$. The person prefers the gamble.

Exercise 7.10 (**a**) Relative to a \$1 reference point, \$1 is coded as ± 0, \$2 as +1, and \$5 as +4. The value of the sure thing is $v(+1) = \sqrt{+1/2} \approx 0.71$ and the value of the gamble $1/2 * v(\pm 0) + 1/2 * v(+4) = 1/2 * \sqrt{\pm 0/2} + 1/2 * \sqrt{+4/2} = 1/2 * \sqrt{2} \approx 0.71$. The person is indifferent. (**b**) Relative to a \$5 reference point, \$1 is coded as –4, \$2 as –3, and \$5 as ± 0. The value of the sure thing is $v(-3) = -2\sqrt{|-3|} = -2\sqrt{3} \approx -3.46$. The value of the gamble is $1/2 * v(-4) + 1/2 * v(\pm 0) = 1/2 * (-2)\sqrt{|-4|} + 1/2 * (-2)\sqrt{|\pm 0|} = 1/2 * (-2)\sqrt{4} = -2$. The person prefers the gamble.

Exercise 7.11 The analysis in this section says that people are more risk prone when they are in the realm of losses. Assuming people who are made to feel poor end up feeling like they are in the realm of losses, the analysis suggests that they should be more likely to choose lottery tickets – which is exactly what the researchers found.

Exercise 7.12 (**a**) $v(+48 + 27) = v(+75) = \sqrt{75/3} = 5$. (**b**) $v(+48) + v(+27) = \sqrt{48/3} + \sqrt{27/3} = 7$. (**c**) It is better to segregate.

Exercise 7.14 (**a**) Chances are you would make fewer purchases. (**b**) The arrangement would encourage you to segregate your losses but integrate your gains, which would simultaneously increase the pain of paying for your stuff and reduce the enjoyment you would derive from it.

Exercise 7.15 The procedure encourages travelers to segregate the costs, which is likely to make them feel worse about the expenditure than they would if they were integrated.

Exercise 7.16 (**a**) $v(-144 - 25) = v(-169) = -3\sqrt{169} = -39$. (**b**) $v(-144) + v(-25) = -3\sqrt{144} + (-3)\sqrt{25} = -51$. (**c**) It is better to integrate.

Exercise 7.17 The advertisement is supposed to make you buy more books by encouraging you to integrate the costs. Because the books are different, you are less likely to integrate them.

Exercise 7.18 (**a**) You should encourage voters to integrate: "You're still taking home \$900k!" (**b**) You should encourage voters to segregate: "You made \$1M! That money is yours! The government is taking \$100k of your money!"

Exercise 7.19 (**a**) $v(-9 + 2) = v(-7) \approx -5.29$. (**b**) $v(-9) + v(+2) = -5$. (**c**) It is better to segregate.

Exercise 7.21 (**a**) You must ignore the column marked B. (**b**) You must ignore the columns marked P and R.

Exercise 7.22 The choice pattern (1a) and (2a) is excluded, as is the choice pattern (1b) and (2b).

Exercise 7.24 A strict preference for A over B entails that $EU(\text{A}) > EU(\text{B})$, which means that $1 * u(30) > 0.80 * u(45)$. Divide each side by four, and you get $0.25 * u(30) > 0.20 * u(45)$. A strict preference for D over C entails that $EU(\text{D}) > EU(\text{C})$, which means that $0.20 * u(45) > 0.25 * u(30)$. But this is inconsistent.

Exercise 7.26 Being ambiguity averse, you would rather bet on the fair coin.

Exercise 7.27 Since the probabilities are *most* ambiguous in game 3, you would be *least* likely to bet on that game.

Exercise 7.28 (**a**) Donner is ambiguity prone. (**b**) Assuming he considers himself competent at real-estate investing, his attitudes are compatible with the competence hypothesis.

Exercise 7.33 The value of a dollar for sure is $\pi(1)v(1) = v(1)$, given that $\pi(1) = 1$. The value of the lottery is $\pi(1/1000)v(1000)$. If $v(\cdot)$ is *S*-shaped, $v(1000) < 1000 * v(1)$. But given that $\pi(x) > x$ for low probabilities, $\pi(1/1000) > 1/1000$. If $\pi(1/1000)$ is *sufficiently* greater than 1/1000, the value of the lottery will exceed the value of the dollar.

Exercise 7.34 See Figure A.16.

Exercise 7.35 (**a**) $\sqrt{1{,}000{,}000} = 1000$. (**b**) $0.90 * \sqrt{(1.04 * 1{,}000{,}000)} + 0.10 * \sqrt{1{,}000{,}000} \approx 1018$. (**c**) $0.40 * \sqrt{(1.21 * 1{,}000{,}000)} + 0.40 * \sqrt{1{,}000{,}000} + 0.20 * \sqrt{(0.90 * 1{,}000{,}000)} \approx 1030$. (**d**) Invest in stocks. (**e**) Invest in bonds.

Exercise 7.36 The value of the status quo is $v(0) = 0$. The value of the gamble is $v(\text{G}) = 1/2 * \sqrt{10/2} + 1/2 * (-2)\sqrt{|-10|} = 1/2\sqrt{5} - \sqrt{10} \approx -2.04$. You would reject the gamble in favor of the status quo.

Exercise 7.37 (**a**) By segregating expenditures, you will feel the loss of money more intensely, which can lead to reduced spending. (**b**) The term is *mental accounting*. (**c**) One problem is that you might overspend in one category and underspend in another, violating fungibility.

Exercise 7.38 (**a**) Silver lining. (**b**) Mental accounting. (**c**) Competence hypothesis. (**d**) Certainty effect or possibly ambiguity aversion. (**e**) Ambiguity aversion.

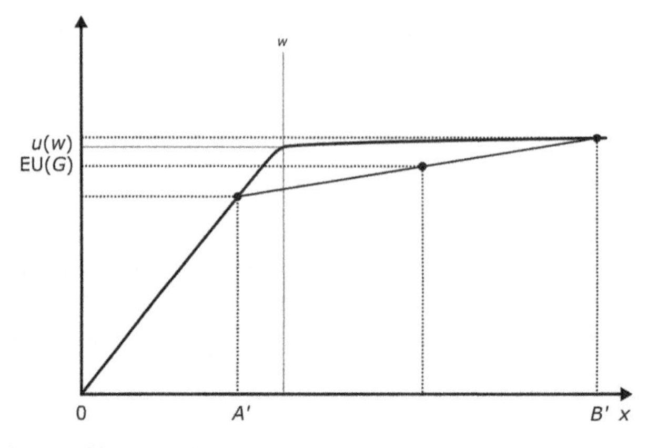

Figure A.16 Rabin's calibration theorem, cont.

Chapter 8

Exercise 8.3 See Table A.4.

Table A.4 Cost of credit

Credit-card offer	$1000	$100	$10,000
Silver Axxess Visa Card	$247.20	$67.92	$2,040.00
Finance Gold MasterCard	$387.50	$263.75	$1,625.00
Continental Platinum MasterCard	$248.20	$68.92	$2,041.00
Gold Image Visa Card	$213.50	$53.75	$1,811.00
Archer Gold American Express	$296.50	$118.75	$2,074.00
Total Tribute American Express	$332.50	$168.25	$1,975.00
Splendid Credit Eurocard	$294.50	$94.25	$2,297.00

Exercise 8.5 $r = 0.20 = 20$ percent.

Exercise 8.8 (**a**) $105. (**b**) $162.89. (**c**) $1146.74.

Exercise 8.9 (**a**) $8664.62. (**b**) 8603.62. (**c**) 14,104 percent. (**d**) Just don't do it.

Exercise 8.12 (**a**) 1.00, 0.30, 0.04, and 1.34. (**b**) Choose **d.** (**c**) Choose **a.**

Exercise 8.13 The grasshopper's delta is lower than the ant's.

Exercise 8.14 (**a**) Low. (**b**) High. (**c**) High. (**d**) Low. (**e**) High.

Exercise 8.15 (**a**) The impartial spectator's delta is one. (**b**) Ours is much lower.

Exercise 8.17 To figure out when you would be indifferent, set up this equation $-1 = \delta * (-9)$ and solve for $\delta = 1/9$. You would (weakly) prefer one stitch just in case $\delta \geq 1/9$.

Exercise 8.18 (**a**) The curve would be steeper. (**b**) The curve would be flatter.

Exercise 8.19 (**a**) $\delta = 1/3$. (**b**) $\delta = 3/4$. (**c**) $\delta = 1/2$. (**d**) $\delta = 3/4$.

Exercise 8.20 (**a**) 2/3. (**b**) 1. (**c**) 1/2.

Exercise 8.21 (**a**) $\delta = 1$. (**b**) $\delta = 0$. (**c**) $\delta = 0.5$.

Exercise 8.23 The table would look like Table A.5, and $\delta = 80/609$.

Table A.5 Time discounting

	$t = 0$	$t = 1$
a	81	16
b	1	625

Exercise 8.24 Young people discount the future too much – that is, their delta is too low – to put much weight on what happens 30–40 years hence.

Exercise 8.25 (**a**) People with higher discount factors tend to have higher credit scores. (**b**) People with higher discount factors are more likely to save for the future and better at managing debt, which means that they are more creditworthy.

Exercise 8.26 The researchers predicted that Calvinists would discount the future the least and Catholics the most, and that atheists would be somewhere in the middle.

Exercise 8.27 We should expect them not to discount the future very much.

Exercise 8.28 (**a**) $\delta = 1/(1 + i)$, meaning that $r = i$. (**b**) $\delta = 1/\sqrt{(1+i)}$.

Chapter 9

Exercise 9.2 (**a**) $1 + 1/3 * 1 * 3 + 1/3 * 1^2 * 9 = 5$. (**b**) $1 + 1 * 2/3 * 3 + 1 * (2/3)^2 * 9 = 7$. (**c**) $1 + 1/3 * 2/3 * 3 + 1/3 * (2/3)^2 * 9 = 3$.

Exercise 9.4 (**a**) $U^{\text{Thu}}(\mathbf{a}) = 8$ and $U^{\text{Thu}}(\mathbf{b}) = 10$; on Thursday, you would choose **b**. $U^{\text{Wed}}(\mathbf{a}) = 6.67$ and $U^{\text{Wed}}(\mathbf{b}) = 8.33$; on Wednesday, you would choose **b**.
(**b**) $U^{\text{Thu}}(\mathbf{a}) = 8$ and $U^{\text{Thu}}(\mathbf{b}) = 2$; on Thursday, you would choose **a**. $U^{\text{Wed}}(\mathbf{a}) = 1.33$ and $U^{\text{Wed}}(\mathbf{b}) = 0.33$; on Wednesday, you would choose **a**.

(**c**) $U^{\text{Thu}}(\mathbf{a}) = 8$ and $U^{\text{Thu}}(\mathbf{b}) = 6$; on Thursday, you would choose **a.** $U^{\text{Wed}}(\mathbf{a}) = 4$ and $U^{\text{Wed}}(\mathbf{b}) = 6$; on Wednesday, you would choose **b.**
(**d**) $U^{\text{Thu}}(\mathbf{a}) = 8$ and $U^{\text{Thu}}(\mathbf{b}) = 4$; on Thursday, you would choose **a.** $U^{\text{Wed}}(\mathbf{a}) = 2.67$ and $U^{\text{Wed}}(\mathbf{b}) = 2.67$; on Wednesday, you would be indifferent between **a** and **b.**

Exercise 9.5 (**a**) 8 and 4. (**b**) 12 and 6. (**c**) 3 and 1. (**d**) 3 and 6. (**e**) Benny. (**f**) Benny.

Exercise 9.6 (**a**) 2/3. (**b**) 3/4.

Exercise 9.7 (**a**) $\beta = 3/4$ and $\delta = 2/3$. (**b**) $x = 4.5$.

Exercise 9.8 $\beta = 4/5$ and $\delta = 1/2$.

Exercise 9.9 Wicksteed is indifferent at $\beta = 1/8$. If he does *not* reach for the blanket, β can be no greater than that, so $\beta \leq 1/8$.

Exercise 9.11 (**a**) If you are an exponential discounter, from the point of view of $t = 0$, you choose between $U^0(\mathbf{a}) = 3$, $U^0(\mathbf{b}) = 5$, $U^0(\mathbf{c}) = 8$, and $U^0(\mathbf{d}) = 13$. Obviously you prefer **d**, and because you are time consistent, that is the movie you will watch.
(**b**) If you are a naive hyperbolic discounter, from the point of view of $t = 0$, you choose between $U^0(\mathbf{a}) = 3$, $U^0(\mathbf{b}) = 1/2 * 5 = 2.5$, $U^0(\mathbf{c}) = 1/2 * 8 = 4$, and $U^0(\mathbf{d}) = 1/2 * 13 = 6.5$. Thus, you will skip the mediocre movie and plan to see the fantastic one. From the point of view of $t = 1$, you choose between $U^1(\mathbf{b}) = 5$, $U^1(\mathbf{c}) = 1/2 * 8 = 4$, and $U^1(\mathbf{d}) = 1/2 * 13 = 6.5$. You will skip the good movie, still planning to see the fantastic one. From the point of view of $t = 2$, you choose between $U^2(\mathbf{c}) = 8$, and $U^2(\mathbf{d}) = 1/2 * 13 = 6.5$. You watch the great movie, forgoing the opportunity to see the fantastic one.
(**c**) If you are a sophisticated hyperbolic discounter, you know that you would be unable to skip the great movie at $t = 2$ and that you consequently will not get to watch the fantastic movie. You also know that from the point of view of $t = 1$, your only realistic options would be $U^1(\mathbf{b}) = 5$ and $U^1(\mathbf{c}) = 1/2 * 8 = 4$. Consequently, you would watch the good movie. From the point of view of $t = 0$, then, your only realistic options are $U^0(\mathbf{a}) = 3$ and $U^0(\mathbf{b}) = 1/2 * 5 = 2.5$, meaning that you will watch the mediocre movie.

Exercise 9.12 (**a**) 8. (**b**) 13. (**c**) 4. (**d**) 12.33.

Exercise 9.15 Less pleasant.

Exercise 9.16 Given that the episode represented by a solid line has higher peak utility than the episode represented by a dashed line, and that the two have the same end utility, the person would favor the former over the latter.

Exercise 9.17 Assuming the college years contain great peak experiences and end on a high note, e.g., with a wonderful graduation ceremony, the peak–end rule will make people remember those years with great fondness. This is true even if there are long periods of tedium or worse. Not the courses in behavioral economics, of course, which are sheer joy throughout :)

Exercise 9.19 Projection bias.

Exercise 9.20 (**a**) People overestimate the degree of variety they will want when eating the food, and consequently diversify too much. (**b**) People tend to be hungry when selecting their foods, and project their current hunger onto their future self, although the future selves will get progressively less hungry as they eat. (**c**) People in a hot, hungry state cannot fully empathize with their future, less hot and hungry state.

Exercise 9.21 Gilbert could question the validity of the numbers. But assuming they are correct, he would say (1) principal, then (2) executive chef, and then (3) loan officer. Unless you are yourself Gilbert, you know more about you than Gilbert does. But he thinks you are so bad at "simulating" future experiences that your knowledge does not translate into any kind of advantage.

Exercise 9.22 (**a**) 16/3 and 4. (**b**) 8 and 6. (**c**) 4 and 2. (**d**) 4 and 6. (**e**) Yves. (**f**) Ximena.

Exercise 9.23 (**a**) Refrain in youth and in middle age, but hit in old age. (**b**) Refrain in youth but hit in middle and old age. (**c**) Hit throughout.

Exercise 9.24 In all likelihood, Orpheus was a naive hyperbolic discounter. An exponential discounter who goes off to rescue Eurydice is time consistent and so will follow through on the plan not to look back. A sophisticated hyperbolic discounter who is unable not to look back would correctly anticipate this and (presumably) cancel the rescue operation.

Exercise 9.25 (**a**) No. This person's problem is not that he changes his mind about how to weight current vs. future consumption. (**b**) No. A person following the peak–end rule would never voluntarily emerge from the "sea of happiness." (**c**) Yes. A person showered with every earthly blessing is arguably not living a very interesting story.

Exercise 9.26 Several answers might be correct, but underprediction of adaptation certainly is one: we think being a prisoner, etc., is much worse than it is because we fail to anticipate the extent to which prisoners adapt to their conditions. In passing, it may be interesting to know that the Stoics were read and cited approvingly by Adam Smith.

Exercise 9.27 (**a**) Hot–cold empathy gaps will make it hard to imagine what it will be like to no longer be young, in love, or sexually aroused. Impact bias will exaggerate the effect of marriage on happiness. Projection bias will conceal the degree to which preferences will change over the long term. (**b**) People could ask their parents, or the parents of the prospective spouse, for advice.

Exercise 9.28 (**a**) Hyperbolic discounting. (**b**) Choosing not to choose. (**c**) Preference over profiles. (**d**) Hyperbolic discounting. (**e**) Hyperbolic discounting. (**f**) Preference over profiles. (**g**) Choosing not to choose. (**h**) Misprediction/miswanting.

Chapter 10

Exercise 10.5 (**a**) ⟨U, L⟩ and ⟨D, R⟩. (**b**) ⟨U, L⟩. (**c**) ⟨U, R⟩.

Exercise 10.10 Suppose that Player I plays U with probability p and Player II plays L with probability q. (**a**) There is an equilibrium in which $p = q = 1/3$. (**b**) There is an equilibrium in which $p = 1/2$ and $q = 1$.

Exercise 10.11 Suppose that Player I plays U with probability p and Player II plays L with probability q. (**a**) There are two equilibria in pure strategies, ⟨U, L⟩ and ⟨D, R⟩, and a mixed equilibrium in which $p = 4/5$ and $q = 1/5$. (**b**) There are no equilibria in pure strategies but a mixed equilibrium in which $p = q = 1/2$. (**c**) There are two equilibria in pure strategies, ⟨U, L⟩ and ⟨D, R⟩, but no equilibria in mixed strategies.

Exercise 10.13 (**a**) The payoff matrix is given in Table A.6. (**b**) In the unique Nash equilibrium, both players randomize with probability 1/3, 1/3, and 1/3 (cf. Example 11.7 on page 249).

Table A.6 Rock-paper-scissor payoff matrix

	R	P	S
R	0, 0	−1, 1	1, −1
P	1, −1	0, 0	−1, 1
S	−1, 1	1, −1	0, 0

Exercise 10.16 This game has two Nash equilibria in pure strategies, ⟨S, ¬S⟩ and ⟨¬S, S⟩, and a mixed equilibrium in which each player plays S with probability 1/3.

Exercise 10.17 This game has two Nash equilibria in pure strategies, ⟨D, D⟩ and ⟨H, H⟩, and a mixed equilibrium in which each player plays D with probability 1/3.

Exercise 10.20 Yes.

Exercise 10.21 (**a**) This game has three Nash equilibria in pure strategies: ⟨U, L⟩, ⟨M, M⟩, and ⟨D, R⟩. (**b**) ⟨U, L⟩ and ⟨M, M⟩, but not ⟨D, R⟩, are trembling-hand perfect.

Exercise 10.25 The unique subgame-perfect equilibrium is ⟨D, RL⟩. That is, Player II plays R at the left node and L and the right node, and (anticipating this) Player I plays D.

Exercise 10.26 (**a**) In the unique subgame-perfect equilibrium, players always Take. (**b**) No.

Exercise 10.28 (**a**) ⟨U, L⟩ and ⟨D, R⟩. (**b**) ⟨D, L⟩ and ⟨U, R⟩. (**c**) ⟨U, R⟩. (**d**) ⟨U, L⟩, ⟨U, R⟩, and ⟨D, R⟩. (**e**) ⟨U, L⟩ and ⟨D, R⟩.

Exercise 10.29 They all are, except ⟨D, R⟩ in game (e).

Exercise 10.30 (**a**) Only ⟨D, R⟩. (**b**) Both ⟨D, L⟩ and ⟨U, R⟩. (**c**) None. (**d**) All three. (**e**) Only ⟨D, R⟩.

Exercise 10.31 (**a**) $p = q = 3/5$. (**b**) $p = q = 5/9$. (**c**) None. (**d**) None. (**e**) None.

Exercise 10.32 At the third and last stage, Player I will play L. At the second stage, given this, Player II will play R. This means that Player I is indifferent between L and R at the first node. Therefore there are two subgame-perfect equilibria: one that follows path L–R and yields (2,1) and one that follows path R–R and yields (2,3).

Exercise 10.33 You would predict that economics majors would defect more frequently than non-majors, and that the economics majors therefore would do worse when playing against each other than non-majors would. Empirical evidence supports the prediction.

Chapter 11

Exercise 11.3 In a subgame-perfect equilibrium, a utilitarian Player II will accept any offer, since to her any division is better than ($0,$0). Because a utilitarian Player I would actually prefer ($5,$5) to any other outcome, that is the division that he will propose.

Exercise 11.4 See Table A.7 for the actual games played by (**a**) egoists, (**b**) utilitarians, (**c**) enviers, and (**d**) Rawlsians. The answers are: (**a**) ⟨D, D⟩, (**b**) ⟨C, C⟩ and ⟨D, D⟩, (**c**) ⟨D, D⟩, (**d**) ⟨C, C⟩ and ⟨D, D⟩.

Table A.7 Social preferences

	C	D
C	4,4	0,5
D	5,0	3,3

(a) Egoists

	C	D
C	8,8	5,5
D	5,5	6,6

(b) Utilitarians

	C	D
C	0,0	–5,5
D	5,–5	0,0

(c) Enviers

	C	D
C	4,4	0,0
D	0,0	3,3

(d) Rawlsians

Exercise 11.5 Whether this kind of behavior is consistent with standard rationality or not depends on the shape of members' preferences. If everything members cared about was acquiring baked goods, they could easily find them more cheaply elsewhere. But assuming people care about other things – kids learning how to swim and not drowning, their own reputation as generous members of the community, or whatnot – the observed behavior is obviously consistent with standard rationality.

Exercise 11.8 See Section 11.2.

Exercise 11.9 Rawlsian preferences.

Exercise 11.10 See Table A.8 for the games (in utility terms) played in each of these three scenarios. (**a**) There are two equilibria in pure strategies: $\langle U, L\rangle$ and $\langle D, R\rangle$. There is an equilibrium in mixed strategies where Player I plays U with probability 1/2 and Player II plays L with probability 1/2. (**b**) There are two equilibria in pure strategies: $\langle U, L\rangle$ and $\langle D, R\rangle$. There is an equilibrium in mixed strategies where Player I plays U with probability 1/3 and Player II plays L with probability 1/3. (**c**) There are two equilibria in pure strategies: $\langle D, L\rangle$ and $\langle D, R\rangle$.

Table A.8 Egoists, utilitarians, and enviers

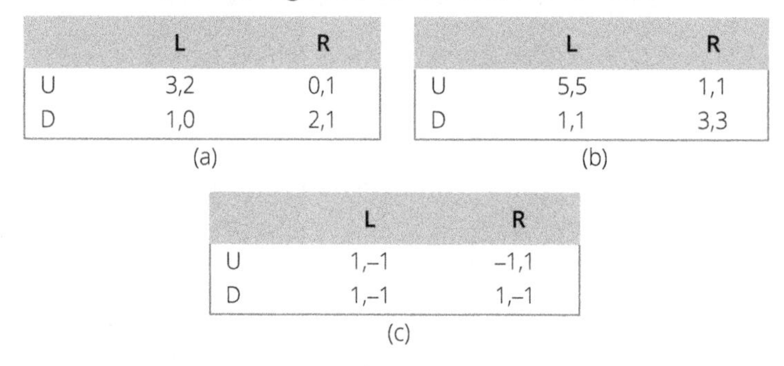

	L	R
U	3,2	0,1
D	1,0	2,1

(a)

	L	R
U	5,5	1,1
D	1,1	3,3

(b)

	L	R
U	1,–1	–1,1
D	1,–1	1,–1

(c)

Exercise 11.11 See Table A.9 for the payoff matrices. (**a**) There are two Nash equilibria in pure strategies, ⟨V, ¬V⟩ and ⟨¬V, V⟩, and a Nash equilibrium in mixed strategies in which both players play V with probability 1/2 and ¬V with probability 1/2. (**b**) There are two Nash equilibria in pure strategies, ⟨V, ¬V⟩ and ⟨¬V, V⟩, and a Nash equilibrium in mixed strategies in which both players play V with probability 3/4 and ¬V with probability 1/4. (**c**) There is only one Nash equilibrium: ⟨V, V⟩. NB: In a mixed equilibrium social preferences make players more likely to volunteer, and when volunteering is beneficial, everyone *always* volunteers.

Table A.9 The volunteer's dilemma

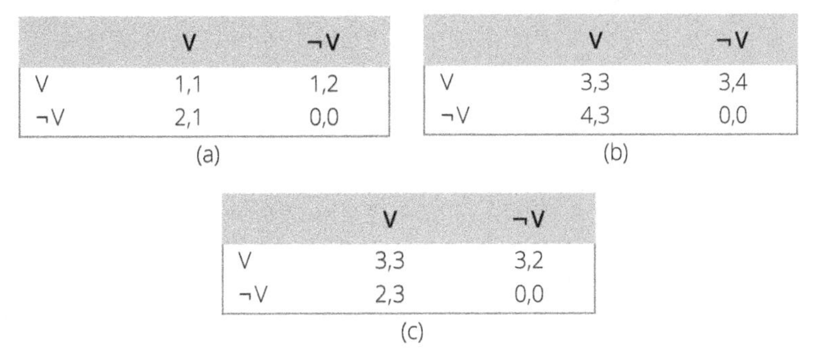

	V	¬V
V	1,1	1,2
¬V	2,1	0,0

(a)

	V	¬V
V	3,3	3,4
¬V	4,3	0,0

(b)

	V	¬V
V	3,3	3,2
¬V	2,3	0,0

(c)

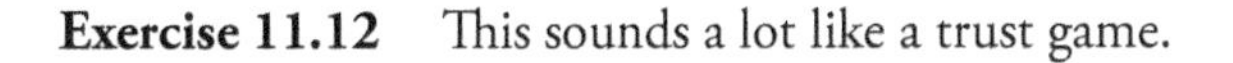

Exercise 11.12 This sounds a lot like a trust game.

Exercise 11.13 This is a paradigmatic beauty contest.

Exercise 11.14 From the point of view of the besieging army, a level-0 strategist would go: "The doors are open! Let's invade!" A level-1 thinker, by contrast, might say: "Why would Kong Ming leave the gates open? They must think we're level-0 strategists and will invade as soon as we are able. The only reason they would open the gates, then, is if this is a trap and they want us to invade. Let's not." A level-2 strategist might say: "They believe we're level-1 strategists, who would not dare invade when the gates are wide open. They have opened the gates to trick us into withdrawing. But we have no reason not to invade, so let us do it." Note that under the circumstances, both level-0 and level-2 strategists would have successfully invaded the city. Level-1 strategists would not, and since Kong Ming accurately predicted the cognitive sophistication of the besieging army, his move was successful. As is often the case, the key is to be exactly one step ahead of your opponent.

Chapter 12

Exercise 12.1 (**a**) ban; (**b**) nudge; (**c**) ban; (**d**) nudge; (**e**) incentive; (**f**) ban.

Exercise 12.2 The intervention simply changes the default option from French fries to apple slices.

Exercise 12.3 The "Bloomberg ban" is not a nudge because it interferes with the freedom of choice of rational and informed customers who want to buy large sodas.

BIBLIOGRAPHY

Adams, Susan (2015), "The happiest and unhappiest jobs in 2015," *Forbes*, February 26, http://www.forbes.com/sites/susanadams/2015/02/26/the-happiest-and-unhappiest-jobs-in-2015/. Accessed April 7, 2015.

Ainslie, George (1975), "Specious reward: A behavioral theory of impulsiveness and impulse control," *Psychological Bulletin, 82* (4), 463–96.

Allais, Maurice (1953), "Le comportement de l'homme rationnel devant le risque: Critique des postulats et axiomes de l'école américaine," *Econometrica, 21* (4), 503–6.

Allingham, Michael (2002), *Choice Theory: A Very Short Introduction*, Oxford: Oxford University Press.

Anand, Easha (2008), "Payday lenders back measures to unwind state restrictions," *Wall Street Journal*, October 28, p. A6.

Angner, Erik (2006), "Economists as experts: Overconfidence in theory and practice," *Journal of Economic Methodology, 13* (1), 1–24.

——— (2007), *Hayek and Natural Law*, London: Routledge.

——— (2015a), "'To navigate safely in the vast sea of empirical facts': Ontology and methodology in behavioral economics," *Synthese, 192* (11), 3557–75.

——— (2015b), "Well-being and economics," in Guy Fletcher, ed., The *Routledge Handbook of the Philosophy of Well-Being*, London: Routledge, pp. 492–503.

——— (2019), "We're all behavioral economists now," *Journal of Economic Methodology, 26* (3), 195–207.

Angner, Erik and George Loewenstein (2012), "Behavioral economics," in Uskali Mäki, ed., *Handbook of the Philosophy of Science: Philosophy of Economics*, Amsterdam: Elsevier, pp. 641–90.

Ariely, Dan (2008), *Predictably Irrational: The Hidden Forces That Shape our Decisions*, New York, NY: Harper.

Ariely, Dan and George Loewenstein (2006), "The heat of the moment: The effect of sexual arousal on sexual decision making," *Journal of Behavioral Decision Making, 19* (2), 87–98.

Ariely, Dan, George Loewenstein, and Drazen Prelec (2003), "'Coherent arbitrariness': Stable demand curves without stable preferences," *The Quarterly Journal of Economics, 118* (1), 73–105.

Ariely, Dan and Klaus Wertenbroch (2002), "Procrastination, deadlines, and performance: Self-control by precommitment," *Psychological Science, 13* (3), 219–24.

Aristotle (1999 [*c* 350 BCE]), *Nicomachean Ethics*, Terence Irwin, trans., Indianapolis, IN: Hackett Publishing Co.

Arkes, Hal R. and Catherine Blumer (1985), "The psychology of sunk cost," *Organizational Behavior and Human Decision Processes, 35* (1), 124–40.

Associated Press (2007), "Ireland: Another metric system fault," *New York Times*, November 1.

Bar-Hillel, Maya (1980), "The base-rate fallacy in probability judgments," *Acta Psychologica, 44* (3), 211–33.

Baron, Jonathan and John C. Hershey (1988), "Outcome bias in decision evaluation," *Journal of Personality and Social Psychology, 54* (4), 569–79.

Becker, Gary S. (1976), *The Economic Approach to Human Behavior*, Chicago, IL: University of Chicago Press.

Beckett, Samuel (1989), *Nohow On*, London: Calder.

Bentham, Jeremy (1996 [1789]), *An Introduction to the Principles of Morals and Legislation*, Oxford: Clarendon Press.

Bicchieri, Cristina (2005), *The Grammar of Society: The Nature and Dynamics of Social Norms*, Cambridge: Cambridge University Press.

Binmore, Ken (1999), "Why experiment in economics?," *The Economic Journal, 109* (453), F16–24.

——— (2007), *Game Theory: A Very Short Introduction*, New York, NY: Oxford University Press.

Blackburn, Simon (2001), *Being Good: A Short Introduction to Ethics*, Oxford: Oxford University Press.

Boethius, (1999 [*c* 524]), *The Consolations of Philosophy*, Rev. ed., Victor Watts, trans., London: Penguin Books.

Brooks, David (2008), "The behavioral revolution," *New York Times*, October 28, p. A31.

Bruine de Bruin, Wändi, Andrew M. Parker, and Baruch Fischhoff (2007), "Individual differences in adult decision-making competence," *Journal of Personality and Social Psychology, 92* (5), 938–56.

Buehler, Roger, Dale Griffin, and Michael Ross (1994), "Exploring the 'planning fallacy': Why people underestimate their task completion times," *Journal of Personality and Social Psychology, 67* (3), 366–81.

Burroughs, William S. (1977 [1953]), *Junky*, Harmondsworth, Middlesex: Penguin Books.

Camerer, Colin F. (2003), *Behavioral Game Theory: Experiments in Strategic Interaction*, New York, NY: Russell Sage Foundation.

Camerer, Colin F., Anna Dreber, Eskil Forsell, Teck-Hua Ho, Jürgen Huber, et al. (2016), "Evaluating replicability of laboratory experiments in economics," *Science, 351* (6280), 1433–36.

Camerer, Colin F., George Loewenstein, and Drazen Prelec (2005), "Neuroeconomics: How neuroscience can inform economics," *Journal of Economic Literature, 43* (1), 9–64.

Camerer, Colin F., Linda Babcock, George Loewenstein, and Richard H. Thaler (1997), "Labor supply of New York City cabdrivers: One day at a time," *The Quarterly Journal of Economics, 112* (2), 407–41.

Caplan, Bryan (2013), "Nudge, policy, and the endowment effect," http://econlog.econlib.org/archives/2013/07/nudge_policy_an.html. Accessed February 9, 2015.

Chetty, Raj (2015), "Behavioral economics and public policy: A pragmatic perspective," *American Economic Review, 105* (5), 1–33.

Clark, Andrew E. (2018), "Four decades of the economics of happiness: Where next?," *Review of Income and Wealth 64* (2), 245–69.

Consumer Federation of America (2006), "Press Release: How Americans view personal wealth vs. how financial planners view this wealth," January 9.

Coupland, Douglas (2008), *JPod*, New York, NY: Bloomsbury USA.

Cowper, William (1785), *The Task: A Poem, in Six Books*, London: J. Johnson.

Davis, John B. (2011), *Individuals and Identity in Economics*, Cambridge: Cambridge University Press.

Dawes, Robyn M. and Richard H. Thaler (1988), "Anomalies: Cooperation," *The Journal of Economic Perspectives, 2* (3), 187–97.

de Sade, Donatien Alphonse François, Marquis (1889 [1791]), *Opus Sadicum: A Philosophical Romance* (Paris: Isidore Liseux). Originally published as *Justine*.

Dixit, Avinash K., Susan Skeath, and David Reiley (2009), *Games of Strategy*, 3rd ed., New York, NY: W. W. Norton & Co.

Dostoyevsky, Fyodor (2009 [1864]), *Notes from the Underground*, Constance Garnett, trans., Indianapolis, IN: Hackett.

Durlauf, Steven N. and Lawrence Blume (2010), *Behavioural and Experimental Economics*, New York, NY: Palgrave Macmillan.

Earman, John and Wesley C. Salmon (1992), "The confirmation of scientific hypotheses," in Merrilee H. Salmon, John Earman, Clark Glymour, James G. Lennox, Peter Machamer, J. E. McGuire, John D. Norton, Wesley C. Salmon, and Kenneth F. Schaffner, eds., *Introduction to the Philosophy of Science*, Englewood Cliffs, NJ: Prentice Hall, pp. 7–41.

Ellingsen, Tore, Magnus Johannesson, Johanna Möllerstrom, and Sara Munkhammar (2012), "Social framing effects: Preferences or beliefs?," *Games and Economic Behavior, 76* (1), 117–30.

Ellsberg, Daniel (1961), "Risk, ambiguity, and the Savage axioms," *The Quarterly Journal of Economics, 75* (4), 643–69.

Englich, Birthe, Thomas Mussweiler, and Fritz Strack (2006), "Playing dice with criminal sentences: The influence of irrelevant anchors on experts' judicial decision making," *Personality and Social Psychology Bulletin, 32* (2), 188–200.

Epicurus (2012 [*c* 300 BCE]), *The Art of Happiness*, George K. Strodach, trans., London: Penguin Books.

Farhi, Paul (2010), "Juan Williams at odds with NPR over dismissal," *The Washington Post*, October 22, p. C1.

Finucane, Melissa L., Ali Alhakami, Paul Slovic, and Stephen M. Johnson (2000), "The affect heuristic in judgments of risks and benefits," *Journal of Behavioral Decision Making*, 13 (1), 1–17.

Fischhoff, Baruch (1975), "Hindsight is not equal to foresight: The effect of outcome knowledge on judgment under uncertainty," *Journal of Experimental Psychology: Human Perception and Performance, 1* (3), 288–99.

Fischhoff, Baruch, Paul Slovic, and Sarah Lichtenstein (1977), "Knowing with certainty: The appropriateness of extreme confidence," *Journal of Experimental Psychology: Human Perception and Performance, 3* (4), 552–64.

FOX6 WBRC (2009), "Tension builds around courthouses' reopening," October 8.

Francis, David (2014), "DOD is stuck with a flawed $1.5 trillion fighter jet," *Fiscal Times*, February 18, http://www.thefiscaltimes.com/Articles/2014/02/18/DOD-Stuck-Flawed-15-Trillion-Fighter-Jet. Accessed February 20, 2015.

Frank, Robert H. (2005), "The opportunity cost of economics education," *New York Times*, September 1, p. C2.

Frank, Robert H., Thomas Gilovich, and Dennis T. Regan (1993), "Does studying economics inhibit cooperation?," *The Journal of Economic Perspectives, 7* (2), 159–71.

Frank, Thomas (2007), "Security arsenal adds behavior detection," *USA Today*, September 25, p. B1.

Frederick, Shane, George Loewenstein, and Ted O'Donoghue (2002), "Time discounting and time preference: A critical review," *Journal of Economic Literature, 40* (2), 351–401.

Friedman, Milton and Rose D. Friedman (1984), *Tyranny of the Status Quo*, San Diego, CA: Harcourt Brace Jovanovich.

Gardner, Sarah (2012) "Nevada's boom and bust economy," January 31, http://www.marketplace.org/topics/elections/real-economy/nevada%E2%80%99s-boom-and-bust-economy. Accessed March 23, 2015.

Gigerenzer, Gerd and Daniel G. Goldstein (1996), "Reasoning the fast and frugal way: Models of bounded rationality," *Psychological Review, 103* (4), 650–69.

Gilbert, Daniel (2006), *Stumbling on Happiness*, New York, NY: Alfred A. Knopf.

Goncharov, Ivan Aleksandrovich (1915 [1859]) *Oblomov*, C. J. Hogarth, trans., New York, NY: The Macmillan Co.

Goodreads.com (2013), "What makes you put down a book?," July 9, http://www.goodreads.com/blog/show/424-what-makes-you-put-down-a-book. Accessed February 20, 2015.

Hafner, Katie (2006), "In Web world, rich now envy the superrich," *New York Times*, November 21, p. A5.

Haisley, Emily, Romel Mostafa, and George Loewenstein (2008), "Subjective relative income and lottery ticket purchases," *Journal of Behavioral Decision Making, 21* (3), 283–95.

Hampton, Isaac (2012), *The Black Officer Corps: A History of Black Military Advancement from Integration Through Vietnam*, New York, NY: Routledge.

Harsanyi, John C. (1975), "Can the maximin principle serve as a basis for morality? A critique of John Rawls's theory," *The American Political Science Review*, *69* (2), 594–606.

Hastie, Reid and Robyn M. Dawes (2010), *Rational Choice in an Uncertain World: The Psychology of Judgment and Decision Making*, 2nd ed., Los Angeles, CA: Sage Publications.

Hayek, Friedrich A. (1933), "The trend of economic thinking," *Economica*, (40), 121–37.

Heath, Chip and Amos Tversky (1991), "Preference and belief: Ambiguity and competence in choice under uncertainty," *Journal of Risk and Uncertainty*, *4*(1), 5–28.

Heuer, Richards J. (1999), *Psychology of Intelligence Analysis*, Washington, DC: Central Intelligence Agency Center for the Study of Intelligence.

Heukelom, Floris (2014), *Behavioral Economics: A History*, New York, NY: Cambridge University Press.

Hobbes, Thomas (1994 [1651]), *Leviathan: With Selected Variants from the Latin Edition of 1668*, Indianapolis, IN: Hackett Pub. Co.

Holt, Charles A. (2019), *Markets, Games, and Strategic Behavior: An Introduction to Experimental Economics*, 2nd ed., Princeton, NJ: Princeton University Press.

Huber, Joel, John W. Payne, and Christopher Puto (1982), "Adding asymmetrically dominated alternatives: Violations of regularity and the similarity hypothesis," *The Journal of Consumer Research*, *9* (1), 90–8.

Hume, David (2000 [1739–40]), *A Treatise of Human Nature*. Oxford: Oxford University Press.

International Ergonomics Association (2015), "What is ergonomics," http://www.iea.cc/whats/index.html. Accessed February 9, 2015.

Jevons, W. Stanley (1965 [1871]), *The Theory of Political Economy*, 5th ed., New York, NY: A. M. Kelley.

Jobs, Steve (2005), "Commencement address," Stanford University, June 14, http://news.stanford.edu/news/2005/june15/jobs-061505.html. Accessed March 30, 2015.

Kagel, John H. and Alvin E. Roth, eds. (1995), *The Handbook of Experimental Economics*, Princeton, NJ: Princeton University Press.

Kahneman, Daniel (2011), *Thinking, Fast and Slow*, New York, NY: Farrar, Straus and Giroux.

Kahneman, Daniel and Amos Tversky (1979), "Prospect theory: An analysis of decision under risk," *Econometrica*, *47* (2), 263–91.

Kahneman, Daniel, Jack L. Knetsch, and Richard H. Thaler (1991), "Anomalies: The endowment effect, loss aversion, and status quo bias," *The Journal of Economic Perspectives*, *5* (1), 193–206.

Kahneman, Daniel, Peter P. Wakker, and Rakesh Sarin (1997), "Back to Bentham? Explorations of experienced utility," *The Quarterly Journal of Economics*, *112* (2), 375–405.

Keynes, John Maynard (1936), *The General Theory of Employment, Interest and Money*, New York, NY: Harcourt, Brace.

Kierkegaard, Søren (2000 [1843]), "Either/or, a fragment of life", in Howard V. Hong and Edna H. Hong, eds., *The Essential Kierkegaard*, Princeton, NJ: Princeton University Press, pp. 37–83.

Kruger, Justin and David Dunning (1999), "Unskilled and unaware of it: How difficulties in recognizing one's own incompetence lead to inflated self-assessments," *Journal of Personality and Social Psychology*, *77* (6), 1121–34.

Krugman, Paul (2009), "How did economists get it so wrong?," *New York Times Magazine*, September 6, pp. 36–43.

Kuang, Cliff (2012), "The Google diet," *Fast Company*, *164* (April), p. 48.

Lambert, Craig (2006), "The marketplace of perceptions," *Harvard Magazine*, March–April, 50–57, 93–95

Layard, P. Richard G. (2005), *Happiness: Lessons from a New Science*, New York, NY: Penguin Press.

Levitt, Steven D. and Stephen J. Dubner (2005), *Freakonomics: A Rogue Economist Explores the Hidden Side of Everything*, New York, NY: William Morrow.

Lichtenstein, Sarah and Paul Slovic (1973), "Response-induced reversals of preference in gambling: An extended replication in Las Vegas," *Journal of Experimental Psychology, 101* (1), 16–20.

Loewenstein, George and Daniel Adler (1995), "A bias in the prediction of tastes," *The Economic Journal 105* (431), 929–37.

Loewenstein, George and Erik Angner (2003), "Predicting and indulging changing preferences," in George Loewenstein, Daniel Read, and Roy F. Baumeister, eds., *Time and Decision: Economic and Psychological Perspectives on Intertemporal Choice*, New York, NY: Russell Sage Foundation, pp. 351–91.

Loewenstein, George and Nachum Sicherman (1991), "Do workers prefer increasing wage profiles?," *Journal of Labor Economics, 9* (1), 67–84.

Loewenstein, George and Peter Ubel (2010), "Economists behaving badly," *New York Times*, July 15, p. A31.

Loewenstein, George, Daniel Read, and Roy F. Baumeister, eds. (2003), *Time and Decision: Economic and Psychological Perspectives on Intertemporal Choice*, New York, NY: Russell Sage Foundation.

Lord, Charles G., Lee Ross, and Mark R. Lepper (1979), "Biased assimilation and attitude polarization: The effects of prior theories on subsequently considered evidence," *Journal of Personality and Social Psychology, 37* (11), 2098–109.

Luce, R. Duncan and Howard Raiffa (1957), *Games and Decisions: Introduction and Critical Survey*, New York, NY: Wiley.

Lyubomirsky, Sonja (2013), *The Myths of Happiness*, New York, NY: Penguin Books.

Mas-Colell, Andreu, Michael D. Whinston, and Jerry R. Green (1995), *Microeconomic Theory*, New York, NY: Oxford University Press.

McKinley, Jesse (2009), "Schwarzenegger statement contains not-so-secret message," *New York Times*, October 29, p. A16.

Meier, Stephan and Charles D. Sprenger (2012), "Time discounting predicts creditworthiness," *Psychological Science, 23* (1), 56–8.

Mischel, Walter (2014), *The Marshmallow Test: Mastering Self-Control*, New York, NY: Little, Brown, and Co.

Myers, David G. (1992), *The Pursuit of Happiness: Who Is Happy—and Why*, New York, NY: W. Morrow.

Nagourney, Adam (2011), "California bullet train project advances amid cries of boondoggle," *New York Times*, November 27, p. A18.

Nickerson, Raymond S. (1998), "Confirmation bias: A ubiquitous phenomenon in many guises," *Review of General Psychology, 2* (2), 175–220.

O'Brien, Miles (2004), "Apollo 11 crew recalls giant leap 35 years later," *CNN*, July 21, http://www.cnn.com/2004/TECH/space/07/21/apollo.crew/. Accessed March 17, 2015

O'Donoghue, Ted and Matthew Rabin (2000), "The economics of immediate gratification," *Journal of Behavioral Decision Making, 13* (2), 233–50.

OECD (2017), *Behavioural Insights and Public Policy: Lessons from Around the World*, Paris: OECD Publishing.

Osborne, Martin J. and Ariel Rubinstein (1994), *A Course in Game Theory*, Cambridge, MA: MIT Press.

Oskamp, Stuart (1982), "Overconfidence in case-study judgments," in Daniel Kahneman, Paul Slovic, and A. Tversky, eds., *Judgment Under Uncertainty: Heuristics and Biases*, Cambridge: Cambridge University Press, pp. 287–93.

Paglieri, Fabio, Anna M. Borghi, Lorenza S. Colzato, Bernhard Hommel, and Claudia Scorolli (2013), "Heaven can wait: How religion modulates temporal discounting," *Psychological Research, 77* (6), 738–47.

Paul, L. A. (2014), *Transformative Experience*, Oxford: Oxford University Press.

Perry, John (1996), "How to procrastinate and still get things done," *The Chronicle of Higher Education*, February 23.

Peterson, Martin (2009), *An Introduction to Decision Theory*, New York, NY: Cambridge University Press.

Pigou, Arthur C. (1952 [1920]), *The Economics of Welfare*, 4th ed., London: Macmillan.

Popper, Karl (2002 [1963]), *Conjectures and Refutations: The Growth of Scientific Knowledge*, London: Routledge.

Proust, Marcel (2002 [1925]), *The Fugitive*, Vol. 5 of *In Search of Lost Time*, New York, NY: Allen Lane.

Rabin, Matthew and Richard H. Thaler (2001), "Anomalies: Risk aversion," *Journal of Economic Perspectives, 15* (1), 219–32.

Ramsey, Frank P. (1928), "A mathematical theory of saving," *The Economic Journal, 38* (152), 543–59.

Rawls, John (1971), *A Theory of Justice*, Cambridge, MA: Belknap Press.

Read, Daniel and Barbara van Leeuwen (1998), "Predicting hunger: The effects of appetite and delay on choice," *Organizational Behavior and Human Decision Processes, 76* (2), 189–205.

Redelmeier, Donald A. and Daniel Kahneman (1996), "Patients' memories of painful medical treatments: Real-time and retrospective evaluations of two minimally invasive procedures," *Pain, 66* (1), 3–8.

Ross, Don (2005), *Economic Theory and Cognitive Science: Microexplanation*, Cambridge, MA: MIT Press.

Russell, Bertrand (1959), *Common Sense and Nuclear Warfare*, New York, NY: Simon and Schuster.

Schelling, Thomas C. (1960), *The Strategy of Conflict*, Cambridge, MA: Harvard University Press.

Schwartz, Barry (2004), *The Paradox of Choice: Why More Is Less*, New York, NY: Ecco.

Seneca (2007 [*c* 49]), *Seneca: Dialogues and Essays*, John Davie, trans., Oxford: Oxford University Press.

Shafir, Eldar, Itamar Simonson, and Amos Tversky (1993), "Reason-based choice," *Cognition, 49* (1–2), 11–36.

Shang, Jen and Rachel Croson (2009), "A field experiment in charitable contribution: The impact of social information on the voluntary provision of public goods," *The Economic Journal*, 119 (540), 1422–39.

Shelburne, Ramona (2014), "Kobe, Bill Clinton talk youth sports," *ESPN*, January 14, http://espn.go.com/los-angeles/nba/story/_/id/10291171/kobe-bryant-says-healthy-competition-keyyouth-sports. Accessed March 30, 2015.

Shiller, Robert J. (2019), *Narrative economics: How stories go viral and drive major economic events*, Princeton, NJ: Princeton University Press.

Sidgwick, Henry (2012 [1874]), *The Methods of Ethics*, Cambridge: Cambridge University Press.

Simon, Herbert A. (1996), *The Sciences of the Artificial*, 3rd ed., Cambridge, MA: MIT Press.

Simonson, Itamar (1990), "The effect of purchase quantity and timing on variety-seeking behavior," *Journal of Marketing Research, 27* (2), 150–62.

Skyrms, Brian (1996), *Evolution of the Social Contract*, Cambridge: Cambridge University Press.

Smith, Adam (1976 [1776]), *An Inquiry into the Nature and Causes of the Wealth of Nations*, 5th ed., Chicago, IL: University of Chicago Press.

——— (2002 [1759]), *The Theory of Moral Sentiments*, 6th ed., Cambridge: Cambridge University Press.

Smith, James P., John J. McArdle, and Robert Willis (2010), "Financial decision making and cognition in a family context," *The Economic Journal, 120* (548), F363–80.

Staw, Barry M. and Jerry Ross (1989), "Understanding behavior in escalation situations," *Science, 246* (4927), 216–20.

St. Petersburg Times (2001), "Dream car is a 'Toy Yoda,'" July 28, http://www.sptimes.com/News/072801/State/Dream_car_is_a__toy_Y.shtml. Accessed March 23, 2015.

Sunstein, Cass R. (2014), "Nudging: A very short guide," *Journal of Consumer Policy, 37* (4), 583–588.

Szuchman, Paula and Jenny Anderson (2011), *Spousonomics: Using Economics to Master Love, Marriage and Dirty Dishes,* New York, NY: Random House.

Tabarrok, Alex (2013), "Stayaway from layaway," October 24, http://marginal-revolution.com/marginalrevolution/2013/10/stayaway-from-layaway.html. Accessed April 1, 2015.

Thaler, Richard H. (1980), "Toward a positive theory of consumer choice," *Journal of Economic Behavior & Organization, 1* (1), 39–60.

——— (1985), "Mental accounting and consumer choice," *Marketing Science, 4* (3), 199–214.

——— (2015), *Misbehaving: The Making of Behavioral Economics,* New York, NY: W. W. Norton & Co.

Thaler, Richard H. and Eric J. Johnson (1990), "Gambling with the house money and trying to break even: The effects of prior outcomes on risky choice," *Management Science, 36* (6), 643–60.

Thaler, Richard H. and Cass R. Sunstein (2008), *Nudge: Improving Decisions About Health, Wealth, and Happiness,* New Haven, CT: Yale University Press.

Thompson, Derek (2013), "Money buys happiness and you can never have too much, new research says," *The Atlantic,* April 29, http://www.theatlantic.com/business/archive/2013/04/money-buys-happiness-and-you-can-never-havetoo-much-new-research-says/275380. Accessed March 17, 2015.

Todd, Peter M. and Gerd Gigerenzer (2000), "Précis of *Simple heuristics that make us smart,*" *Behavioral and Brain Sciences, 23* (5), 727–41.

Tomberlin, Michael (2009), "3rd lawsuit claims rigged jackpot," *The Birmingham News,* October 8, p. B1.

Tversky, Amos and Daniel Kahneman (1971), "Belief in the law of small numbers," *Psychological Bulletin, 76* (2), 105–10.

——— (1974), "Judgment under uncertainty: Heuristics and biases," *Science, 185* (4157), 1124–31.

——— (1981), "The framing of decisions and the psychology of choice," *Science, 211* (4481), 453–58.

——— (1983), "Extensional versus intuitive reasoning: The conjunction fallacy in probability judgment," *Psychological Review, 90* (4), 293–315.

——— (1986), "Rational choice and the framing of decisions," *The Journal of Business, 59* (4), S251–78.

Tversky, Amos and Eldar Shafir (1992), "The disjunction effect in choice under uncertainty," *Psychological Science, 3* (5), 305–9.

US Department of Defense (2002), "DoD news briefing: Secretary Rumsfeld and Gen. Myers," February 21, http://www.defense.gov/transcripts/transcript.aspx?transcriptid=2636. Accessed March 23, 2015

Velleman, J. David (1991), "Well-Being and Time." *Pacific Philosophical Quarterly, 72* (1), 48–77.

Vonnegut, Kurt. (2006), *A Man Without a Country,* Daniel Simon, ed., London: Bloomsbury.

Ware, Bronnie (2012), *The Top Five Regrets of the Dying: A Life Transformed by the Dearly Departing,* Carlsbad, CA: Hay House.

Watts, Tyler W., Greg J. Duncan, and Haonan Quan (2018), "Revisiting the marshmallow test: A conceptual replication investigating links between early delay of gratification and later outcomes," *Psychological Science, 29* (7), 1159–77.

Wicksteed, Philip H. (2003 [1933]), *The Common Sense of Political Economy,* London: Routledge.

Wilde, Oscar (1998 [1890]), *The Picture of Dorian Gray,* Oxford: Oxford University Press.

Wilkinson, Nick and Matthias Klaes (2017), *An Introduction to Behavioral Economics,* 3rd ed., London: Red Globe Press.

World RPS Society (2011), "How to beat anyone at Rock Paper Scissors," http://www.worldrps.com/how-to-beat-anyone-at-rock-paper-scissors/. Accessed April 13, 2015.

INDEX